ARMED
BLUEJACKETS
ASHORE

ARMED BLUEJACKETS ASHORE

US Navy Landing Guns
1850-1942

NELSON H. LAWRY

FONTHILL

Fonthill Media Language Policy

Fonthill Media publishes in the international English language market. One language edition is published worldwide. As there are minor differences in spelling and presentation, especially with regard to American English and British English, a policy is necessary to define which form of English to use. The Fonthill Policy is to use the form of English native to the author. Nelson H. Lawry was born and educated in the USA and now lives in that country, therefore American English has been adopted in this publication.

www.fonthill.media
office@fonthillmedia.com

First published in the United Kingdom
and the United States of America 2024

British Library Cataloguing in Publication Data:
A catalogue record for this book is available from the British Library

Copyright © Nelson H. Lawry 2024

ISBN 978-1-62545-082-1

The right of Nelson H. Lawry to be identified as the author of this work has been asserted by him in accordance with the Copyright, Designs and Patents Act 1988

All rights reserved. No part of this publication may be reproduced, stored in a retrieval system or transmitted in any form or by any means, electronic, mechanical, photocopying, recording or otherwise, without prior permission in writing from Fonthill Media Limited

Typeset in 10pt on13pt Sabon
Printed and bound in England

For my late wife and helpmate, Wynn S. Lawry, who sat by my side in more libraries and archives than I can number, accumulated dust from old records up to her elbows, but who will not be with me when the result of those labors sees the light of day. I miss you, dear.

Today we expect all of you—in fact, you must, of necessity—be prepared not only to handle a ship in a storm or a landing party on a beach, but to make great determinations, which affect the survival of the country.

—John F. Kennedy, former naval officer.
Commencement remarks upon the graduation of
the class of 1961, United States Naval Academy.

Contents

Acknowledgments   10
Glossary   13
Introduction   19
Prefatory: An Armed Landing on a Far and Remote Shore   23

1	The US Navy's First Landing Gun: The Dahlgren Boat Howitzer	27
2	Dahlgren Howitzers in Action During the 1850s and the Civil War	42
3	Shinmiyangyo, June 1871	69
4	The 1870s–1880s and a New Landing Howitzer	85
5	Smaller Caliber Additions for Landing Parties	113
6	Into the New Age	129
7	American Imperial Adventures	147
8	Landing Guns for the Twentieth Century	171
9	The Marines Get Their Share	197
10	Veracruz, April 1914	219
11	World War I and Intervention in Russia	241
12	US Naval Landing Guns in the 1920s	260
13	US Naval Landing Guns in the 1930s	272
14	The Beginning of World War II and the End of US Navy Landing Guns	284

Retrospective   302
Appendix: US Naval Landing Gun Survivors   304
Index   309

Acknowledgments

First and foremost, I wish to express my gratitude to my late wife, Wynn Shugarts Lawry, who not only accompanied me on numerous expeditions, both field and archival, but also rolled up her sleeves to peruse many, many old documents. To her credit, although initially she found the information therein bewildering, she quickly caught on to what was wheat and what was chaff. It is a pity that she did not live to see the fruits of our labors. Praise is due no less to Glen M. Williford, who was there at the beginning and the end and in between. More than generously, he served not only as a sidekick during our numerous raids on the National Archives, but also as a sounding board, when not infrequently I needed someone to bounce my ideas off—those notions at times being a deal less inspired than others. I want to thank Kenneth E. Thompson for being at home, seemingly whenever I telephoned, and dropping everything to check his voluminous library for otherwise difficult-to-find details on long-gone naval, marine, and military officers who had a part in this narrative. And not least to Martha Couture, for her forbearance and at one critical juncture, the loan of more modern technology than I had at hand.

Other friends or colleagues who provided, loaned, or shared valuable information or otherwise deserve my appreciation include Ralph Rottet, Karl Schmidt, Donald M. Kehn, Jr., Thomas Batha, Kenneth Bondurant, Rick and Dave Coughlin, Kurt House, Bruce Canfield, Tyler Sciarrino, Stephen Jendrysik, Ray Trygstad, Joel W. Eastman, Kim Dolce, Donald Radcliffe, and Spencer Tucker in the United States; Harold Skaarup, Douglas Knight, Terry Honour, Maxwell Toms, and Warren Noble in Canada. And special nods to Peter Stubbs in England, Thomas Duvernay in South Korea, and Damien Allan in Australia.

Although not a collector of ordnance, I received helpful assistance from several members of the collecting world: James Schoenung, Donald Lutz (deceased), Bernie Paulson, Matthew Switlik, Douglas Houser, and John Morris.

At Fonthill, I wish to thank Alan Sutton, Jasper Hadman, Jamie Hardwick, Josh Greenland, Jay Slater, and Kena Smith. Also a friendly smile to Heather Martino, who has since moved on to other endeavors, for her lively and stimulating chats.

Acknowledgments

For the remainder of the persons whom I found wonderfully helpful, I shall list them by institution. Given professional transience these days, some of them may not still be at the institution specified, because in those instances it has been several years since our last contact.

American Legion (Henlopen) Post 5, Rehoboth Beach, DE: Robert Pruchniewski and Kathy Stebbing
American Legion Post 6, Chapel Hill, NC: Edward H. Gill
American Legion Post 139, Reese, MI: Robert and Joyce Spiekerman
American Legion (Tidewater) Post 327, Norfolk, VA: Alonzo M. Scott
Arizona State Library: Peter Grant, Andrew Bourgeois, and Amelia Raines
Arizona State University, Hayden Library: Shirley Whitehouse
Arizona State University West, Fletcher Library: Lynn Duron
Boston Athenæum: Mary Warnement
Central Massachusetts Veterans Service District: Adam Costello
Chicopee, MA, Emily L. Partyka Central Library: Brigitte Bisaillon
Clark's Trading Post, Lincoln, NH: Robert Henderson
Dover, NH, Public Library: Denise LaFrance, Carrie Tremblay, and Emily Ainaire
Grand Canyon University Library: Kelly O'Neill
Heard Museum: Marcus Monenerkit
Marine Corps Association: Patricia Everett
Massachusetts National Guard Armory, Concord, MA: Leonid E. Kondratuik and Keith T. Vezeau (the latter now at the Massachusetts State Archives)
Michigan Department of Military and Veterans Affairs: Walter Sullivan
Mystic Seaport Museum, CT: Paul O'Pecko
Nashua, NH, Public Library: Courtney Caisse
National Museum of the Marine Corps: Alfred V. Houde, Bruce Allen, and Joan C. Thomas
National Museum of the Royal Navy: Heather Johnson
National Museum of the US Navy: Jim Bruns (retired)
National Register of Historic Places Archives: Jeff Joeckel
New Hampshire Division of Historical Resources: Peter Michaud (now at the Portsmouth Naval Shipyard)
New Hampshire State Library: Donna Gilbreth
Patriots Point Naval & Maritime Museum, SC: David A. Clark
Peabody Essex Museum, Phillips Library: Kathy Flynn and Catherine Robertson
Royal Armouries, Fort Nelson Museum: Philip Magrath
Royal Marines Museum: Ian Maine
University of New Hampshire, Dimond Library: Louise Buckley; Physics Department: Michael Briggs
US Army Field Artillery Museum, Fort Sill: Gordon A. Blaker
US Army Picatinny Arsenal: Robert D. Frutchey
US Marine Corps History Division: Kara Newcomer and Alisa Whitley
US Marine Corps Library: Nancy K. Whitfield

US National Archives, Archives I: Special nods to Mark Mollan (now deputy historian, US Coast Guard) and Chris Killillay; also Kim McKeithan, Susan Abbott, Damani Davis, and Juliette Arai; Archives II: Nathaniel Patch and Daryl D. Bottoms; Boston (Regional) Archives: Nathaniel Wiltzen
US Naval Academy Archives: David D'Onofrio, Adam Minakowski, and Dr. Jennifer Bryan
US Naval Academy Museum: James Cheevers
US Naval Support Activity Crane: Jeffrey Nagan
US Naval War College Historical Collection: Dr. Evelyn Cherpak (retired)
US Naval War College Library: Gina Brown
US Naval War College Museum: John Pentangelo (now director, Hampton Roads Naval Museum), John W. Kennedy, and Robert Doane
US Navy Art Collection: Pam Overmann
US Navy Department Library: Steven J. Lynch and Dennis S. Wilson
US Navy, Norfolk Naval Shipyard: Jeffrey R. Cunningham and Marcus Robbins
Veterans of Foreign Wars Post 3438, West Carrollton, OH: Wesley L. Johnson (deceased)
Westborough, MA, Public Library: Nancy Odell

If in this chronicle of those folks providing kindly assistance to me, I have omitted someone, please believe it was not intended and accept my profuse apology for the oversight. And, it should go without saying, none of the folks on the above muster roll is in any way responsible for errors appearing in this book. Those mistakes are mine only.

Glossary

Abatis: Felled trees with their sharpened limbs pointing outward and acting as a defensive barricade (the somewhat more refined variant is a *cheval de frise*).

Accles drum feed: The third magazine feed for the Gatling gun, consisting of a circular drum that sat perpendicular to the piece, its doughnut hole in line with the barrel, and worked by mechanical means to feed bullets into the gun. Although it allowed the greatest gun elevation, the device proved troublesome to load, and was replaced by an improved Bruce feed.

BAR: The Model 1918 Browning automatic rifle, loading a magazine of twenty .30-caliber rounds. It was introduced as a squad automatic weapon near the end of World War I, and served throughout World War II and the Korean conflict.

Becket: A rope handle, for example on an early ammunition chest.

BL: Breechloading or breechloader, that is, a cannon or small arm loading from its breech end, the rounds either fixed, with the projectile attached to the propellant case, or separate, with the projectile and propellant loaded discretely.

Bluejacket: A sailor seeing duty ashore, typically in an armed landing party.

Bouche: A bushing, most often of copper, that lined the vent of a muzzleloading cannon in order to extend its service life.

Breechblock (or Breech plug): A steel member, whether interrupted screw or sliding wedge (either horizontal or vertical), that closes off the breech of an artillery piece and contains the detonation of the propellant. See the individual types for further clarity.

Broadwell drum feed: The earliest magazine feed for the Gatling gun, consisting of a circular drum that sat flat atop the gun, held in place only by two ribs and slots and the force of gravity, and thus limiting the gun in elevation.

Broadwell ring: The circular gas check initially used with bagged powder in the 1870s 3-inch landing howitzer. At the end of that decade it was supplanted by the cup gas check of smaller diameter that put less stress on the breech mechanism.

Bruce feed: The second magazine feed for the Gatling gun, in which a bronze frame having two tracks fed the rounds into the gun by gravity. After one track had emptied, the loading device shifted automatically to the other track.

Caisson: Though differing somewhat, often another term for a limber.

Caliber: As a primary dimension, the diameter of the bore of the barrel, whether a .45-caliber pistol, a .30-caliber rifle, or a 3-inch landing gun. As a secondary dimension, the length of a heavy gun barrel, in particular a naval one, as expressed in multiples of *that* bore caliber; thus a 3-inch/50-caliber (or 3-inch/50cal) deck gun is 150 inches long, whereas a 14-inch/50-caliber (or 14-inch/50cal) main battery gun is 700 inches or 58+ feet long.

Canister: A number of lead, iron, or steel balls encased in a metal canister, which split apart upon the detonation of the propellant and thus did not require a burster charge; otherwise put, a shotgun cartridge for a cannon and limited to close-range work.

Capsquare: One of a pair of hinged pieces or caps on a gun carriage that swung down and locked the trunnions in their seats, or any member having a similar function.

Cascabel: The knob on the breech end of a muzzleloading cannon. On the Dahlgren landing howitzers, it was pierced and threaded as part of the elevating mechanism.

Chase: Loosely, that part of a nineteenth-century gun between the trunnions and the muzzle.

Cole cart: A two-wheeled handcart equipped with a drawbar designed for pulling by marines and later bluejackets ashore. It became available in several types and sizes, depending upon what it was intended to carry, whether a weapon of a specific type, ammunition, or other essentials.

Commodore (Commo.): A former naval rank next senior to captain, now replaced in the US Navy by the rank of rear admiral of the lower half.

Crenelation: The top of the wall of a protective battlement, characterized by a linear array of indentations for firing, whether by archers or riflemen, at attacking or besieging hostile forces.

CSS: Confederate States Ship.

Drift: Much simplified, for a projectile fired from a rifled gun, the gradual divergence from its aimed trajectory in the direction of its rotation. The greater the range, the more the drift, and this factor must be corrected for when firing.

En bloc clip: A metal charger holding the number of rounds that will fit into a rifle's magazine and be inserted as a whole. The clip will drop from the rifle upon chambering the first round (e.g. the Winchester-Lee M1895) or be ejected with the final round (e.g. the M1).

Ensign: The lowest commissioned rank in several navies. Until 1912, except during wartime, the achieving of this rank in the regular US Navy required four years at the naval academy, followed by two years of fleet service. Since 1912, midshipmen are commissioned such after successfully completing four years at the academy. The rank was established in 1862 and is equivalent in the army and marines to the commissioned rank of second lieutenant.

Field gun: Specifically among US Navy shipboard ordnance intended for use on shore, both of the Fletcher guns, Mark I and Mark I mod 1, introduced respectively to fleet service shortly before and after 1900, which generated a muzzle velocity of 1,150 feet per second. That definition is in contradistinction to the landing guns of higher marks firing a larger round at 1,650 feet per second.

Glossary

Fixed round: The projectile and propellant joined together for greater convenience in loading. Such joining was accomplished in different ways in different generations.

Flag officer: In the US Navy, a temporary rank for a squadron commander until the change in naval rank structure in 1862, which introduced the rank of rear admiral. It is presently an encompassing term for any admiral or general.

Fuse: The device within the projectile, acting either from concussion or burning-time interval from a powder train, that serves to explode the burster charge. The Bormann, Boxer, and Hotchkiss fuses were innovative types seeing use at different times in different nineteenth-century landing guns.

Gingal (Jingal): An ancient firearm otherwise known as a wall gun, usually a matchlock, ranging in size from a large musket fired from a rest to one mounted in some fashion atop the parapet.

Gomer chamber: The namesake of its designer, Louis-Gabriel de Gomer, a partly conical powder chamber well suited to howitzers and mortars of the nineteenth century.

High Explosive or HE shell: A relatively thin-walled round leaving a greater space for its highly explosive filler, such as Explosive D (aka dunnite or ammonium picrate), soon superseded by TNT (trinitrotoluene) in the twentieth century. Such rounds continue to depend upon concussion and fragmentation to neutralize enemy personnel and emplacements.

HMS: Her or His Majesty's Ship in the Royal Navy.

Howitzer: A cannon having a short barrel and the ability to fire at a significant elevation in order to provide a suitably curved trajectory. The Dahlgren howitzer was so designated because of its short barrel length and Gomer powder chamber.

Interrupted screw breech mechanism: A hinged steel plug or breechblock with alternating threaded and blank sections, which is pushed into the countering blank and threaded sections of the breechblock recess (or screw box), then turned so that the threaded sections mesh and lock to form a barrier to the detonation of the propellant. Although often coupled with bagged powder, in some instances in the past it found use with metal cartridge cases.

Lieutenant commanding (Lt. Cmdg.): A lieutenant who commanded a small warship or prize vessel and was, temporarily, superior in rank to a regular lieutenant. In 1862 in the US Navy, the rank became the permanent one of lieutenant commander.

Limber: A two-wheeled cart attached to the trail of the field carriage and carrying ammunition for the gun. Through the years, the increasing number of rounds carried made the combined weight of gun, carriage, and limber too heavy for man-hauling over more than short distances.

Master: An historic rank that in several navies had evolved to commissioned navigator. In March 1883 the US Navy replaced the rank with that of lieutenant (junior grade), abbreviated Lt. (j.g.).

Midshipman: Typically, a student at a naval academy. Between 1902 and 1912, the rank persisted for two years of fleet service after four years at the US Naval Academy, before the young officer qualified for his commission as ensign. The rank commanded every courtesy deserving of a warrant officer.

Minié ball: The namesake of its inventor, Claude-Étienne Minié, a conical-nosed, hollow-based bullet used in mid-nineteenth-century ML rifled muskets in service in Europe and the United States.

ML: Muzzleloading or muzzleloader, that is, a cannon or small arm loading from its muzzle end, with the rounds either fixed or separate.

Naval cadet: Between 1882 and 1902, the rank that persisted for two years of fleet service after four years at the US Naval Academy, before the young officer qualified for his commission as ensign. The rank commanded every courtesy deserving of a warrant officer.

Obturation: The reliable and safe closing-off of the breech of the gun behind the powder charge achieved by various means, thus protecting the gunners.

Pounder: When preceded by an arabic number—e.g. 12-pounder—the alphanumeric specifies a gun defined by the weight of its projectile. In the Dahlgren 12-pounder SB howitzer, this projectile weight correlated with a bore diameter of 4.62 inches, whereas the more modern 1-pounder has a 37mm bore.

Primer: An explosive device that worked by friction or percussion to set off the main propellant charge in the gun, either way by means of a pull on a lanyard. Also an explosive cap on a fixed metallic cartridge that performs the same function when struck by the firing pin.

QF: Quick-firing, loosely a fast-operating breech mechanism of British design, in guns of small to intermediate size.

Reinforce: Loosely, the thick part of a nineteenth-century gun between the breech and the trunnions.

RF: Rapid fire, the US designation roughly equivalent to the British QF.

Rifling: The spiral groove cut into the gun bore to impart a spin to the projectiles fired.

RML: Rifled muzzleloader.

RN: Royal Navy.

RRSC: Russian Railway Service Corps, comprised of uniformed but otherwise civilian American railroad men who oversaw and operated the eastern portions of the Trans-Siberian and Chinese Eastern Railways, from mid-1918 to mid-1920.

Sabot: A device, often made of wood, attached to the base of a projectile in order to center it in the bore of a muzzleloading gun. Upon firing the piece, the sabot fell away when the projectile left the muzzle.

SB: Smoothbore, that is, a gun having an unrifled bore.

Shrapnel (Spherical case): As originally defined, a projectile loaded with lead, iron, or steel balls to achieve its destructive effect upon the detonation of the burster charge; distinct from shell, which contained a greater payload of explosive and generated metal shards. It was named for its inventor, Lt. (eventually Lt. Gen.) Henry Shrapnel, Royal Artillery.

Sliding wedge breech mechanism: A steel barrier sliding horizontally or vertically within a mortice, and coupled with a metallic cartridge case that provides its own obturation.

Stripper clip: A metal charger holding the number of rounds that will fit into a rifle's magazine and be inserted as a whole. The clip is manually removed, stripping off the rounds before the first round is chambered (e.g. the Springfield M1903).

Tinclad: A pejorative, at times jocular, applied to lightly armored Union gunboats deployed in inland waters during the American Civil War. The name was also applied to certain heavy cruisers, laid down during the mid- to late 1920s, having overly light armor in order to satisfy the weight restrictions of the Washington Naval Treaty signed in 1922. They were known as treaty cruisers, or worse, "treaty tinclads."

Trunnions: The pair of cylindrical "ears" projecting from each side of a nineteenth-century cannon, which fitted into pockets on the carriage and then were locked into place by capsquares, securely mounting the piece. Trunnions could be cast as part of the gun or a part of a ring slipped on during the gunmaking process.

USAT: United States Army Transport, i.e. an army transport or cargo vessel.

USCGC: United States Coast Guard Cutter.

USMC: United States Marine Corps.

USN: United States Navy.

USNA: United States Naval Academy.

USNRF: United States Navy Reserve Force, created in 1916 to replace the naval militia, and from which the navy drew additional members for its wartime needs.

USS: United States Ship.

Vent: The channel into which the primer fitted, and when exploded, allowed the flame to reach the powder chamber. It could either be radial, passing through the gun reinforce directly into the chamber, or axial, along the centerline of the bore, through the breechblock into the chamber.

Introduction

During the course of my research for this book, I learned that a 3-inch naval landing gun—a Fletcher type Mark I discussed at length in the main text—stood in front of an American Legion post in rural Michigan. After some effort, I reached the post commander, and it turned out, most fortuitously, that he was very interested in the gun, had rebuilt its wood-spoked wheels, and had repainted it as best he could figure out. Trouble was, despite a deal of research on his part, he didn't know what the gun was. I was able to tell him, and sent him photos and drawings to erase any doubt (after all, at that point I was a complete stranger). In return, the delighted commander and his wife, the Spiekermans, obligingly sent me a series of photographs exactly as I had requested them. Those photos in turn enabled me to answer a question that had eluded me despite hours perusing naval ordnance textbooks, manuals, and pamphlets. This is only one instance of many happy collaborations that I enjoyed—to be sure, often stumbled onto—while researching and writing this book.

Indeed, the Legion commander's experience mirrored my own of years before, when I had my first "what-the-devil-is-that?" encounter. As with most military historians, I was rather familiar with the artillery pieces that had served during World War II and the Korean conflict. And of course I could recognize *la bonne Soixante-quinze*, the famous French 75, as well as other animal-drawn pieces of similar vintage of German or Austrian origin, brought to these shores as trophy guns following World War I. My first look at a breechloading naval landing gun, however, registered naught, and soon led to visits to libraries and archives, including national ones, not to mention hours (and hours) on the road to access these wonderful repositories. There was no book then available pulling together US Navy landing guns, their technical details, and histories, combat or otherwise. At that precise point, I determined to write such a book.

Too soon I learned that such endeavors were not measured by hours on the clock, but rather in the amount of dust and grime transferred from these old records to oneself, especially under the fingernails. Along the way, I met a lot of friendly souls happy to assist me, more than once taking measuring tape and kitchen magnet to a nearby landing gun. In the latter case, how else do you ensure if some telling component is made of steel or bronze?

For nonfiction works such as this, some words of introduction and explanation are regarded as essential for the reader's better understanding of both the material considered in the book and the author's foibles as a penman. The period covered is from 1850, with the introduction of the US Navy's first purpose-built landing howitzers, designed by Lt. John Dahlgren, to November 1942, the execution of Operation Torch, the end of the first year of World War II for the United States, and essentially the end of naval landing parties in their long-recognized form.

To the distress of some and the delight of others—I suspect very few occupy the middle ground on this one—I have decided against endnotes. Before anyone draws the wrong conclusions, I fully appreciate the value of such specific reference markers. In my previous life, I worked as a laboratory biologist in the study of cyanobacteria (aka blue-green algae), and my published papers include numerous such markers citing previous publications. After all, in the interpretation of one's own empirical data, the earlier works of others, whether in concurrence or opposition, merit such consideration and citation. On the other hand, I think frequent and/or extensive endnotes in a book of even moderate length can be awful distractions to the attentive reading of the main text. They may cause the reader with perhaps too much fondness for detail to flip back and forth in an attempt to keep up with every jot and tittle provided, and thus miss the forest for the trees. Perhaps my furnishing each chapter with a bibliography of all sources consulted will suffice, although making not a few readers of history unhappy. As much as possible, I have cited the most relevant pages of each source, except in those instances where such numbers of pages are absurdly large, where then I have cited the entire work.

Because the events I describe in this book ended eighty years ago and extend back much farther than that, I use the English units of weights and measures common at the time (e.g. tons, pounds, ounces, and grains; miles, yards, feet, and inches). In those cases where formally designated, I specify millimeters for the bores of gun barrels (e.g. 6mm, 37mm, 75mm, etc.). Places are denoted as they were recognized at the time in the English-speaking world. For the rest of my usages, I have relied on the consensus of contemporary style manuals, with my favorite go-tos *The Chicago Manual of Style*, *The Oxford Guide to English Usage*, and Strunk and White's little gem, *The Elements of Style*.

As with then-current practice, the names of warships and merchant vessels are not preceded by "the" unless an intervening modifier is present, thus *New Hampshire* (a battleship) stands alone, but the heavy cruiser *Houston* is so modified. The navy sometimes changed the names of its warships, usually to free up the name of an existing obsolescent vessel for one of modern construction. The armored cruiser *New York*, launched in 1891, became *Saratoga* two decades later, but a mere six years thereafter, she assumed the name *Rochester*. Ships' names appear in italics, whereas the names of classes of warships are denoted in roman type, as exemplified in the following paragraph.

In fleet units of the New Navy of the 1880s and thereafter, I have followed the name of the ship with the number and caliber of its main battery—so defining its relative importance—with the specification whether m number of individual gun mounts (e.g. six 4-inch guns) or n number of turrets mounting the number of those guns (e.g. two twin 13-inch gun turrets) were on board. In the instances of 3-inch and 5-inch guns, I have specified them additionally by their bore length in calibers—e.g. 5-inch/51-caliber or more simply 5-inch/51cal—because those guns existing in different lengths so varied in characteristics and capabilities. Such specification is also useful in showing the escalation of main guns, particular in battleships as the international race to build such capital ships progressed. Of course, numerous intermediate, secondary, and tertiary guns were also mounted, particularly onerous in pre-dreadnought battleships, where intermediate and secondary size ordnance differed by as little as 1 inch in bore diameter, and thus made precise fire control for such batteries very difficult to impossible. I have strayed from the main-battery-only designation in the peculiar Kearsarge- and Virginia-class battleships, where a twin 8-inch gun turret was superimposed atop each of the two main gun turrets, with consequent—and one would have thought readily predicted—problems arising from that configuration.

As was done during the period when such ammunition remained in use, differentiation is made between shrapnel and shell. Shrapnel was an exploding round whose payload consisted of a considerable number of lead, iron, or steel balls, their diameter a function of the size of the gun. Shell was either common or high explosive, after fuse detonation breaking up into metal shards and splinters. It is currently popular to call these metal shards shrapnel, but in the nineteenth and early twentieth centuries, men of the ordnance branch and of the field artillery firmly distinguished between these injurious shell fragments and shrapnel's metal balls.

Squads, sections, platoons, companies, batteries, and battalions are referred to as units; regiments, brigades, divisions, and corps as formations. An individual serving in a particular armed service is uniformly in lower case: sailor or seaman, marine, soldier, ranger, infantryman, artilleryman, or gunner. Likewise, when not used as a title, captain, corporal, or ensign are ordinary words. Rank is abbreviated if it precedes the bearer's full name, but the unabbreviated rank is applied if just the surname appears: Lt. Cmdr. Arthur Smith, but Staff Sergeant Greene.

I have cited the authors of published works devoid of naval or military rank, despite in many instances that rank having been indicated in the original publication. In line with current convention, I have cited the volume numbers of all periodicals as arabic numbers, even in those instances where such volumes originally appeared as roman numerals; the issue number for that volume immediately follows within parentheses, for example, 38(7).

With a few isolated deviations, I have given minimal attention to the monetary aspects of the navy's acquisition of its various landing pieces. That avoidance of dollar particulars includes most instances where a manufacturing company failed in its contractual obligations to the US government. Only with breakthrough technology, such as the navy's first breechloading gun, the 3-inch landing howitzer of the 1870s, and the navy's first attempt at recoil absorption on a field carriage, the Fletcher Mark I 3-inch field gun of the 1890s, have I provided a lot of fine ordnance detail. This book is not necessarily an introduction to field artillery, for I assume the reader is already familiar with the decided advantages, for example, of the sliding wedge breech mechanism, the hydropneumatic recuperator, and the split-trail field carriage.

As both expected and feared, two official sources sometimes provide different and largely contradictory data. If I was fortunate, I stumbled upon a third source and now had two sources in agreement, and thus could ignore the odd datum out. If, however, I did not so stumble or worse yet, now had three disparate data, I had either to make an astute guess, or inform the reader of the awful dilemma and thus dump the responsibility for coming up with the right answer into his lap.

Some readers, I hope, will be in full agreement with the decisions I have made in writing this book. Others will think I have been altogether arbitrary. I expect there are solid arguments to be made for either point of view.

PREFATORY

An Armed Landing on a Far and Remote Shore

As the cutter neared the beach, a large swell lifted it and the naval cadet in charge, thirteen months out of the academy at Annapolis, could now see that the breaking surf was heavier than it had appeared from aboard the armored cruiser on which he served. The next swell began to curl into a wave and in so doing, forced the stern of the cutter around to the left. Alarmed at the chance they would broach to in the next wave, the cadet ordered the starboard oarsmen to bend to their work and their counterparts on the port side to slack off. The experienced coxswain hardly required the young officer's order to shove the tiller hard over to the right. By the time the following wave began to break behind them, the boat had assumed a more assuring heading and they glided safely toward shore. The oarsmen on both sides near the bow scrambled out to pull the cutter onto the beach, some losing their footing because of the heavy undertow. But in the next moment they were fortuitously assisted in their task by a larger than usual wave that acted to shove the boat an extra few feet in the desired direction.

Once the cutter had been firmly beached, the young officer sent two men, armed with rifles and bandoliers, trotting toward the line of trees bordering the stretch of sand, instructing them to keep a close watch. Meanwhile, the men still on board and those in the water together lifted the field carriage with its pair of wooden spoked wheels off the somewhat more protected port side and rolled it ashore. It was followed by the pair of laden ammunition chests. At the same time, other men removed the lashings close to the bow, lifted the timber spar made fast to the gun barrel, and passed it to several pairs of waiting hands outside the boat. Having freed the 3-inch gun from the spar, the men quickly and expertly reunited it with the wheeled carriage waiting on the beach. Next they turned the piece completely around so its trail pointed forward and its muzzle rearward, and then they unpinned the smaller trail wheel and dropped it into place. That accomplished, the men placed the ammunition chests on their racks, one on each side, outboard of the gun and inboard of the respective carriage wheel.

Several of the sailors donned the harnesses of the drag ropes, while those not in the hauling crew struggled into their packs and picked up sacks, each carrying an additional 3-inch shell or shrapnel round. In addition, most of the men were accoutered with rifle, bayonet, and bandoliers of 6mm ammunition; the cadet and senior petty officer wore .38-caliber revolvers. The men carried a canteen full of fresh water and previously prepared food to sustain them, as well as a medical pouch. At the cadet's order, the men in harness began to pull the piece toward the line of green vegetation where the men sent ahead waited.

The cadet glanced somewhat apprehensively toward his right, because contrary to plan—likely due to the unexpectedly heavy surf—some boats transporting only riflemen had landed between his cutter and the launch carrying both the other landing gun belonging to the battery and the lieutenant junior grade who served as the battery commander. Unlike the cadet's gun, that on the launch had been mounted on a boat carriage in the bow to fire forward in the event such was needed, which in this instance it had not been. To his relief, the cadet spotted his commanding officer some distance down the beach, who turned at that moment and gave him a hasty smile and reassuring nod.

Farther on, the youthful officer saw the company of marines assigned as the advance unit move quickly across the beach, form a skirmish line, fix bayonets, and disappear into the scrub. Behind him, the three steam launches and single steam cutter that had towed the strings of pulling boats from the three warships lying more distantly seaward remained close inshore, the gun crews of the 1-pounders or 3-pounders arming those boats casting a wary eye at the activities on the beach, ready to provide supporting fire when needed.

This depiction of an early 1890s landing in strength shows the officer commanding, his aide, and a squad of marines for security being taken ashore in a pulling cutter. At the bow three sailors man an encased short Gatling gun with an Accles feed drum. Beyond, a steam launch armed with a 3-inch boat howitzer tows a string of pulling boats. The national ensign having a thirteen-star union or canton was then traditionally flown on US Navy small boats. Drawn by Thure de Thulstrup for *Harper's Weekly,* August 1892.

Just at that moment, the gun crew ran onto a sizable area of soft sand. Despite the men's best efforts, the gun, with its carriage weighing considerably more from the heavy ammunition chests aboard, sank nearly to its wheel hubs, alarmingly near to bogging down. This obstacle had been anticipated, and at least the other gun carriage had been equipped beforehand with wheels having wider metal tires. Because there had been only the one replacement set aboard, however, the cadet's gun still had the narrower tires that made negotiating deep beach sand more difficult.

Fortunately, the cruiser's pioneer platoon was just then passing to their left, commanded by an ensign well known to the cadet. The latter called out, "Mister Phillips, lend a hand here if you please!" The ensign had been a year ahead at the academy and had given the cadet an awful time when a fourth classman—a lowly plebe—but now they were close friends. His formal address to his messmate was solely for the good order and discipline of the enlisted men they both commanded. Indeed, the cadet awaited his own commission as ensign eleven months hence with much impatience, assuming all went well in the meantime, which included surviving the present landing.

The pioneer platoon leader uttered but few words, made some hand signals that would have proved mystifying to the uninformed observer, followed by the sweep of his right arm, and a file of men commanded by a petty officer peeled off toward the motionless landing gun. As it served as more of a hindrance in such sand, the cadet ordered the men on the trail to unpin and raise the trail wheel. Pushing and pulling, the reinforced gun crew soon had the piece beyond the soft patch, and with a word of thanks from the cadet, the pioneers returned to their platoon.

Almost immediately thereafter, the gun section penetrated the line of trees and brush at the edge of the beach. One of the men sent ahead came back to report finding a trodden path through the vegetation, but still they moved more cautiously now. The cadet sent one of his advance guard as a runner in the direction of the other section to ask for instructions. The cadet's boat had landed in the correct place, and he expected that the other gun would move leftward to join him, rather than the other way around, but as he was the more junior, he needed to be certain of his commanding officer's intention.

For the moment, his gun needed to be in a better position to support the infantry and so the young officer instructed the gun crew to move toward some higher ground he could see through the thinning vegetation. As the gun crew did so, all became aware of an odd humming sound amidst them, accompanied by the spatter of the surrounding leaves. The men noticed several of those leaves now had large holes in them that hadn't been there before, and there came to their ears a distant pop-popping. One of the pioneers to their left grunted and then exclaimed, "I'm hit!" Although at some point expected, actually being taken under fire came as a rude and unpleasant shock. The naval cadet drew his revolver and after another careful look to his front, motioned his crew to haul the landing gun forward toward the knoll where it could be deployed to better advantage.

With his gun emplaced, the young officer put away his sidearm, pulled out his binoculars, and peered at the distant line where blue smoke was evident. He could make out earthworks there and hostile riflemen atop them, surprisingly foregoing the cover offered by the defenses. Given that circumstance, the cadet quickly decided that shrapnel would serve better than shell, at least until the distant riflemen went to ground there. Between that line of resistance and his own position, he saw the friendly infantry, both sailors and marines, seeking what cover was afforded by the scrub, and heard them returning the enemy fire. He needed his commander's order to open fire, however, and waited impatiently for that word.

After what seemed an eternity, his runner returned and said the other gun had emplaced a distance to their right. To the youthful officer's satisfaction, the lieutenant had also given permission to commence firing. At that moment the blast of the other gun was heard. Taking another hasty observation through his field glasses, the naval cadet made a quick calculation, directed a specific elevation of the barrel and timing of the fuse, and ordered the gun captain to open fire. He had displayed a distinct aptitude for such fire direction at the academy, and he was confident in his reckoning. Indeed, the first round exploded just under the line of earthworks in the distance, and it required little adjustment to put the successive rounds among the riflemen firing from the top of the earthworks, whereupon they scattered and disappeared. Given that change in disposition, the cadet ordered his gun crew to shift to firing shell.

The cadet reflected that after four years at the naval academy and a year on active service with the fleet—and at barely twenty years of age—he was at last in action. Moreover, he had accomplished his first and perhaps most essential objective: he had gotten his detachment intact onto and then off the beach. He allowed himself the smallest of self-satisfied smiles, only to resume his stern demeanor and turn once more to the task at hand.

Note of explanation: Between 1882 and 1902, students at the US Naval Academy bore the designation of naval cadet. Four years of instruction—in the classroom, on the drill and athletic fields, on the Severn River, and at sea during the summers—were followed by two years with the fleet. The US Navy required those six years of such varying service before considering whether the young man was qualified to hold the commission of ensign (indeed, the fitness reports written by his commanding or supervising officer went back to the naval academy). In 1902, after the navy had declared its intention four years previous, the designation of naval cadet reverted to midshipman, but the six-year term of service continued to be mandatory. During his final two years, the naval cadet/midshipman was considered equivalent to a warrant officer, meriting the address of "sir" and a hand salute from enlisted men. It took another decade, in 1912, before midshipmen were commissioned as ensigns upon completing four years satisfactorily at the academy. The landing described above is envisioned to have taken place during the mid-1890s.

1

The US Navy's First Landing Gun: The Dahlgren Boat Howitzer

Origin

The United States Navy had long been aware of the need for a small artillery piece that would be used in a multitude of roles, but mostly to be mounted in ships' boats or on sturdy field carriages to accompany naval landing parties ashore. Until 1850, little more than a decade before the onset of the American Civil War, a variety of makeshift pieces saw such service. Most commonly, carronades of British design, and small field guns and mountain howitzers acquired from the US Army filled this need. The lightest of them, mounted on special carriages in the fighting tops of sailing and steam warships, were intended to fire down on the gun crews of the enemy warship when two ships were engaged at close quarters. None proved entirely satisfactory for these various functions, particularly if constructed of iron, which made the weapon too heavy for the task.

The amphibious operations undertaken during the Mexican-American War, 1846–48, served to remind the navy even more acutely of the need for such dedicated guns in the antipersonnel and landing role, as well as in cutting out small coastal vessels that had taken refuge in the shallow waters of coves and river mouths. As a result of intensive work by Lt. John A. B. Dahlgren, an ordnance officer assigned to the Washington Navy Yard under Commo. Lewis Warrington, the chief of the Bureau of Ordnance and Hydrography until his death in 1851, the US Navy got its long-desired boat gun. The small howitzer that resulted evermore bore Dahlgren's name. The development of this eventually excellent muzzleloading (ML) piece would need to overcome not a few obstacles, both metallurgical and political, in the latter instance from conservative naval officers opposed to the adoption of such a light howitzer. Their resistance very likely had to do as much with it being so clearly identified as a junior officer's brainchild as anything else, and it hardly helped that Dahlgren too often proved as jealous and petty as he did technologically inventive.

Nonetheless, in 1849 such a gun, with ammunition and accoutrements, was issued for trial purposes to the second-rate sailing frigate USS *John Adams*, soon to depart

Left, a stern, fashionably muttonchopped John A. B. Dahlgren (1809–1870) wears the two wide stripes of a commander, so specified during the 1850s. Dahlgren designed a generation of naval ordnance and served twice as chief of the US Navy's Bureau of Ordnance. Right, a bearded seaman holds a Colt Model 1860 .44-caliber cap-and-ball revolver as he stands by a Dahlgren heavy 12-pounder landing howitzer. *Frank Leslie's Illustrated Newspaper,* April 20, 1861. (*Naval History and Heritage Command*)

for African waters to assist in disrupting the slave trade. The warship would not return to the United States for another four years. During the intervening period, however, in December 1850, Secretary of the Navy William Graham authorized the fruit of Dahlgren's labors for adoption by the navy.

In the eleven years between 1850 and the firing on Fort Sumter, a number of different types of Dahlgren boat and landing howitzers were developed and adopted. Except for a relatively small number of wartime rifled pieces made from ferrous metal, all were cast of bronze in the ordnance shops at the Washington Navy Yard, using mostly the cupola air furnace newly installed in the foundry there. Although the standard gun formulation of this alloy contained nine parts copper to one part tin, Dahlgren specified 9.5 to 9.8 parts copper to 1.2 parts tin for his howitzers, without "a particle of zinc." The use of bronze achieved a weight saving over iron, though decidedly not a cost saving, the latter factor providing yet another cause for criticism of Dahlgren's howitzer. Because of the limitations in facilities and personnel assigned, production remained slow and supply limited, such that few warships within the fleet ever had more than a single such piece aboard. In 1854 the navy erected a new ordnance building at the Washington yard, and two years later a new foundry as well, mainly for the casting and production of the Dahlgren howitzers.

Initially, three types of smoothbore ML howitzer were made—light and medium 12-pounders and a 24-pounder—a function of the rate (size) of the warship for which each was intended, though more specifically the size of the ships' boats carried in those classes of vessels, their launches in particular. The designations of the 12-pounders led to confusion (which persists today), so the medium 12-pounder became the heavy 12-pounder. In time, three more versions appeared: a small smoothbore (SB) 12-pounder and rifled 12- and 20-pounders—though because of its weight and design, the 20-pounder rifle saw exclusive use as a deck gun. In practice, the navy restricted the 24-pounder SB, although designated a boat gun, to shipboard use as well because of its heaviness.

Some Specifications of the Six Types of Dahlgren Howitzer

Type	Bore diam. (inches)	Length (inches)	Weight (pounds)	Trunnions (yes/no)	Powder ch. (pounds)	Number made
small 12-pdr SB	4.62	42.5	300	no	0.5	23
light 12-pdr SB	4.62	50.95	430	no	0.625	188
heavy 12-pdr SB	4.62	63.5	760	no	1.0	456
12-pdr rifle	3.4	63.5	870	no	1.0	423
24-pdr SB	5.82	68.2	1,310	no	2.0	1,012
20-pdr rifle	4.0	68.2	1,350	yes	2.0	100

Ranges differed as a function of the size of howitzer, the type of round, and the elevation of the gun (in degrees above the horizontal). By design, the Dahlgren smoothbore howitzers fired three types of projectile: shell, shrapnel, and canister. Their intended use obviated the provision of both solid shot and grape, although there occurred wartime exceptions in shipboard firing. The most extensively manufactured and used smoothbore pieces, the 24-pounder and the heavy 12-pounder, at 5 degrees elevation threw their projectiles about 1,300 yards and 1,100 yards respectively (because heavier rounds of otherwise equal size are less retarded by air resistance, the somewhat heavier shrapnel generally reached farther than the shell from the same gun). At the same elevation, the rifled 20-pounder (or 4-inch rifle) fired its shell to about 1,960 yards and the rifled 12-pounder (or 3.4-inch rifle) to 1,770 yards.

Although the original pattern of rifling in both pieces consisted of three wide grooves, the production version of the 3.4-inch rifle had twelve grooves and was thus an example of the more satisfactory polygroove rifling. Not only did the rifles fire to greater range than their smoothbore counterparts, they did so more accurately. These rifled pieces, however, had an Achilles heel: rifling in bronze did not wear well from repeated firings. Because the 12-pounder heavy SB and the 12-pounder rifle were made from the same initial castings, in later years, ordnance machine shops bored out more than one worn 3.4-inch rifle, it thereby becoming a 12-pounder heavy SB of 4.62-inch bore and losing substantial mass in the process.

Description of Internal and External Design Details

All six types of light Dahlgren ordnance incorporated a powder chamber on the breech end in the shape of a frustum of a cone, known as a Gomer chamber from its inventor, French artillery officer Louis-Gabriel de Gomer. His innovation was particularly suitable for mortars and howitzers, which generally fired at higher angles than guns. Whereas the Gomer chambers of all the smoothbore howitzers and the 3.4-inch rifle took the form of a flat-bottomed cone (i.e. flat-ended at the breech), that for the 4-inch rifle was round-bottomed.

Given other features of the larger rifle that strayed from those of the five true Dahlgren howitzers—it was cast with trunnions, they were not; its barrel displayed a distinctive traditional taper, theirs did not—but mainly because of Dahlgren's own intention that the piece serve entirely as a light shipboard piece, further discussion of the Dahlgren 4-inch rifle or light 20-pounder will be limited to immediate relevance.

In external design, the Dahlgren howitzer's breech end formed part of a sphere, melding briefly into a cylinder, and then transforming into a mildly and nearly imperceptible tapering barrel. The barrel was devoid of muzzle swell, reinforcing rings or hoops, or indeed adornment of any kind. When in service it was never supposed to be polished, in order to prevent as much as possible the reflection of the sun from ruining the gunner's aim or revealing to the enemy its presence while on the march. In fact, Dahlgren considered the verdigris of normal copper oxidation to be the ideal hue. As mentioned before, notably absent were trunnions, which he eschewed because of the increased weight of the gun carriage thus required. Instead, he devised a loop under the gun barrel through which to run a pin in order to secure the piece to its carriage, similar to the manner of attaching the older naval carronade.

The cascabel knob was pierced and threaded to take the elevation screw, which bore a disk below the threaded portion having a coarsely milled rim, in order to turn more readily to elevate or depress the breech, and thus have the opposite and intended effect on the muzzle. Many howitzers that survive today have their front sight cast at top dead center of the muzzle, although a certain number possess an adjustable front sight above the mounting loop on the howitzer's left side. The design called for an adjustable rear sight to be attached to a casting on the left side of the breech. Also atop the breech, usually but not always on the right side of the howitzer, projected the casting for the hammer of the firing mechanism.

When in action, the gunner fired the piece by pulling the lanyard attached to the hammer lock. On the tug of the lanyard, the hammer fell, exploding a percussion primer, whose flame flashed through the copper-bouched vent to detonate the propellant charge. The perforated hammer head remained in place, with the hot gas discharged back through the vent thereupon escaping through the aperture and thus not displacing or damaging the hammer head.

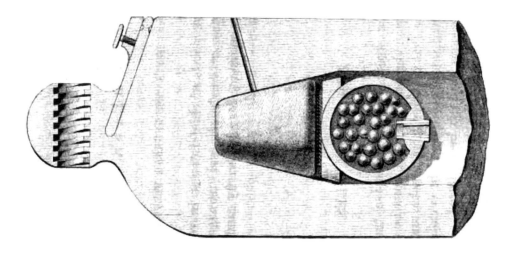

A conical Gomer chamber designed for SB mortars and howitzers is shown loaded with a fixed spherical case round. From right to left, note the fuse (will be set alight by the propellant discharge, in turn to ignite the burster charge), the shrapnel balls within the projectile, the wooden sabot to stabilize the round while in the gun, and the powder bag. Farther left is the angled vent opening into the powder chamber and fitting the primer, and lastly the cascabel threaded for the elevating screw. John A. Dahlgren, *Boat Armament of the U.S. Navy*, 1856.

Two images of the heavy 12-pounder standing guard at the Ashburnham (MA) Historical Society hall reveal the principal external features of the Dahlgren landing howitzer. The barrel is imperceptibly tapered and unpolished, with an unadorned muzzle. The barrel lacks trunnions, but rather an underpin attaches gun to carriage. The firing lock or hammer is poised above the vent opening, and the left-side casting will hold the rear sight. The iron field carriage bears wheels with wooden spokes and felloes. The terminal socket for the trail stave to move the gun in traverse extends rearward of the trail wheel. (*Nelson H. Lawry*)

The Dahlgren Howitzer Carriages

There were two entirely different carriages designed for the Dahlgren 12-pounder howitzer, reflecting the different roles envisioned for the piece. To fire against small vessels of hostile intent or warlike persons ashore required a strong boat carriage. This carriage consisted of a wooden bed mounting the howitzer, which in turn sat on a pivoting wooden slide resting on a yellow pine (hard pine) scantling running the length of a ship's launch or cutter—the launch being the larger and better choice—morticed into one or two cross pieces of identical wood. On a standard howitzer launch, there existed six different metal pivot plates, three forward and three aft, placed as two distinct equilateral triangles to direct the fire of the piece, each allowing the gun to fire through a 120-degree arc, in theory without the need to change the direction of the boat. In practice, however, small course changes by the coxswain to point the howitzer seem to have been the norm. Of course, the smaller the warship, the smaller its boats, thus explaining the need for the lighter howitzers (and occasioning the legitimate question by those ever-dubious officers: When was a light howitzer simply too light to accomplish its job?).

The boat carriage also incorporated a compressor that operated by two screw clamps to dampen the howitzer's recoil when fired. The device was tightened or loosened by the two large handscrews atop the bed that acted down through the slide to the metal plate under it. Members of the gun crew set the compressor prior to firing to reduce the recoil and that force's resulting action on the boat. After the piece had recoiled and had been reloaded, the crewmen slackened the compressor to permit the gun being run out again.

Upon coming ashore, the 12-pounder howitzer was readily shifted from being a boat gun to a landing gun in order to support the assault of naval infantry and/or marines, and required but a few men and a few minutes to accomplish the changeover. The means of underpinning the howitzer to either the boat or field carriage particularly lent itself to the quick shift. Gun crews exercised regularly in this maneuver, which required a minimum of specialty gear such as a muzzle block, a shifting spar, and a pair of stout skids to get the gun out of the boat and back into it after the action ashore had ended. The crews were cautioned that the howitzer itself was *never* to be treated as an independent entity, but must always be mounted on the boat carriage or the field carriage.

The wrought-iron field carriage for the 12-pounder howitzer weighed a bit less than 500 pounds, so light that a rod brace was incorporated from each side of the trail end, the pair splaying outward to the axle housing just inside of each wheel. Initially the wheels bore wooden spokes and felloes, but during the war years, to make the landing piece more durable and reliable, thin iron wheels resembling those on agricultural implements supplanted the wooden ones. In his teachings, Dahlgren stressed that both the gun *and* carriage must not be too heavy, lest they become burdensome to the gun crew while bringing them to shore in the surf, a task made even more dangerous when a heavy sea was running.

The field carriage incorporated a third wheel, located near the end of the trail, for easier mobility by sailors hauling the guns on shore. They used a drag rig—essentially

The US Navy's First Landing Gun: The Dahlgren Boat Howitzer 33

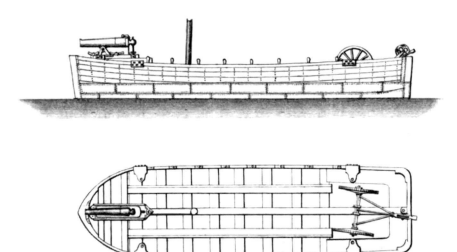

FRIGATES LAUNCH.

These gull's-eye views from both afloat and aloft of the large launch carried by frigates reveal the six swivel points for mounting the Dahlgren boat howitzer, viz. at the apex of the bow, starboard and port bow, starboard and port quarter, and mid-transom. At the stern, the launch also carries the field carriage to be married with the gun upon reaching shore. The accompanying ammunition chests are not shown. Dahlgren, *Boat Armament*, 1856.

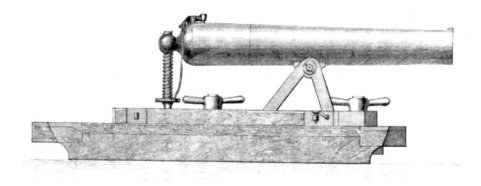

12 PDR. AND BOAT CARRIAGE.
BY J. A. DAHLGREN

A Dahlgren medium 12-pounder bronze howitzer is shown mounted on a boat carriage. Note the two large handscrews that tightened the bed onto the slide, acting as a compressor to diminish the recoil when the gun fired, then slacked off to run out the piece again. Around the time of the Civil War, the medium 12-pounder was redesignated the heavy 12-pounder. Dahlgren, *Boat Armament*, 1856.

Two Dahlgren heavy 12-pounder howitzers on display at the historic Fairfax County (VA) Courthouse are mounted on the expedient light field carriages whose iron wheels resemble those of then-contemporary farm equipment. The right-hand gun mounts its firing lock on the left side of the breech, whereas the left-hand gun mounts its lock on the breech's right side. (*Nelson H. Lawry*)

a harness woven of hemp 20-some feet long—to get the landing gun from one place to another, including a pair of guide ropes to enable steering. Crew members acted to brake the piece using two short drag ropes attached to the front of the carriage. When the piece deployed for action, a member of the gun crew pulled a pin to enable the trail wheel to be lifted up out of the way, allowing the trail to dig into the ground and better absorb the recoil. A slim socket was affixed to the upwardly curving terminus of the trail in order to insert the end of the trail stave, which acted as a handy lever in training (traversing) the piece.

If possible, sailors in the landing force other than the gun crew hauled the landing guns when on shore in order that the gun crews not be fatigued when going into action. That ideal was not always practicable, so gun crews had to cope with the existing conditions ashore. When mounted on the field carriage, the Dahlgren howitzer was operated by roughly a twelve-man crew, a chief of piece, and a quarter-gunner. The latter, typically a petty officer, had the responsibility for the ammunition and ancillary items positioned at the gun's rear.

As previously stated, because of their weight and bulk, Dahlgren's 4-inch rifle and 24-pounder SB howitzer saw use—the former exclusively and the latter almost exclusively—as shipboard pieces. Both were mounted in broadside on wooden pivot carriages that possessed a bed sliding in recoil, similar in principle to the boat carriage for the 12-pounder howitzer described earlier.

Ammunition and Ammunition Service

The Dahlgren howitzer was first and foremost an antipersonnel weapon, or to be brutally frank, a mankiller. In consequence, the most important factors lay in the types of projectiles it fired and the rapidity in which it fired them. All three types of ammunition fired by Dahlgren smoothbore howitzers were fixed, that is, the projectile and bagged propellant were joined as one, ensuring a rapid rate of fire. A fresh crew adequately trained could get off between seven and eight rounds per minute with the gun on its field carriage, and perhaps three to five with the piece mounted in the boat. Because these guns were muzzleloaders, a wooden sabot attached to the base of the projectile held the round properly centered within the chamber of the gun. The cartridge bag had an inner and outer sleeve, with the inner sleeve tied and closing off the powder, and then the outer sleeve pulled up over the groove cut into the sabot, securing the propellant to the sabot and the projectile (both the smoothbore shell and shrapnel round being spherical). Proper stowage of the ammunition, with the weight resting on the sabot, diminished the chance of the propellant bag separating from the remainder of the fixed round.

Close-range work required the use of canister. The projectile consisted of thirty-nine cast-iron or lead balls packed in sawdust within a cylindrical tin canister—thus the name of the round—the upper end closed by a wrought-iron plate and the lower end by a wooden plug also serving as the sabot. Because the canister came apart within the bore of the gun upon detonating the propellant, no burster charge was required. The most effective range of canister lay between 200 and 350 yards from the muzzle of the piece. In both the 12-pounder and 24-pounder sizes, the projectiles were layered from the base of the canister in four courses of eight balls—in each instance seven balls surrounding a single center ball, with that single ball offset somewhat above the surrounding ones—and one upper course of seven peripheral balls with no center one. Needless to write, these sizable metal balls erupting from the muzzle of a Dahlgren howitzer had a lethal effect.

Canister Used in Dahlgren Smoothbore Howitzers

Type	No./Diam. of balls	Weight of round, iron balls*	Weight of round, lead balls*
12-pounder	39 / 1 inch	9 pounds	11 pounds
24-pounder	39 / 1.3 inches	17 pounds	22 pounds

*The weight specified is that of the complete fixed round.

In order to achieve the same shotgun effect on enemy troops assembled in the open beyond the range of canister, the gun crew made use of shrapnel, also known as spherical case shot. The name derived from its inventor, then Lt. Henry Shrapnel, who served in the British Army during the eighteenth and nineteenth centuries and eventually rose to the rank of lieutenant general. The fuse set off a burster charge, which released a large number of .65-caliber lead balls, packed in a sulfur resin matrix to prevent their deformation from the first of the two detonations they would endure. Although the balls dispersed minimally upon the detonation of the small burster charge, collectively they continued the forward momentum of the hitherto intact round. Otherwise explained, the projectile body was thus essentially a vehicle having a superior ballistic coefficient to transport the lethal payload, the metal balls, to a greater distance than they could have traveled independently, and enabled them upon release to retain sufficient velocity for their deadly effect. Only sufficient practice and experience achieved the close estimation of the intervening distance, and the correct cutting of the fuse to explode the round just in front of the body of hostile troops.

Shrapnel Used in Dahlgren Smoothbore Howitzers

Type	No./Diam. of lead balls	Weight of burster	Weight of propellant	Weight of round*
12-pounder	80 / 0.65 inch	0.8 ounce	1 pound	12/13 pounds
24-pounder	175 / 0.65 inch	1+ ounce	2 pounds	24/26 pounds

*The weights specified are those of the round, with and without the propellant charge included.

Shell, with its greater explosive payload, saw use against enemy troops either concealed or entrenched, from several hundred to well more than 1,000 yards distant. The heavy 12-pounder fired to an effective maximum range of 1,100 yards, while its 24-pounder bigger brother did so to nearly 1,300 yards. The fuse ignited when the propellant detonated and in turn set off the shell. The body of the shell fragmented with the exploding charge, scattering metal shards and splinters in every direction, though they still possessed a forward momentum, particularly if the round exploded in air.

Shell Used in Dahlgren Smoothbore Howitzers

Type	Explosive payload	Weight of round	Weight of propellant	Weight complete
12-pounder, heavy	0.5 pound	10 pounds	1 pound	11 pounds
24-pounder	1 pound	18 pounds	2 pounds	20 pounds

The shell and shrapnel for the smoothbore light 12-pounder weighed 1 pound less than their counterparts for the heavy 12-pounder, with six tenths of the weight of the propellant charge, providing a proportionately diminished payload and range. Because of that reality, the heavier piece was preferred during the Civil War and later encounters.

As in other pieces of ordnance, rifling brought both greater range and accuracy to naval boat and landing guns. The 3.4-inch or 12-pounder rifle could fire shell accurately to 1,700 yards. By design, the new Dahlgren rifled howitzers were for the most part shell guns, with shrapnel seemingly not provided. Their shells did not

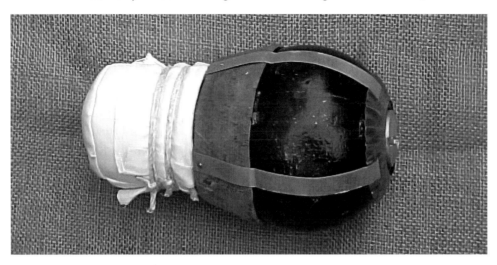

This spherical projectile is a reproduction of that fired from the army's Model 1841 12-pounder smoothbore mountain howitzer, similar to that used in the Dahlgren SB boat howitzer. It could be either shell or shrapnel, with each type painted a distinctive color. Shrapnel, or spherical case, held metal balls, whereas shell was filled with a greater gunpowder payload. (*Courtesy of Donald Radcliffe, AAA Munitions*)

require sabots, although lead cups or sleeves intended to expand upon the detonation of the propellant charge, allowing the rifling to bite into them, did adorn the elongate and ogival-pointed projectiles. There survives ample evidence that most of the shells provided by private manufacturers, particularly the Shenkl shell with its papier-mâché collar (confusingly termed a sabot), required separate loading.

Shell Used in Dahlgren Rifled Howitzers

Type	Explosive payload	Weight of round	Weight of propellant	Weight complete
3.4-inch*	0.5 pound	11 pounds	1 pound	12 pounds
4-inch*	0.86 pound	18 pounds	2 pounds	20 pounds

*Despite the officially correct distinction that rifled guns be expressed in inches, more often than not the Dahlgren rifled howitzers were designated as pounders, particularly when intermixed in field batteries with their smoothbore counterparts.

Wartime exigencies did see the Dahlgren rifled howitzers firing non-standard projectiles. On occasion they fired solid shot, particularly against enemy timber fortifications and wooden ships. Even though the use of canister was eschewed in these pieces as mutually deleterious—the balls damaged the rifling and the rifling imparted a rotary motion to the balls that degraded their effect—on occasions where the circumstance merited, canister and even grape were fired. For example, during the action at Wilson's Wharf, Virginia, on May 24, 1864, USS *Dawn* fired three rounds of canister and two stands of grape from her Dahlgren 3.4-inch rifle.

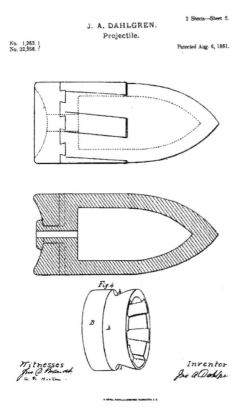

This drawing from John Dahlgren's patent application for the elongate pointed shell for use in his rifled howitzers shows the hard iron body separated from the soft lead collar or sabot (labeled Fig. 4) to be fixed to the shell's base. The collar's slightly greater diameter permitted the rifling to bite into it to impart spin. Rounds from rifled guns possessed greater range and accuracy than those fired by smoothbore pieces. John A. Dahlgren, August 6, 1861.

Ashore, shell, shrapnel, and canister rounds were transported in two ammunition chests lashed to or slung from the carriage, carrying either nine or eighteen rounds apiece; the chests were typically furnished with beckets (rope handles). The advantage of the lesser number of rounds lay in easier handling of the chest when the gun came into action. In addition, most members of the gun crew carried a round in a special leather pouch. This could be augmented to two rounds each for particularly able-bodied men when the officer in command expected the action to be prolonged, but the marching distance inland not excessive. Of course, a greater distance from the beach required a greater amount of ammunition to sustain the landing force, but under those circumstances, other means had to be found to carry the additional rounds. Gun crew performance as a function of fatigue had to be kept in mind at all times.

Primers and Fuses

The percussion primer saw the most use in firing Dahlgren howitzers. It consisted of a small goose quill tube filled with fine-grained gunpowder, capped by an explosive wafer of cartridge paper filled with fulminate of mercury mixed with a small amount of mealed powder. The tube was inserted into the vent through the reinforce. Upon being struck by the hammer, which fell upon the pull of the lanyard, the cap exploded and ignited the powder in the tube, in turn firing the main charge in the gun. Not infrequently, the army friction primer served as a back-up and in many instances the primer of choice. Its brass tube, also to be inserted into the vent, contained gunpowder. At its top, a short spur soldered on at a right angle contained a mixture of two to one parts antimony sulfide and potassium chlorate. Through this mixture ran a wire roughened on one end and twisted into a loop on the other. Tugging smartly on the lanyard attached to the loop pulled the roughened end of the slider wire through, causing sparks and igniting the chemical mixture, thereupon setting off the powder train in the shaft.

Spherical case (or shrapnel) rounds in general and shell discharged from the Dahlgren howitzers differed from most other spherical shells provided for naval service, which used the standard navy time fuse. Explosive rounds fired from the Dahlgren howitzers typically used either the Bormann time fuse or the percussion fuse.

The Bormann time fuse, invented by Belgian army officer Charles Bormann, consisted of a circular metal disk, three to one parts tin to lead, threaded to permit it to be screwed into the top of the projectile. It contained a powder arc of five seconds maximum duration for time of flight (Confederate copies, when they worked, permitted five and a half seconds), that finite time being intentionally limited to offset the natural tendency to overestimate the range. The thin metal covering over this train of powder was marked into seconds and quarter-seconds. The gunner firmly punctured the thin cover with a knife, punch, or more specialized fuse cutter at the desired time interval, commensurate with the projectile range required. The cut exposed the powder at that point to the flash from the propellant, which started the fuse burning. The zero end of the horseshoe-shaped powder train communicated with the fuse's center core of powder, which once ignited, blew off the protective bottom plate and exploded the burster charge of the projectile. Depending upon preference and circumstance, either the chief of piece or the quarter-gunner, as directed by the chief, cut these time fuses.

The percussion fuse commonly saw use when the gun commander wished the shell to explode on impact. It too was screwed into the projectile just before firing. In one form, the impact on the target thrust the metal nose cap containing fulminate of mercury back against the nipple of the detonator, fixed in place, exploding it and blowing out the fuse's bottom plug, made of lead or even cork. Most percussion fuses used inertia, in some fashion holding back a floating (movable) plunger with a cap containing mercury fulminate while the projectile remained in flight, and then upon impact causing the cap to slam forward against an anvil, detonating it. Under most circumstances, the percussion fuse proved sufficient to set off the burster charge of the round. The angle on the face of the target and the angle of the projectile path relative to the target, however, were always critical factors in successfully exploding the detonator and in turn the shell. Many a round failed to explode because it merely glanced off the target.

In summary, range, flight time of projectile, and burning time of fuse were inextricably mixed, with the best arbiters of those factors being natural aptitude, experience, and not least of all, cool nerve whenever there ensued a hot return of one's fire.

Although the navy issued the goose quill percussion primer, set off by the howitzer's firing lock striking the explosive cap, many gun crews preferred the army friction primer. This primer was inserted into the vent and then detonated by a smart tug on the lanyard attached to the loop of the horizontal spur. The roughened wire pulled through the ignition mixture, causing sparks and setting off the powder train in the vertical tube and, in turn, the propellant charge. (*From the author's collection*)

At an estimated success rate of 75 percent, the Bormann time fuse was likely the most reliable of the navy's fuses of the type. A cut was made just to the right of the number of seconds reckoned, to a maximum of five seconds. The cut exposed the powder at that desired burning time, which corresponded to the time of flight and the range desired. When the crew fired the gun, the exploding propellant flamed the fuse's powder train, which burned to the powder-filled core of the fuse and thereupon blasted through to the burster charge. Dahlgren, *Boat Armament*, 1856.

References Consulted

Barrett, E., *Naval Howitzer. Instructions Condensed for the Volunteer Officers of the U.S. Navy* (New York: Barrett, 1863), pp. 57–61, 63, 65–71.

Brandt, J. D., *Gunnery Catechism, As Applied to the Service of Naval Ordnance* (New York: Van Nostrand, 1864), pp. 79–86, 91, 93–95, 98, 133–41, 149–52, 182–85, 192, 194, 196–97.

Coggins, J., *Arms and Equipment of the Civil War* (Garden City, NY: Doubleday, 1962), pp. 62, 67, 74, 76–77, 80–83, 144–45.

Dahlgren, J. A., *Boat Armament of the U.S. Navy*, 2nd edit. (Philadelphia: King and Baird, 1856).

General Order, December 17, 1850, issued by William A. Graham, Secretary of the Navy, *in* US National Archives, Record Group 80, General Records of the US Navy Dept., Washington, DC.

Luce, S. B., *Instruction for Naval Light Artillery Afloat and Ashore* (New York: Van Nostrand, 1862), pp. 12–15, 19–21, 23.

Manucy, A., *Artillery through the Ages: A Short Illustrated History of Cannon, Emphasizing Types Used in America*, Interpretive Series History No. 3, National Park Service (Washington, DC: Govt. Print. Office, 1949), pp. 15, 25–27, 66–71.

Olmstead, E., Stark, W. E., and Tucker, S. C., *The Big Guns: Civil War Siege, Seacoast and Naval Cannon* (Alexandria Bay, NY: Museum Restoration Service, 1997), pp. 103–10, 177–78, 181, 190–92, 204–05.

Paulson, B., Personal communications, May–August 2014.

Peterson, H. L., *Round Shot and Rammers* (Harrisburg: Stackpole, 1969), pp. 45, 107, 110–19.

Ripley, W., *Artillery and Ammunition of the Civil War* (New York: Van Nostrand, 1970), pp. 87–91, 107, 379.

Tucker, S. C., "The Dahlgren Boat Howitzer," Technical Report, *Naval History*, Fall 1992, 6(3): 50–54.

US National Archives, Record Group 74, Records of the Bureau of Ordnance, Washington, DC.

2
Dahlgren Howitzers in Action During the 1850s and the Civil War

The 1850s

In the decade before the onset of the American Civil War, the US Navy employed Dahlgren boat and landing howitzers most frequently and aggressively in the coastal waters of Africa and Asia. The period also saw one of the most auspicious accomplishments by the United States Navy, which despite in the end being an entirely peaceful occasion, involved the participation of a substantial number of naval landing howitzers. The years of civil war soon following would demand a great deal more application of Dahlgren's bronze artillery pieces.

On September 4, 1853, Commo. Isaac Mayo, commanding the African Squadron, arrived off the mouth of West Africa's Cavalla (or Cavally) River in the squadron's flagship, the sail frigate *Constitution*. The river runs into the Gulf of Guinea and serves as the boundary between Liberia and Ivory Coast. In 1853 it also separated the territories of two frequently warring tribes, the Barbo and the Gribo. The main function of the American squadron was the suppression of the slave trade, frequently in consort with ships of the Royal Navy. On this occasion, however, Old Ironsides came here at the request of the magistrate of the American colony nearby, because the warfare had much inconvenienced American trade in the area.

Despite the ready acquiescence of the Gribo to Mayo's offer to arrange a peace, the far more warlike Barbo curtly rebuffed it and threatened to kill the messenger. The commodore decided to force the issue, but his options were limited because of the violent surf pounding the shore. On the 5th, he put forth with an armed party in five boats. When a further entreaty under a white flag also failed, Mayo had several signal rockets fired as a warning. While promising, the pyrotechnics still did not produce the desired end. Mayo thereupon ordered Lt. John De Camp, commanding the launch and its 12-pounder heavy smoothbore (SB) howitzer, to fire shrapnel rounds above the Barbo town, about 600 yards distant. After thirty such projectiles had been fired in rapid succession, though with no blood yet shed, a white flag appeared and peace talks were thereafter held aboard *Constitution*.

In his action report, De Camp praised the accuracy and range of the Dahlgren howitzer, which by his own admission had exceeded his expectations. Apparently the compressors on the boat carriage had not been fastidiously tightened during the rapid firing, and Mister De Camp urged that officers engaged thereafter in such undertakings ensure that those necessary devices be thoroughly screwed down before each shot, particularly if a heavy sea was running.

Much has been written about Commo. Matthew C. Perry's two uninvited visits to Japan in the mid-nineteenth century and what enormous changes the Convention of Kanagawa wrought on the history of Japan and that of the civilized world in general. Although both Perry and the government of the United States wished a peaceful resolution so that commerce could be carried on with that insular and isolationist nation, Perry's East India Squadron was fully prepared to use force in opening up Japan to the West. The ships of the squadron were heavily armed, as were the sailors and marines that accompanied Perry and his officers whenever they came ashore. On those occasions, their armament included a number of Dahlgren bronze boat and landing howitzers.

Commodore Perry brought four warships to Edo (or Yedo) Bay—now Tokyo Bay—when he first visited Japan in July 1853: sidewheel frigates *Mississippi* and *Susquehanna*, and sail sloops-of-war *Saratoga* and *Plymouth*. After much negotiation, the initial conference between the Americans and high-ranking representatives of the shogun—the Tokugawa shogunate ruled Japan until overthrown in favor of the emperor in 1868—took place on the 14th. Not knowing if the meeting would be a hostile one, Perry had the two steam frigates move as close to shore as their draft allowed, with their decks cleared for action and their Dahlgren howitzers mounted in boats alongside, ready to move shoreward in case of treachery. This conference proved inconclusive, but Perry promised to return in the spring.

When the larger squadron of nine ships made its reappearance in early March 1854, it brought a dozen Dahlgren SB howitzers: one 24-pounder and the remainder 12-pounders. A substantial armed escort accompanied the American party landing at Yokohama on the 8th, backed up by a few Dahlgren howitzers on field carriages. Others mounted in the bows of the ships' boats close inshore fired two salutes: twenty-one guns in honor of the emperor and seventeen guns for the Japanese high commissioner. Nearly three weeks later, the visitors from the United States entertained a large Japanese diplomatic party aboard. In addition to the copious food and spirit offered up, the American hosts lowered a boat mounting a Dahlgren howitzer, which to the delight of their guests was repeatedly discharged.

Close to the end of Perry's visit, as a result of seeing the various evolutions of the boat and landing howitzers, the Japanese asked for the gift of three of the bronze guns and three launches on which to mount them. The Americans gave them one gun. In a letter to the secretary of the navy written on April 4, 1854, the commodore declared that the sloop-of-war *Saratoga* was about to return home—she sailed that morning—carrying the signed but not yet ratified treaty. He indicated that the howitzer would be taken from her. But in his history of the voyage published in 1856, Perry specified the source as the frigate *Mississippi*. Likely he chose that alternative so *Saratoga*'s

Commo. Matthew Calbraith Perry (1794–1858) is shown in this well-known photograph by Matthew Brady not long before the commodore's death. Perry's fame extended far beyond his mid-1850s expeditions that opened Japan to the West. Long a proponent for improved naval education and modernization of the fleet, he became "the father of the steam navy." Perry was not simply a thinker, however, for as a young officer he had seen action in the War of 1812 and the Second Barbary War. Photograph by Matthew B. Brady, *c.* 1857–1858.

This 1850s lithograph depicts the meeting between American and Japanese diplomats at Yokohama in early March 1854. Three Dahlgren landing howitzers are evident: one standing on the meeting ground, one being hauled by its crew up the embankment, and one being unloaded by sailors from a ship's launch. Frontispiece of Dahlgren, *Boat Armament*, 1856.

departure would not be delayed. Townsend Harris, the successful and influential American minister to Japan in the late 1850s and early 1860s, reported during his tenure that the Japanese had fabricated a thousand copies of the Dahlgren howitzer. That seemingly fantastic number in so short a time may, of course, be apocryphal.

Soon after his second visit to Japan, Perry signed a similar convention with the then-autonomous kingdom of Loo-Choo, part of the Ryukyu chain of islands. Two Dahlgren howitzers, one each from the steam warships *Mississippi* and *Powhatan*, accompanied Perry and his inevitably well-armed escort while ashore at Great Loo-Choo (now Okinawa) on July 11, 1854.

When most of the East India Squadron sailed for Japan the second time, sloop-of-war *Plymouth* stayed behind in Chinese waters, but hardly remained idle. During the Taiping Civil War, serious misbehavior by imperial troops near the international settlement at Shanghai became increasingly aggressive. In early April 1854, the consuls of Great Britain and the United States, along with the commanding officers of screw sloop HMS *Encounter* and USS *Plymouth*, Capt. George O'Callaghan, RN, and Cmdr. John Kelly, USN, made a formal representation to the Chinese general in command that he pull his troops back in order to stop the annoyances and depredations. If he failed to comply, the Anglo-American officers promised they would force the issue. Upon the general's refusal to withdraw, the naval commanders landed a force of more than 200 bluejackets and marines (150 British, 60 American) from the two ships already named, as well as from brig-rigged sloop HMS *Grecian*. A trainband from the European quarter of the city equipped with two privately owned fieldpieces and reinforced by thirty American merchant seamen armed with a bronze howitzer, thereupon strengthened the Anglo-American naval force. There soon followed the action at Muddy Flat, adjacent to a tributary of Soochow Creek.

Lt. John Guest commanded the *Plymouth*'s Dahlgren landing howitzer, firing forty to fifty rounds of shell, shrapnel, and canister. In his action report, he wrote, "The fixed ammunition was perfect; not a single shell failed to burst, not a fuze or a tube disappointed us; and, consequently, the officers and men were inspired with perfect confidence in the gun, both as a means of assailing the enemy and of defence when attacked." Guest used Bormann time fuses in the shells and shrapnel projectiles and marveled at their precision. He also expressed great satisfaction in the performance of canister in both raking the top of the Chinese breastworks and repelling an attack by a large force of infantry in the field. After the light artillery had fired for fifteen or twenty minutes, the Anglo-American naval infantry moved forward under moderate fire to capture the Chinese positions, which they leveled after the withdrawal of the imperial troops. Among the several casualties suffered by the attackers was R. H. Pearson of Newburyport, Massachusetts, master of the merchant vessel *Rose Standish*, mortally wounded.

Showing the flag remained the far more typical duty for the US Navy on the China coast during the mid-1850s, although some minor actions took place against bandits ashore and pirates afloat. More often than not, these episodes developed in concert with vessels of the Royal Navy, one service calling upon the other in about equal measure. Always appreciated by the British seamen were the Dahlgren boat howitzers the Americans almost invariably brought along, and with which they had by then become pretty straight shooters. The minor nature of these altercations ended in early August 1855 with a hair-raising fight in Tai-O Bay off Kowloon.

The crews of the sidewheel frigate USS *Powhatan* (Capt. William McCluney) and screw sloop HMS *Rattler* (Cmdr. William Fellowes) engaged a large fleet of "piratical junks" that had been committing depredations against coastal commerce. In addition, the paddle steamer HMS *Eaglet*, a chartered auxiliary warship, was employed to tow the pulling boats (also termed rowing boats) into action in the shallow shoal water. As the battle unfolded, boats of both services chased a number of large armed junks, with the Americans banging away with their howitzers in the bows, and the sailors and marines thereafter boarding the hostile craft. During the course of those events, the crew of one junk blew up their vessel, and killed or wounded a number of the boarders, both American and British. At the end of the day, much of the pirate fleet had been destroyed or captured, with hundreds of their crews dead. Casualties to the Anglo-American naval force numbered fewer than twenty.

Lt. Robert Pegram, who commanded the American boat contingent, had the occasion to write, "In connection with this report, I deem it not inappropriate to express my great satisfaction at the performance of the 12-pdr. boat howitzer of Lieutenant Dahlgren's construction. For rapidity and precision of fire, facility of working, certainty of execution, and completeness of arrangement of ammunition, I have never seen any thing to compare with them." In their respective reports on the action, Pegram and Fellowes had complimentary things to write about the other's coolness and courage under fire. Pegram would win later fame in the Confederate States Navy.

The sidewheel frigate USS *Powhatan* is pictured at anchor in an Asian port, flying her commissioning pennant from the mainmast. She took part in four episodes recounted in this book: Perry's second visit to Japan, March 1854; the battle against pirates near Kowloon, China, August 1855; Lt. David Porter's secret mission to Pensacola, FL, April 1861; and the second attack on Fort Fisher, NC, January 1865. The artist/engraver is unknown. (*NavSource Naval History Photo Archives*)

Unrest continued in China, with the next major eruption occurring in October 1856, in Canton. Responding to an urgent request from the American consul, the sail sloop-of-war *Portsmouth* proceeded there from Whampoa, and her captain, Cmdr. Andrew Foote, led eighty-three men ashore. His landing force included a Dahlgren howitzer. By early November, the original force had been supplemented by men from the sail sloop *Levant* and the screw frigate *San Jacinto*, the latter having Commo. James Armstrong aboard. As Foote returned up the Pearl River in mid-November, with plans to withdraw the American landing force, one of the Chinese barrier forts fired on his boat flying the national colors, and the situation escalated from there. Attempts to negotiate with the Chinese government in Canton failed, so on the 20th, *Portsmouth* and *Levant* commenced a heavy bombardment of the barrier forts, which was returned in kind by the forts' guns. Foote again commanded the landing force, this time of nearly 300 bluejackets and marines, backed up by three Dahlgren howitzers. The landing was supported by the ships' guns.

The Americans readily captured the first barrier fort, but soon thereafter 3,000 Imperial Chinese troops made repeated attempts to retake it. All failed due to accurate musketry and howitzer fire. By the 22nd, the Americans had seized all four of the forts defending the harbor. From the following day through early December, the sailors and marines systematically destroyed those forts, blowing up their granite walls, and first spiking their guns, then rolling them into the river. The Americans withdrew on the 6th. In this violent episode, they had suffered six men killed and between twenty and thirty wounded; the Chinese casualties numbered in the hundreds.

In response to John Brown's insurrectionist raid on the federal arsenal at Harpers Ferry, then Virginia, in October 1859, a detachment of marines under the direct command of 1st Lt. Israel Greene and the overall command of Bvt. Col. Robert E. Lee, US Army, departed in two trains for that place. Another officer who would win fame wearing grey in the impending war, 1st Lt. James E. B. ("Jeb") Stuart, accompanied Lee as his aide. Not knowing what size force they would face or how well armed, the marines took two Dahlgren howitzers, along with some rounds of shrapnel. [Marine Corps historian Bernard C. Nalty declares that they were 3-inch howitzers, but because there was no such thing in either navy or army service, and the marines brought them from the Washington Navy Yard, they were almost certainly Dahlgrens.] Greene did not deploy the howitzers, but instead, he and his men stormed the locomotive house in which Brown and his followers had taken refuge with some hostages. After a short hand-to-hand struggle, the marines killed or captured the insurrectionists. With the coming of the war, Greene, too, would go South.

The War of Rebellion

If the Dahlgren boat and landing howitzers had vividly suggested their potential during the 1850s, they proved their real worth during the four years of civil war that followed. Throughout the course of that war, Union and Confederate navy and army units armed with these pieces expressed great appreciation for their worth.

In addition to the four smoothbore types already in service during the previous decade, a 3.4-inch rifled howitzer utilizing the same casting as the heavy 12-pounder smoothbore and a trunnioned 4-inch rifle intended strictly for shipboard use came into being during the war. Demand for Dahlgren's howitzers of all types so lagged behind the production capability of the gun shops at the Washington Navy Yard, however, that two commercial armaments firms in Massachusetts received contracts to fabricate these handy weapons: Cyrus Alger in Boston and the Ames Manufacturing Company in Chicopee Falls. Although each company made a relative few of other types, they concentrated on the production of the heavy 12-pounder boat howitzer (Alger made fifty-seven and Ames 202) and the 24-pounder shipboard piece (Alger constructed 302 and Ames 673). Dahlgren's rage over what he perceived as poaching on his personal domain may be fairly described as apoplectic. He declared in writing, "My private rights have been most villainously invaded." Indeed, he claimed the US Navy complicit in that invasion and described outright the founders at Alger as "rasculs."

A limited number of a puddled wrought-iron 3.4-inch rifled howitzer of otherwise identical pattern to the Dahlgren rifled howitzer would be provided to the Union army by the Wiard Iron Works in Trenton, New Jersey. At Richmond, Virginia, the Tredegar Iron Works, now in Confederate hands, manufactured an indeterminate number of 12-pounder bronze boat howitzers of both light and heavy patterns, and a smaller number of 24-pounders. How they performed relative to those of original manufacture remains uncertain.

With the secession of the Southern states in general, and the capture almost intact of the Gosport Navy Yard, Norfolk, Virginia, with its enormous store of naval matériel in particular, the US Navy lay in a diminished state. The vastly weakened service—one fifth of its officers had gone South—owed its resurrection to the enormously capable Secretary of the Navy Gideon Welles and his energetic deputy, Assistant Secretary of the Navy Gustavus Fox, a former navy officer. Not only did Welles order new warships from both naval and commercial shipyards, but he and Fox began to buy up practically everything that would float and carry cannon. By war's end in 1865, the Union fleet consisted of more than 500 warships, most of them specialty craft, from sleek propeller-driven cruisers on the high seas to double-ended, paddle-wheeled, ex-ferryboats providing the army supporting fire from shallow creeks. For sheer diversity, such an assortment of wartime naval vessels would not be seen again by the US Navy for another eighty years.

Opposing Welles and Fox was the Confederacy's Secretary of the Navy Stephen R. Mallory, not only a capable administrator but a master of improvisation. By juggling his often meager resources, Mallory contracted to build in local yards or purchase from abroad more than three score of ironclads and numerous commerce raiders. But, the South had few such natural resources, few shipyards, and with time, few dollars. However gallant his officers and brave his crews, what Mallory could *not* do was build a fleet.

A century after the event, author Jack Coggins neatly summarized the naval war against the Confederacy as "a story of continuous and growing harassment and encroachment from the sea: seaports blockaded, fortresses reduced, river barriers

With one of their number assuming a martial stance, these midshipmen repose beside a Dahlgren 12-pounder landing howitzer on the lawn of the Atlantic Hotel, Newport, RI, at that time quarters of the US Naval Academy. Two weeks after Fort Sumter fell on April 13, 1861, with the uncertainty of whether Maryland would secede, the navy relocated the academy from Annapolis to Fort Adams, a harbor defense work at Newport. In August, the service leased the hotel, with its more adequate spaces for rooming, messing, and teaching. (*US Naval Academy Archives*)

forced, and the great inland water highways taken and used by the enemy." To that one may only add the many lesser inland waterways—the creeks, branches, small rivers, and inlets—perhaps not permanently taken or even for long, but nonetheless allowing the Union navy penetration for the damaging raids that taken as a whole sapped the strength of the South.

In carrying out these forays, the US Navy frequently made use of Dahlgren howitzers to great advantage. Although the navy regularly undertook landing and cutting-out parties where the presence of a small artillery piece or two was not only a useful adjunct, but perhaps the deciding factor, Dahlgren howitzers of all sizes saw far more naval combat as shipboard guns. Particularly aboard the gunboats of the riverine flotillas, such guns fired almost weekly against hostile sharpshooters, and infantry and cavalry units, as well as provided counterbattery fire against enemy artillery. Indeed, during the early months of the war, these howitzers aboard the barely satisfactory gunboats put into emergency service often constituted their sole armament.

As secessionist sentiment heated up in the South and then boiled over into outright war, Dahlgren howitzers were put to immediate use. During the uneasy peace existing even after the bombardment and surrender of Fort Sumter, the fate of Fort Pickens, a large masonry work on Santa Rosa Island that commanded the entrance into Pensacola Bay, Florida, hung in the balance. On the night of April 12, 1861, 100 marines from the naval vessels lying offshore, commanded by Lt. John Cash, reinforced the troops

Previously serving as a New York harbor ferryboat, double-ended sidewheel gunboat USS *Hunchback* is photographed up the James River in Virginia. The walking beam aft of her funnel is typical for paddlewheel steamers. Her maximum wartime armament included four 9-inch Dahlgren smoothbore cannon, one 100-pounder Parrott rifle, and two Dahlgren landing howitzers: a 3.4-inch rifle and a 12-pounder SB. Three of the Dahlgren nines and the Parrott rifle are visible in this image. (*NavSource Naval History Photo Archives*)

The officers and enlisted crewmen of river gunboat *Hunchback* posture shamelessly before Matthew Brady's camera. The breech ends of two of her 9-inch Dahlgrens may be seen on the lower deck, and her two Dahlgren landing howitzers appear on the upper deck, the 12-pounder SB on the right and the 3.4-inch rifle on the left. Note the difference in mass surrounding the howitzers' bores. Photograph by Matthew B. Brady. (*Library of Congress*)

of the fort's garrison. Bluejackets also went ashore to bolster the marines, but soon returned to their ships.

In mid-April, Lt. Cmdg. David Dixon Porter, who would attain the rank of rear admiral less than two years later, arrived at Pensacola Bay temporarily commanding the sidewheel steam warship *Powhatan* (she had seen service in Asian waters). His was "a secret and desperate expedition" on the direct orders of President Abraham Lincoln to enter the harbor and prevent a Confederate naval attack on Fort Pickens. He was dissuaded from that hazardous mission, however, by Capt. Montgomery Meigs, acting on the behalf of Bvt. Col. Harvey Brown, who commanded the army contingent defending the fort and the island.

These officers convinced Porter instead to support the army, and on the 23rd, he ordered a Dahlgren howitzer mounted in one of the ship's launches to investigate heavily manned rebel boats that were reconnoitering the island's beaches. If they appeared to be preparing to land, Porter instructed the boat crew to fire on them. Two of the other warships present also lowered armed boats for that purpose. A 12-pounder howitzer was returned to the ship on May 1, after serving ashore with a naval landing party. Twelve days later, *Powhatan* transferred two heavy 12-pounder howitzers, along with their boat and field carriages, to the 500-ton, sidewheel steamship *Philadelphia*, commandeered the previous month. In the first month of the war, therefore, *Powhatan*'s Dahlgren howitzers played the two principal roles for which they had been intended. Equally so, the episode typifies the excellent cooperation between the Union navy and army during the War of Rebellion, with Dahlgren 12-pounder howitzers often having a substantial part in joint operations.

Also of relevance, Porter or any other lieutenant so assigned held the rank of lieutenant commanding only as long as he commanded a vessel. In February 1862 the rank structure of the US Navy changed markedly, thereupon to include among others, the regular rank of lieutenant commander.

Gunners in the Union and Confederate armies found the Dahlgren howitzers useful, too. Although unclear whether a gift or a loan, two such pieces found their way to the 71st New York State Volunteer Infantry Regiment not long before the First Battle of Bull Run (First Manassas). That regiment mustered for three months' service in mid-April 1861 and arrived in Washington the following month. An artillery unit assigned to the regiment in early June became I Company, commanded by Capt. Augustus Van Horne Ellis. Some of the companies bivouacked at the Washington Navy Yard, which they had responsibility for guarding, along with the Long Bridge spanning the Potomac River to Alexandria, Virginia. While at the yard, the members of I Company received training on the Dahlgren howitzer, and in early July, Cmdr. John A. Dahlgren issued two of his 12-pounders to the unit. When the unit departed from the capital city later in the month, the howitzers accompanied it.

The 71st New York became part of Col. Ambrose E. Burnside's 2nd Brigade, 2nd Division, and set off on the march to Centreville, Virginia, with I Company drawing its howitzers sailor-fashion by drag ropes. When the brigade deployed to fight on July 21, I Company was the second artillery battery to go into action, ultimately becoming part of twenty-four fieldpieces in that sector of the Union lines, of which three quarters were rifled

It is tempting to think these Union soldiers are members of I Company, 71st New York State Volunteer Infantry, posing with one of their Dahlgren 12-pounder SB howitzers not long before the battle of First Bull Run (First Manassas). The fifth man from the left is a sailor, and the second man from the left may be a naval officer, both perhaps gunnery instructors. No information is available on the date or source.

guns. When two units armed mostly with rifled cannon were ordered forward—ultimately to their destruction—the howitzer company remained to support the 2nd Brigade infantry as the seesaw battle progressed. Upon the retreat of the federal army, I Company retired in good order with its guns. But coming upon the choke point where the Warrenton Turnpike bridge spanning Cub Run had been blocked, and indeed remained under heavy enemy fire, I Company was forced to abandon its pieces unspiked.

Along with other artillery pieces previously belonging to the Union army, the Confederates captured the two Dahlgren howitzers. Although the field artillery contingent of South Carolina's Hampton Legion cast covetous eye on them and offered in exchange two pieces of similar size made by the Tredegar Iron Works of Richmond, persons having the final say issued the Dahlgren howitzers instead to the Rowan Artillery (Reilly's Battery) of North Carolina.

As previously implied, the instances of Dahlgren boat howitzer service in various roles during the American Civil War are too numerous to chronicle. Several additional examples will be provided, with an eye to giving an equal balance to joint navy-army, navy-only, and army-only operations, and in the first two instances, whether as landing, boat, or shipboard guns. In accomplishing that end, however, the examples are taken solely from east coast operations. To be sure, other theaters saw plentiful use of Dahlgren bronze howitzers, including in operations on that long and broad south-running river, the command of which remained so critical to both sides.

Shortly after Rear Adm. David Porter took command of the Mississippi River Squadron in October 1862, he requested 200 Dahlgren howitzers for service on his vessels. Indeed, the most common shipboard gun on the lightly armored or "tinclad" riverine gunboats became the Dahlgren 24-pounder howitzer, because of its weight seeing almost exclusive use as a shipboard gun. One month after he had assumed command, Porter wrote to Rear Adm. Andrew Foote, who had previously led the squadron.

> I am now hard at work fitting out a light-draft semi-ironclad set of steamers, drawing not more than 30 inches when deep, and armed with six and eight 24-pounder Dahlgren howitzers. They are really formidable little craft.... Think how serviceable they would have been to you on the Tennessee and Cumberland Rivers. I have one little fellow up the Tennessee, and she has it all her own way.

However prone to exaggeration and self-inflation Porter could be, and however feigned the sincerity of his last remarks were, he remained truly proud of the naval force he had accumulated to win the vast Mississippi River and its often sizable tributaries. Meanwhile, on the southeastern seaboard, Dahlgren's howitzers were taking the war to the rebels.

In early 1862, the naval barricade of Southern ports still suffered a multitude of holes for the opportunistic blockade runner to slip through. After discussion at the highest levels, the services agreed jointly to deprive the Confederacy of the refuge inside the North Carolina Outer Banks, which included Pamlico, Roanoke, Croatan, and Albemarle Sounds, and the ports fronting on them. The army handed the job of capturing these assets to Ambrose Burnside, now a brigadier general, and his Coast Division of three infantry brigades from the Northeast totaling about 13,000 men. Burnside considered this force, although assigned minimal artillery, more than sufficient. Beginning in late January, a small fleet of transports, supported by both navy and army gunboats, and commanded by Flag Officer Louis Goldsborough, passed with great difficulty across the swash, the sea-washed sandbar obstructing the outer barrier west of Cape Hatteras, into Pamlico Sound. The army gunboat *Zouave*, the floating battery *Grapeshot*, and the steamers *City of New York* and *Pocahontas* went down during a winter gale, for which these waters are infamous.

For the amphibious landing, the senior officers selected the western side of Roanoke Island, facing Croatan Sound. On February 7, in just hours, 12,000 men came ashore at Ashby Harbor. Steam launches pulled long strings of rowing boats laden with troops, which they cast off in a manner identical in principle to the children's game of snap the whip, thus propelling those boats toward the landing. Offshore, the gunboats provided supporting fire. In addition to the naval gun crews, members of the 1st New York Marine Artillery aboard them fired "semi-steel"—actually puddled wrought-iron—Wiard ordnance, including that manufacturer's Dahlgren-look-alike boat howitzers. It rained all night, but the federal soldiers took it in good stride.

On the following day, as the Union regiments pushed north, they soon encountered a large redoubt—termed by the invaders the Three-Gun Battery and by the defenders Fort Defiance—fortified in front by a wet ditch and an abatis of felled trees with sharpened branches, and more difficult yet, protected on both flanks by an extensive cypress swamp. As several New England regiments entered the often waist-deep

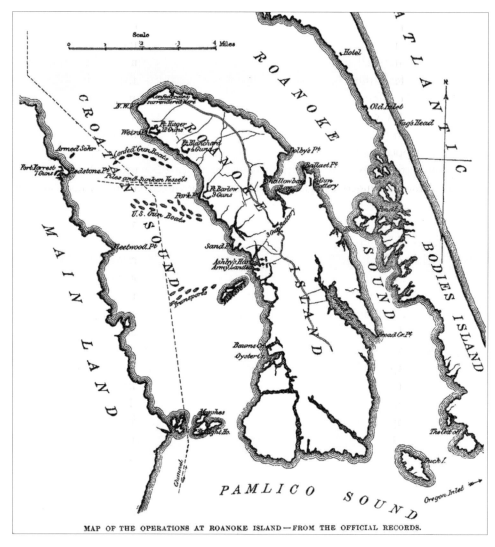

MAP OF THE OPERATIONS AT ROANOKE ISLAND—FROM THE OFFICIAL RECORDS.

Roanoke Island lies at the confluence of four sounds inside North Carolina's Outer Banks. In addition to the three shown in this chart, Albemarle Sound is immediately to the north. On February 7, 1862, Brig. Gen. Ambrose Burnside landed his three brigades, entirely unopposed, at Ashby Harbor on the island's west side, facing Croatan Sound. *Official Records of the Union and Confederate Navies in the War of Rebellion*, Series I, Vol. 6.

morass on either side of the causeway in order to flank the rebel stronghold, Brig. Gen. John Foster, commanding the division's 1st Brigade, called up the battery of six Dahlgren boat howitzers taken from the launches involved in the landing, in order to support the infantry assault.

The naval battery, commanded by seventeen-year-old Midshipman Benjamin Porter, consisted of both smoothbore and rifled pieces. The gun captain in most instances was a naval noncommissioned officer, including Acting Master's Mate J. B. Hammond. The gun crews, however, came from the Union Coast Guard (later redesignated the 99th New York Volunteer Infantry) and the 9th New Jersey Volunteer Infantry. Under heavy counterfire, the battery continued shooting until it was low on ammunition and thereupon was instructed to cease fire and husband what remained. The smoothbore howitzers had fired shell, shrapnel, and canister, the rifled pieces only shell and grapeshot. The battery suffered three men killed and five wounded, two of them seriously.

Thereafter the infantry attack proceeded, with the final winning assault made from both flanks, followed by a frontal bayonet attack by Hawkins' Zouaves (9th New York Volunteer Infantry), shouting their battle cry, "Zou! Zou! Zou!" and giving General Burnside the appearance of "a sea of red caps" because of the narrow front. (More will be written shortly on this colorfully attired regiment.) The three unspiked cannon in the captured redoubt included a single 24-pounder boat howitzer on a heavy field carriage. It would seem the rebels, too, appreciated Dahlgren's handy little demons.

The Confederate island garrison of a few thousand men surrendered shortly afterward. In their reports on the action, Generals Burnside and Foster and Flag Officer Goldsborough were lavish in their praise of the seemingly fearless Porter and Hammond in enduring heavy hostile fire. Burnside went on to capture his other essential objectives of the expedition, most of which stayed in Union hands for the duration of the war. Indeed, his star would never burn more brightly in that conflict. Most notably, the US Army, with essential naval support, had carried out what would be the most successful amphibious operation of the Civil War.

What had happened on the coast of North Carolina, however, at least for the moment, brought little effect on the war being prosecuted elsewhere. After yet another federal defeat on the banks of Bull Run, in early September 1862, Gen. Robert E. Lee led the Army of Northern Virginia on the first major Confederate incursion into Union territory. His being turned back after the battle of Antietam, near Sharpsburg, Maryland, with terrible losses to both sides—made all the more vivid by the battlefield

Seen here as a major general, Ambrose E. Burnside (1824–1881) never enjoyed odds more in his favor than during his successful amphibious assault on Roanoke Island, NC, in early February 1862. Burnside's skill during the battle of Antietam seven months later proved less stellar, and his brief tenure thereafter as commander of the Army of the Potomac led to the disaster at Fredericksburg. Following the war, he served as governor of and then US senator from Rhode Island. His impressive facial hair became the eponymous "burnsides," in later years transposed to "sideburns." *Biographical Cyclopedia of Representative Men of Rhode Island*, 1881, engraving made from an 1863 photograph.

photographs of Alexander Gardner and James Gibson that shocked the sensibilities—remains well known. Perhaps hidden in the carnage of that one-day affray on September 17 is the role of Dahlgren boat howitzers at the hands of K Company, 9th New York Volunteer Infantry, and on the opposing side, of Grimes' Battery from Portsmouth, Virginia.

Whiting's Battery, or K Company, of Hawkins' Zouaves, had been issued both rifled and smoothbore Dahlgren 12-pounder boat howitzers and divided into two sections, one with three smoothbore pieces under the battery commander, Capt. James Whiting, and the other with two rifled howitzers under his second-in-command, Lt. Richard Morris. From the lessons learned at First Bull Run, the guns were now drawn by horses. With the greater part of Maj. Gen. Ambrose Burnside's IX Corps having encountered terrible difficulty in seizing Rohrbach Bridge—the southernmost of three such structures spanning Antietam Creek near Sharpsburg—Brig. Gen. Isaac Rodman, commanding its 3rd Division, received orders to cross at a ford downstream. Rodman in turn directed Whiting's Battery to support the crossing of his division's two brigades. No small problem lay in the imposing bluff on the west side of the creek, rising to nearly 100 feet at its highest elevation, with Confederate infantry well protected behind large trees and a stone wall atop it. This bluff commanded many of the crossing points.

The precise location of the ford had not been specified, so the first effort took place at an unsuitable one not far down the creek. When the crossing failed at that location, a reconnaissance found the far more practicable Snavely's Ford a mile or so downstream. The battery deployed on a hill just below the ford, and protected by the 8th Connecticut Volunteer Infantry, opened fire against the Georgia troops on the far side of the creek to support the successful crossing. For a while at least, Morris's section of two 3.4-inch rifles fired counterbattery against Capt. Benjamin Franklin Eshleman's battery of four guns, part of the Washington Artillery from Louisiana.

In another part of the field, Grimes' Battery, or more formally the Portsmouth Battery from Virginia under the command of Capt. Cary Grimes, included two Dahlgren 12-pounder howitzers among its four guns. The battery served as part of a Confederate artillery battalion, with Captain Grimes as its acting commander, which was firing at enemy infantry from the vicinity of the Piper Farm's stone barn. After things got a bit hot, the batteries began to withdraw to safer ground, but while so doing, a bullet hit Grimes, who was mounted and offering an obvious target. As his men carried him from the field, another bullet struck, mortally wounding him.

A typically perilous engagement of Union naval warships with hostile forces ashore took place up the Rappahannock River in early December, just prior to the Union defeat at the battle of Fredericksburg. Four gunboats of the Potomac Flotilla's 2nd Division, under the temporary command of Acting Master William Shankland—the screw-driven *Currituck* and *Anacostia* and the sidewheelers *Cœur de Lion* and *Jacob Bell*—while anchored on the north side of the river at Port Conway, came under artillery fire from an eminence above Port Royal named Jack's Hill, on the river's opposite side. Unbeknown to Shankland and the commanding officers of the other gunboats, they would be confronted that day by two of the most capable artillery officers in the Confederate States Army.

The officer on the hilltop was Capt. Robert Hardaway, and his gun a British-made 12-pounder Whitworth breechloading rifle, likely having the greatest accuracy and longest range of any field gun of similar size in either army. At 4:15 p.m., Hardaway opened an accurate fire from his elevated position nearly 3 miles from the anchorage of the Yankee gunboats. The federals replied almost at once with guns of various sizes, including both rifled and smoothbore Dahlgren 12-pounder howitzers. The shipboard ordnance fired shell against the battery and canister against infantry in Port Royal, but their guns lacked both the reach and accuracy of the Whitworth.

With Hardaway making things rather too hot for the gunboats, Shankland in council with the other captains decided to shift to a more protected anchorage 3 miles downriver. While proceeding there, the gunboats came under fire from Confederate batteries behind Pratt's Landing commanded by Maj. John Pelham. In this exchange, *Jacob Bell* was hit once by a 12-pounder rifle shell, which did minimal damage, while one of Pelham's gunners lost a leg to grapeshot, a wound from which he subsequently died. Of interest is that *Cœur de Lion* had both a rifled and a smoothbore 12-pounder Dahlgren howitzer aboard, the former firing seventeen shells and the latter five rounds of canister during the engagement.

The murder of two seamen—one white, one black—from the sailing brig USS *Perry* accompanied the botched attempt in early December 1863 to destroy a blockade runner sheltering in Murrells Inlet, South Carolina. John Dahlgren, promoted to rear admiral and transferred from the Ordnance Bureau to command the South Atlantic Blockading Squadron, wanted retaliation. In consequence, near month's end, six of the blockading vessels, 150 seamen, 100 marines, and four boat howitzers, to fire mostly canister and grape, were to undertake the next expedition. Because the opposing force consisted mostly of cavalry, the intention was to issue signal rockets to the landing boats to fire at the Confederates and frighten their horses.

An end-of-the-year storm, however, put a halt to that plan. The expedition commander, Capt. Joseph Green, at that time commanding the screw sloop *Canandaigua*, but now on board the small screw gunboat *Mary Sanford* for this inshore raid, reduced the remainder of his flotilla to the larger screw gunboat *Nipsic* and the sidewheel tug *Daffodil*. On the first day of 1864, pursuant to orders, Cmdr. James Spotts took *Nipsic* in alone, and disembarked forty seamen in two launches, each armed with a 12-pounder boat howitzer, and thirty marines in two cutters. Spotts's executive officer, Acting Master William Churchill, commanded the landing party. He came ashore with one howitzer on a field carriage, in charge of Acting Ensign Taylor, protected by the marine skirmishers under 2nd Lt. Louis Fagan. Wasting no time, Taylor fired five rounds at the small schooner anchored in the protected inlet. Laden with turpentine, the vessel quickly caught fire and burned to the waterline. Thereupon the landing party withdrew.

In late May, Acting Rear Adm. Samuel P. Lee, commanding the North Atlantic Blockading Squadron, arrayed four Union warships in succession on Virginia's James River. On the downriver end, screw steamer *Dawn* moored just below Wilson's Wharf on the north bank, and on the upriver end, screw gunboat *Pequot* anchored above Fort Powhatan on the south bank. Between them, Lee placed two armed vessels previously

Upon his promotion to rear admiral in July 1863, John A. B. Dahlgren (1809–1870) left the Bureau of Ordnance and assumed command of the South Atlantic Blockading Squadron until war's end. He is shown seated behind one of his 50-pounder rifles, which reveals its characteristic trunnion band and breech strap. In 1868, Dahlgren resumed his post as chief of the navy's Bureau of Ordnance, and then briefly commanded the Washington Navy Yard until his death. Photograph by Matthew B. Brady. (*National Archives and Records Administration*)

captured from the Confederate navy, the ironclad *Atlanta* and the screw tug *Young America*. Brig. Gen. Edward Wise, commanding 1st Brigade, 3rd Division, Army of the James, divided his two black infantry regiments between Fort Powhatan and the supply depot at Wilson's Wharf, subsequently named Fort Pocahontas. By May 24, the latter place had been almost completely fortified by earthworks, ditch, and abatis, and garrisoned by the 1st Regiment and several companies of the 10th Regiment, US Colored Troops. One white field artillery battery supported the garrison.

The widely held belief in the white South that black soldiers, even after training, remained unreliable—expressed all too frequently as "niggers won't fight"—attracted the attention of Confederate forces in the area. The defeat, or better yet the capture, of the negro garrison at Wilson's Wharf was deemed to be an easy task, and one that would serve as a lesson to other black formations. Black troops, both free men and former slaves, were appearing in ever increasing numbers in the Union armies, and the sowing of doubt and apprehension in their ranks was no small part of the attack on Wilson's Wharf.

At midday on the 24th, more than 2,500 men of Maj. Gen. Fitzhugh Lee's cavalry division suddenly appeared and charged the defenses at Wilson's Wharf. Much to their surprise, the black soldiers on the other side of the barricade remained cool and disciplined, and delivered volley after volley of deadly fire. After the initial attack, most of the rebel troopers dismounted to fight on foot. Well before the five-and-a-half-hour battle ended, with rather heavier casualties to the Confederates, the artillery battery used up its ammunition. Behind the defenses on the James River, however, USS *Dawn* fired in support. Her commanding officer, Acting Vol. Lt. John Simmons, and executive officer, Acting Master Joseph Jackaway, worked in concert to shift fire from one flank to the other as the need arose. The army signal section commanded by 2nd Lt. Julius Swain aided these efforts.

The small gunboat fired eighty-one rounds from her two Parrott rifles—a 100-pounder and a 20-pounder—and thirty-seven rounds from her Dahlgren 12-pounder rifled howitzer: twenty-one shells with Bormann fuse, eleven shells with percussion fuse, three rounds of canister, and two stands of grape. Under those desperate circumstances, Simmons could ill afford to be fastidious about firing canister and grape from the rifled howitzer. In his action report, though, he urged the return of the pair of 32-pounder smoothbores, better suited to fire such ammunition, that had long been aboard his ship, but removed shortly before the action at Wilson's Wharf.

If the residents of the nation's capital had become inured to the war, they snapped out of it abruptly during the summer of 1864 and Jubal Early's raid on Washington. In the end, Maj. Gen. Lew Wallace's delaying action on July 9, at the battle of Monocacy on the outskirts of Frederick, Maryland, saved the capital city from capture. Nevertheless, it was a close-run thing, with Lieutenant General Early additionally sending a large cavalry force commanded by Brig. Gen. Bradley Johnson in a wide, sweeping raid around Baltimore. Johnson in turn detached a small body under Maj. Harry Gilmor to do mischief to the rails and wires between Baltimore and points north. Gilmor did just that, and moreover, on July 11 he burned a goodly part of the Philadelphia, Wilmington & Baltimore Railroad bridge spanning the Gunpowder River above the Maryland port city.

Forewarned of his coming and his probable targets, the navy dispatched three screw gunboats to protect tidewater points northeast of Baltimore. One of them, *Fuchsia*, fired thirty-four rounds, both shell and canister, from her Dahlgren howitzers at enemy cavalry threatening the Bush River railroad trestle. The gunboat eventually drove the horsemen off, preventing the repetition of the damage to the Gunpowder River structure. As well, the navy sent a large detachment down from the Philadelphia Navy Yard to ensure the security of the railroad ferry and other assets at Havre de Grace, where the Susquehanna River flows into Chesapeake Bay. This force, commanded by Lt. Cmdr. Thomas Harris, included a company of marines under Capt. James Forney and a battery of Dahlgren howitzers. This force remained at Havre de Grace until the Confederates had recrossed the Potomac and the situation in the Maryland tidewater had returned to normalcy.

In spring 1864, the armored ram CSS *Albemarle*, ably captained by Cmdr. James Cooke, began to torment the Union gunboats in the North Carolina sound for which she was named. She mounted two 6.4-inch Brooke rifles and her eight-sided oaken casemate, sheathed in 4 inches of iron, proved impervious to most of the projectiles the gunboats could hurl. Moreover, she bore a long ram plated with iron that she used with deadly effect. Original sepia wash by R. G. Skerrett, reproduced in the *Official Records of the Union and Confederate Navies*, Series I, Vol. 9.

The ironclad ram CSS *Albemarle*, constructed over the course of sixteen months in a North Carolina cornfield—a site selected by nineteen-year-old shipbuilder and detached army lieutenant Gilbert Elliott—first appeared at the mouth of the Roanoke River on April 18, 1864. Her sloped central, eight-sided casemate consisted of iron 4 inches thick, reinforced by southern oak. Within that casemate were mounted a pair of Brooke double-banded 6.4-inch (100-pounder) rifles, fore and aft, with each piece able to pivot to fire through three separate shuttered gunports. Her tapered bow was heavily plated with iron, enabling her to ram enemy warships built of wood with destructive effect. Cmdr. James Cooke became her first and most successful commanding officer. During the next six months, the ram would cause much grief to the US Navy and put its command of Albemarle Sound in grave doubt.

On the day following her initial appearance, *Albemarle* tangled with a pair of double-ended sidewheel gunboats, *Southfield* and *Miami*. They had been lashed together with hawsers at the direction of Lt. Cmdr. Charles Flusser, commander of the Union gunboat flotilla, in order to ensnare the rebel ram. Flusser went aboard *Miami* to direct the fight, but despite his best laid plans, things went badly wrong. A shell fired by the gunboat deflected off the ram's thick iron hide and exploded near Flusser, killing him. *Albemarle* rammed and sank *Southfield*, then disabled *Miami*. This naval defeat led directly to the Union army upriver surrendering Plymouth, North Carolina.

The next action took place on May 5, with two small flotillas engaged: four Union double-enders, a screw gunboat, a former ferryboat, and two smaller sidewheel gunboats, pitted against *Albemarle* and two consorts. In the ensuing action, the ram and USS *Sassacus* became locked in a near-death struggle, with crewman aloft in the gunboat attempting to toss hand grenades down the ram's deck hatch. *Sassacus* took a round to her starboard boiler, but by some miracle, she did not blow up or otherwise sink. CSS (ex-US Army) *Bombshell*, assisting the ironclad in convoying a troopship (which escaped), surrendered, and *Albemarle* withdrew up the Roanoke, licking wounds that were hardly superficial. Not least, her aft Brooke rifle had been broken at the muzzle and needed replacement. One month later, Cmdr. John Maffitt relieved now Captain Cooke as *Albemarle*'s commanding officer.

Several of the Union gunboats involved had on board Dahlgren boat howitzers of one or more types. Sidewheel double-ender *Wyalusing* mounted the greatest number: four 24-pounders, two 12-pounder smoothbores, and two 3.4-inch rifles. The only vessel confirmed to have fired her Dahlgren howitzers—two 12-pounder smoothbores and a 12-pounder rifle—was army steamer *Bombshell*, which changed hands twice in as many weeks. During her second action, she fired her howitzers at the federal gunboats until her surrender and recapture on the 5th. It is difficult to believe, however, that in the frantic melees taking place on those occasions, at least some of the other warships mounting Dahlgren howitzers did not fire them.

It remained abundantly clear that something had to be done about this troublesome Confederate ram. After failed attempts to sink her by shore party, the Union navy turned to Lt. William Barker Cushing, an officer not only renowned for his adventures (and misadventures), but whose highly perceived—many would argue exaggerated—sense of honor revealed a knife edge on which mad courage and madness itself teetered in precarious balance. The feats of derring-do Cushing had engaged in previously included a commando raid designed to abduct Brig. Gen. Louis Hébert, who commanded the Confederate defenses at the mouth of North Carolina's Cape Fear River. Hébert had absented himself on the night of the raid, but Cushing and his men netted the general's chief engineer, Capt. Patrick Kelly.

Cushing retained a free hand to plan the destruction of *Albemarle*, and in so doing, he traveled to the New York Navy Yard to select and fit out small craft. He decided upon two 30-foot screw picket boats, each fitted with a spar torpedo and a Dahlgren 12-pounder howitzer on a boat mount in the bow. Soon thereafter, the lieutenant brought the wooden boats south. During a storm en route, *Picket Boat No. 2* suffered mechanical difficulty, and in seeking a quiet inlet to effect repairs, her commanding officer, Ens. Andrew Stockholm, blundered onto a hostile shore. After a brief fight, he and his crew surrendered, but they managed to set their boat afire. The intrepid Cushing carried on, determined to make the attack on CSS *Albemarle* with the one torpedo boat remaining.

On the night of October 27–28, 1864, Cushing handed off his boat code signal book to the commanding officer of the sidewheel double-ender *Otsego* in order to keep it secure. Thereupon he and fourteen volunteers manning *Picket Boat No. 1* made their way stealthily up the Roanoke River, to find *Albemarle* protected by a log pen. The raiders had pulled off almost complete surprise, but as they neared their quarry, a sentry detected the unidentified craft and after a peremptory challenge, gave the alarm. In order to approach the log boom head on, Cushing directed the coxswain to circle around and thereupon make full speed for the barrier. *Albemarle*'s by then thoroughly alerted crew, as well as soldiers on the shore, took the steam launch under heavy fire. Cushing's gunners got off a round or two of canister from their Dahlgren howitzer before the boat crashed into the boom. Fortunately for Cushing, the logs were slick from accumulated aquatic growth and the torpedo boat slid well up onto them.

The lieutenant quickly deployed the torpedo's spar under the overhang just astern of the ironclad's port bow, and forcibly pulled the lanyard on the device. The detonation came virtually at the same instant as the ram fired a Brooke rifle. Although the round closely missed the torpedo boat, Cushing and his men were trapped. In the end, only the commanding officer and a seaman escaped after arduous exertions, with one man killed in action and the remainder either captured or drowned, fatigue and the cold water their undoing. Tragically, one of those drowned was Acting Master's Mate John Woodman, who had led three successful reconnaissance missions by boat and by foot to ascertain the ram's state of repair. *Albemarle* meanwhile suffered a six-foot hole in her hull and sank into the mud, never again to steam forth against her foes. After the war had ended, the ram's third commanding officer, Alexander Warley, said of Cushing's raid, "A more gallant thing was not done during the war."

As the conflict wore on through late 1864 and into 1865, the small naval landing party, sent ashore for a short period and a narrowly defined mission, evolved into the large fleet brigade, landed to cooperate with the army and provisioned for the rather longer haul.

In late November–early December 1864, the South Atlantic Blockading Squadron, still commanded by Rear Admiral Dahlgren, sent ashore a brigade of nearly 500 men

On the night of October 27, 1864, Lt. William B. Cushing and his volunteer crew stole up the Roanoke River to destroy the Confederate armored ram *Albemarle* in her lair. Firing canister from their Dahlgren 12-pounder boat gun, they put *Picket Boat No. 1* over the slippery log boom. Cushing quickly positioned the spar torpedo against the enemy vessel and pulled its lanyard, blowing a hole in her hull. *Albemarle* thereupon sank to the mud, her forays done. Only Cushing and one other man escaped death or capture. Rendering by Julian O. Davidson, reproduced in *Battles and Leaders of the Civil War*, Vol. 4, 1888.

to support a considerably larger army formation on the South Carolina coast. This expedition was intended to cut the Charleston & Savannah Railroad and thus divert Confederate forces from opposing the seaward advance on Savannah, Georgia, by Maj. Gen. William T. Sherman. The coastal operation came under the overall command of the army's Brig. Gen. John Hatch, with the naval contingent led by Cmdr. George Preble. The fleet brigade consisted of three battalions, one each of sailor infantry and marines, under Lt. James O'Kane and 1st Lt. George Stoddard respectively, and one of artillery, commanded by Lt. Cmdr. Edmund Matthews. The artillery battalion fielded eight Dahlgren landing howitzers: four heavy and two light 12-pounder smoothbores, and two 3.4-inch rifles. At Honey Hill and Tulifinny Crossroads, the strongly entrenched Confederates, fighting on their own soil, successfully withstood the federal attacks. In the latter instance, the defending units included a battalion from The Citadel's Corps of Cadets. Afterward, Matthews expressed his unhappiness with the accuracy of the rifled bronze howitzers, remarks that portended their boring-out postwar. Despite the repulses of the federals, some Confederate formations had been lured away from Sherman's route of march, and the navy had acquitted itself sufficiently well that it contemplated other such operations.

That opportunity arose during the following month, January 1865, on the second attempt to capture Fort Fisher, protecting the mouth of the Cape Fear River and Wilmington, North Carolina, nearly 30 miles upstream. The port remained the only one still open to blockade runners, and thus offered the Confederacy its last hope for resupply. To deny Wilmington to the rebels, a joint army-navy operation was conceived under the overall command of Maj. Gen. Alfred Terry. The fleet of nearly sixty warships intended to bombard the fort came almost entirely from the North Atlantic Blockading Squadron, commanded by Rear Adm. David Porter, following his October relief of Acting Rear Adm. Samuel Lee. Porter's chief-of-staff, Fleet Captain Kidder Breese, led the naval brigade of about 2,200 sailors and marines.

Facing them, the 1,900-man Confederate garrison commanded by Col. William Lamb was part of the Department of Cape Fear, under the direction of Maj. Gen. William Whiting. When overall commander Gen. Braxton Bragg, ever irascible and unyielding, refused Whiting reinforcements, the latter officer gallantly joined Lamb within the fort. A 6,400-man division of the Army of Northern Virginia, led by the usually competent Maj. Gen. Robert Hoke, stood north on the Federal Point peninsula, but remained mostly unengaged as the battle developed, because of Bragg's refusal to commit this division. When Hoke did attempt to reinforce Fort Fisher, he was unable to get an adequate number of men south to the fort.

Insofar as the Union side's tactical plans went, the lessons provided by Burnside's successful amphibious landings nearly three years before would be roundly ignored, as they had been for the failed amphibious attack on Fort Sumter in Charleston in early September 1863. The Fort Sumter debacle should have served as a strong reminder that amphibious operations required substantial planning, preparation, and training. Despite that hard lesson, the naval assault on the fort's northeast bastion on January 15 was ill-conceived, with the sailors carrying just pistols and cutlasses—only the marines and naval sharpshooters had long arms—and going forward entirely

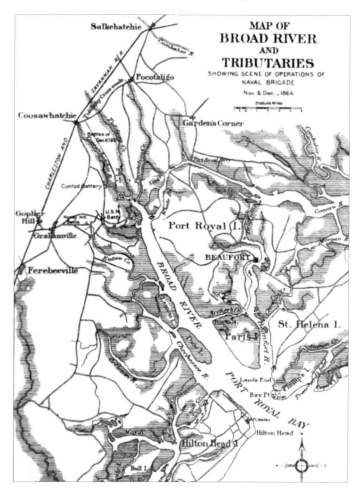

Left: In late 1864, the joint operation up South Carolina's Broad River, which included a 500-man naval brigade from the South Atlantic Blockading Squadron, had two major aims: to cut the Charleston & Savannah Railroad and to draw Confederate forces away from General Sherman's march on Savannah, GA. Battles ensued at Honey Hill, east of Grahamville, and at Tulifinny Crossroads, farther north near Coosawhatchie. The Confederate defenders foiled both Union thrusts. Paris/Parris Island is today the Marine Corps Recruit Depot. *Official Records of the Union and Confederate Navies*, Series I, Vol. 16.

ASSAULT OF THE NAVAL COLUMN ON THE NORTH-EAST SALIENT OF FORT FISHER.

Above: In 1864, Rear Adm. David D. Porter (1813–1891) stands next to a Dahlgren 4-inch (20-pounder) rifled ML howitzer on a pivot carriage, aboard sidewheel steamer USS *Malvern*, flagship of the North Atlantic Blockading Squadron. His distinguished naval family included his father, Capt. David Porter of America's earlier wars, and his adoptive brother, Adm. David G. Farragut, also actively serving during the Civil War. Porter rocketed from lieutenant to rear admiral in fifteen months, commanding the Mississippi River Squadron during that time. A decade after the war ended, the Bureau of Ordnance converted some of the 4-inch ML rifles to breechloaders. Photograph by Alexander Gardner, 1864. (*Naval History and Heritage Command*)

Opposite below: Bluejackets of the naval brigade, armed with only revolvers and cutlasses, are depicted as they sweep forward against the northeast bastion of Fort Fisher, NC. This assault on January 15, 1865, would be repulsed with heavy loss and rout. Best friends Lt. Benjamin Porter and Lt. Samuel Preston were among the eighty-eight sailors and marines killed in the assault. The always-heroic Lt. Cmdr. William Cushing once again survived the ordeal. Drawing by Walton Taber, reproduced in *Battles and Leaders of the Civil War*, Vol. 4, 1888.

In February 1862, while still a seventeen-year-old midshipman, Benjamin Horton Porter (1844–1865) coolly and capably directed his battery of six Dahlgren landing howitzers on Roanoke Island, as members of his gun crews fell from Confederate fire. After a stretch as a prisoner of war in 1863–64, he was exchanged and assigned to command USS *Malvern*, Rear Adm. David Porter's flagship. Customarily brave during the assault on Fort Fisher on January 15, 1865, Lieutenant Porter met his death. Photograph by Matthew B. Brady. (*National Archives and Records Administration*)

unsupported by Dahlgren landing howitzers. Moreover, the army and navy assaults, intended to be closely sequential, lacked coordination.

The army onslaught by Brig. Gen. Adelbert Ames's reinforced division of four brigades from XXIV Corps, closely supported by the inshore gunboat flotilla whose guns, including Dahlgren howitzers, provided heavy, accurate, and deadly fire, did ultimately carry the fort. The attack by the naval brigade began prematurely, well before the marines intended to provide covering fire were properly organized and in place, and it became disordered almost at once. After suffering terrible casualties from the heavy fire, the sailor infantry fled, despite the efforts of their officers to rally them. The habitually indestructible Lt. Cmdr. William Cushing survived the fight, but the equally intrepid Lt. Benjamin Porter, who as a midshipman had coolly stood to his howitzer battery on Roanoke Island in February 1862, did not. On the Confederate side, both Whiting and Lamb had fought like lions but had sustained severe wounds. Whiting died from dysentery within two months while still a prisoner of war, and Lamb suffered from his disability for years afterward. Much to their chagrin, following the capitulation of Fort Fisher, the Confederates blew up Fort Caswell and other defenses across the river mouth, and within the month they withdrew from Wilmington.

In April of that year, other than a few fitful eruptions in distant places, the long bloody war at last ended.

References Consulted

Anderson, B., *By Sea and by River: The Naval History of the Civil War* (New York: Knopf 1962), p. 63.

Bailey, R., *The Bloodiest Day: The Battle of Antietam*, The Civil War, No. 11 (Alexandria, VA: Time-Life Books, 1984), pp. 60, 104–05, 116–17, 120–28, 130.

Byline not specified, "Burnside's Expedition. Its Arrival at Hatteras Inlet, Great Difficulties Encountered, Perpetual Gales of Wind," *New York Times*, January 29, 1862.

Carman, E. A., "The Antietam Manuscript," 1890s, copy in the library, Antietam National Battlefield (also online), pp. 29, 32, 54.

Chaitin, P. M., *The Coastal War: Chesapeake Bay to Rio Grande*, The Civil War, No. 9 (Alexandria, VA: Time-Life Books, 1984), pp. 17–30, 95–97, 163–70.

Coggins, J., *Arms and Equipment of the Civil War* (Garden City, NY: Doubleday, 1962), pp. 125–28, 132–33, 138, 141–42, 144–45, 149.

Cooling, B. F., *Jubal Early's Raid on Washington 1864* (Baltimore: Nautical and Aviation Publishing Co., 1989), pp. 167–69.

Dahlgren, J. A., "Actual Service," in *Boat Armament of the U.S. Navy*, 2nd edit. (Philadelphia: King and Baird, 1856), pp. 158–81.

Graham, M. J., *The Ninth Regiment New York Volunteers (Hawkins' Zouaves)* (New York: E. P. Coby, 1900), pp. 280, 308–309, 315–17, 320.

Kaeuper, R. W., "The Riverine Fleet," undated (online).

Kimball, E. A., commanding 9th New York Vol. Inf. Regt., action report dated Sept. 20, 1862 (online).

McTernan, W., and Kullberg, A. D., "U.S. Marines Face Citadel Cadets at Tulifinny Crossroads," *Leatherneck*, March 2013, 96(3): 44–48.

Nalty, B. C., *United States Marines at Harper's Ferry and in the Civil War*, History and Museums Division, HQ, US Marine Corps (Washington, DC: USMC, 1983), pp. 1–10, 15–21.

Official Records of the Union and Confederate Armies in the War of Rebellion, Series I, Vol. 1; Vol. 2; Vol. 9; Vol. 19, Part I; Vol. 33; Vol. 36, Part II; Vol. 37, Parts I and II; Vol. 44; Vol. 46, Parts I and II.

Official Records of the Union and Confederate Navies in the War of Rebellion, Series I, Vol. 4; Vol. 5; Vol. 6; Vol. 9; Vol. 10; Vol. 11; Vol. 15; Vol. 16; Vol. 23; Series II, Vol. 1.

Perry, M. C., *Narrative of the Expedition of an American Squadron to the China Seas and Japan, Performed in the Years 1852, 1853, and 1854* (Washington, DC: Beverly Tucker, 1856), pp. 238, 248, 252–54, 261–62, 343–46, 375, 393–94, 495–97.

Quarstein, J. V., *A History of Ironclads: The Power of Iron Over Wood* (Charleston: History Press, 2006), several extracts available online.

Schwartz, Z., "'This Insulting Bravado': William B. Cushing and the Martial Culture of 19th Century America." Paper submitted when a midshipman as a requirement of the US Naval Academy, Spring term 2012, pp. 3–7, 12, 14–20.

Sewall, J. S., "The Invincible Armada in Japan," *New Englander and Yale Review*, 1890, 53: 201–12.

Shetter, R. M., "Assaulting the Littorals: The Development and Evolution of a Dedicated American Amphibious Assault Force," master's thesis, Department of History, California State University, Long Beach, 2012, pp. 26–34.

Spalding, J. W., *The Japan Expedition. Japan and Around the World: An Account of Three Visits to the Japanese Empire* (New York: Redfield, 1855), p. 341.

Tucker, S. C., "The Dahlgren Boat Howitzer," Technical Report, *Naval History*, Fall 1992, 6(3): 53.

————, (ed.), *The Civil War Naval Encyclopedia*, Vol. 1 (Santa Barbara: ABC-CLIO, 2011), p.143.

Wingfield, T. C., and Meyen, J. E., (eds.), "Lillich on the Forcible Protection of Nationals Abroad," Appendix I. "A Chronological List of Cases Involving the Landing of United States Forces to Protect the Lives and Property of Nationals Abroad Prior to World War II," *International Law Studies* 77: 120–21, 123–24.

Zilian, F., "After war began, Naval Academy temporarily moved to Newport," *Newport Daily News*, Saturday, April 13, 2013.

3
Shinmiyangyo, June 1871

Background

To Koreans, Shinmiyangyo refers to the disturbance of their nation by a Western power—in this instance the United States—in the year Shinmi, equivalent to 1871 in the Common Era. This incident is distinguished from Pyonginyangyo, the disturbance five years earlier involving another Western power, France. The two intrusions had much the same origin, which was the killing of the foreign power's nationals by Koreans in 1866.

If the Japan opened to the world by Commo. Matthew Perry in 1854 had been isolationist, a dozen years later Korea (or Corea), the 4,000-year-old Hermit Kingdom, remained emphatically xenophobic. The execution of nine French missionaries, not to mention thousands of Roman Catholic converts, led immediately to a punitive expedition mounted under Rear Adm. Pierre-Gustave Roze, commander of France's Far Eastern Squadron. The campaign lasted several weeks and spilled considerable blood, finally ending in November 1866. On the other hand, the less distinct US response to the annihilation of the crew of the American registry *General Sherman* in August 1866 came five years later, with the consequent battle ending after two days.

In January 1866, the rescue and safe passage from Korea of the crew of the storm-wrecked American schooner *Surprise* stood in sharp contrast to the events taking place seven months later. To be sure, the United States government had few illusions about the merchant steamer *General Sherman* or her captain and crew. The latter consisted of Chinese and Malays, who in Korean eyes were the sort of bloodthirsty rogues who for years had been plundering coastal towns and their dwellers. Among the ship's passengers were two or three men of disreputable character, apparently acting in collusion with the captain to further their own ends. Despite being repeatedly warned not to proceed further or to attempt to undertake trading, the ship had sailed up the Taedong River, well into the interior near Pyongyang. When the ship became stranded, fighting soon ensued in which several Koreans died. Korean soldiers thereupon set fire to the upperworks of the iron-hulled vessel, and when the crew abandoned ship, they were shown no mercy.

Soon following the end of its Civil War, the United States made two minor attempts to ascertain the fate of the American flag merchant vessel and its crew, which Korea unequivocally rebuffed. As a consequence, on May 16, 1871, most of the US Asiatic Squadron commanded by Rear Adm. John Rodgers departed from Nagasaki, Japan. His ships consisted of screw frigate (and flagship) *Colorado*, screw sloops *Alaska* and *Benicia*, sidewheel gunboat *Monocacy*, and screw tug *Palos*. The civilian passengers aboard included not only Frederick F. Low, American minister to China, and his staff, but the notable Felice (or Felix) Beato, hired on to be the expedition's official photographer.

Seven years earlier, Beato had photographed the multinational punitive expedition mounted against the anti-shogun forces attempting to close off to all foreign vessels the Strait of Shimonoseki, lying between Honshū and Kyūshū, Japan. Following several bellicose incidents, on September 5–6, 1864, a seventeen-ship squadron of British, Dutch, and French warships first bombarded the forts and batteries along the strait, and then landed a naval brigade to complete their destruction. Almost entirely immersed in the Civil War, the United States Navy was represented by the chartered screw steamer *Ta-Kiang*, carrying a 30-pounder Parrott rife and commanded by Lt. Frederick Pearson, both from the sail sloop-of-war *Jamestown*. Although the American steamer did not land bluejackets, she towed pulling boats, provided supporting fire, and took aboard British wounded for medical treatment. Of equal significance, Maine-born William Henry Harrison Seeley, an ordinary seaman serving in the Royal Navy, won the Victoria Cross for more than one act of gallantry during the action, the first American so awarded.

Unlike the retributive purpose of the multinational squadron, the mission of the American flotilla in 1871 was threefold: to establish political and commercial relations with Korea; to ascertain if any of *General Sherman*'s crew survived in the intervening five years, and if so to bargain for their release; and to negotiate a treaty to prevent such incidents from arising in the future. There was, initially at least, no punitive intent to the mission.

Nonetheless, several hundred seamen and marines served aboard the American warships who could be suitably armed and deployed in the event their presence ashore became necessary. During the week-long voyage west through the East China Sea, and then north through the Yellow Sea to the west coast of Korea, the men detailed for a possible landing exercised daily for infantry service ashore and engaged in firing practice to sharpen their marksmanship.

Such precautions on the part of these ships become all the more comprehensible in the light of an incident four years previous, caused by an almost identical circumstance. In March 1867, the American bark *Rover* foundered during a severe gale off southern Formosa. Although the entire crew, including the captain's wife, reached shore safely, they were soon discovered and thereafter massacred by local tribesmen. Following a period of some months in a vain attempt by the United States government to negotiate just punishment of the murderers, in June of that year, two screw sloops, *Hartford* and *Wyoming*, sailed to Formosa under the command of Rear Adm. Henry Bell to carry out reprisals. Once offshore, the ships landed a party of nearly 200 bluejackets and marines led by Cmdr. George Belknap in order to find and punish the offenders.

A major difficulty became immediately obvious: the garb of the pursuers remained inappropriate to both the climate and the terrain. The uniform of heavy wool best worn at sea against chilly weather hardly suited the hot climate and dense tropical forest of southern Formosa. The cunning tribesmen repeatedly fired from ambush and then quickly melted away, preventing the landing party from coming to grips with them. A number of seamen collapsed from heat exhaustion or heatstroke, and the expedition's second-in-command, Lt. Cmdr. Alexander Slidell MacKenzie, lost his life while leading the final charge against the foe. After a day of it, the Americans had enough.

Admiral Rodgers determined that his landing force would not find itself in the same predicament. Indeed, the American expedition of June 1871 would be a classic example of the successful shore landing from naval vessels. Thorough planning and training, bold leadership, firm resolve, manly strength, splendid courage (indeed, on both sides), encountering and overcoming of unexpected obstacles, and sharply defined but limited aims ensured its success.

The Weapons Ashore

Within six years of the ending of America's Civil War, both the army and navy had significantly reduced the plethora of ordnance types purchased because of the shortages and delays in war production. Gone were the small Dahlgren 12-pounder howitzers, with the heavy and light howitzers remaining in service in the fleet. Gunners recognized that the heavy 12-pounder smoothbore and the 3.4-inch rifle had the most destructive effect, but on the other hand, the light 12-pounder pulled more easily and quickly while on shore. The larger 24-pounder howitzer continued to reign on the decks of gunboats and small auxiliaries, such as armed tugboats, and was considered much too heavy to land.

In the manner of small arms, Remington breechloading carbines and pistols firing metallic cartridges equipped the enlisted members of postwar naval landing parties. These were single-shot weapons using the rolling block breech mechanism, with the familiar Colt and Remington cap and ball revolvers issued to the seamen in the war years now phased out. The justification, however absurd, for adopting single-shot pistols came from the 1866 *Ordnance Instructions for the United States Navy*: "The revolver pistol does not realize in service with seamen the advantages claimed for that description of the arm." Officers, who more often than not provided their own sidearm, of course kept their revolvers, now largely converted to fire metallic cartridges. Although the navy had hoped for the logistical convenience to adopt a pistol and carbine using the same cartridge, that wish turned out to be impossible without either reducing the impact and range of the carbine bullet or making the recoil of the pistol too violent and disrupting successive shots. In the end, the Model 1867 carbine used .50-45 rounds, whereas the Model 1867 pistol fired shorter .50-25 rounds. That is, both rolling block weapons were .50-caliber in bore, but the former cartridge had 45 grains of black powder, and the latter 20 grains less of the propellant. By way of comparison, the somewhat later Model 1870 Remington rolling block infantry rifle fired a .50-70 cartridge and saw use by the navy for the remainder of that decade.

The arming of the marines aboard ship remains odder yet. They retained their wartime Model 1861 or 1863 Springfield muzzleloading rifled musket, firing a .58-caliber Minié bullet by means of an explosive cap and fixing a socket bayonet with an 18-inch triangular blade. Apart from the longer range of the rifle over the carbine, the reason for this inexplicable policy may have been a matter of money or the conservatism of older naval officers. In addition to the Springfields, the marines had some Model 1861 Whitney (or Plymouth) .69-caliber muzzleloaders. By any standard, the marines' shoulder weapons were woefully obsolete. The commanding officer of the expedition's marines, Capt. McLane Tilton—ironically he had been a member of the board selecting the Remington breechloaders—disparaged the older rifles as "muzzle fuzzels."

Whatever the thinking, sound or spurious, these were the weapons that the members of the US Navy and Marine Corps would take ashore in Korea in June 1871. There they would face warriors armed with matchlock muskets, gingals, and cannon—some rifled, some breechloaders—more apt for combat of the sixteenth and early seventeenth centuries. And swords, spears, and arrows from a far earlier age.

The Provocation

On May 23, 1871, the American squadron anchored in Roze Roads, so styled for the admiral who had led the French incursion five years before, seaward of Ganghwado (Ganghwa or Kanghwa Island). Days of dense fog had hindered the voyage and lengthened its duration. A week later, still encountering heavy fog, the ships shifted their anchorage nearer to the mouth of the so-called Salée River, actually the Yeomha or Ganghwa Strait, a narrow north-south passage between the island and the Gimpo Peninsula. On its upper end, the strait connects with the estuary of the Han River, which flows by the capital city of Seoul, then about 40 miles to the southeast.

A small number of Korean vessels arrived on the last day of the month, and three envoys came on board *Colorado*. Minister Low's acting secretary, Edward Drew, and the assistant secretary of the American legation in China, John Cowles, were fluent in Chinese, which remained the language of Korea for formal dealings (indeed, the degree and on what occasions China chose to exercise influence, and even sovereignty, over Korea complicated matters). The secretaries were assisted in the often tricky and delicate linguistics by two Chinese scribes. After conferring briefly with the envoys, Drew and Cowles determined that they were of the third and fifth ranks, and thus Low refused to treat with them, demanding instead to speak with envoys of the first rank representing Korea. Drew assured the envoys, however, that the flotilla had only peaceable intentions and gave them a tour of the ship, almost certainly with the intention of impressing them with her strength.

Despite his assistants' facility in Chinese, the minister misread the intention of the Korean delegation. When Admiral Rodgers requested permission to proceed north in order to survey the strait and make soundings, the envoys said nothing, and Low misinterpreted that response as their acquiescence to the request. The Koreans, however, in being noncommittal intended that no permission to enter the waterway was being granted. This miscommunication between members of the two nations led to provocations by both sides.

In thinking they had received consent to survey the strait, on June 1 the Americans went ahead with that task, using the relatively shallow-draught *Monocacy* and *Palos*, along with three steam launches and a steam cutter, each of the boats having a 12-pounder Dahlgren howitzer in the bow. Interpreted as an armed incursion beyond the limit where foreign vessels were allowed—indeed the part of the Yeomha clearly fortified—the movement outraged the Koreans, who opened fire without warning from a number of forts and hidden batteries. The Americans were in turn outraged by the supposed treachery on the part of the Koreans, and returned the fire. In retrospect, it is eminently clear the Americans, although angered, had not been taken by surprise. Their immediate and closely accurate counterbattery fire strongly suggests their shipboard guns had already been carefully laid and sighted before the Koreans fired.

However actual the treachery the Americans and Koreans believed each other guilty of, the Yeomha proved the most treacherous of all, with its extreme tidal flux, furious current, hazardous shoal waters, and uncharted rocks. The current proved particularly strong at a sharp bend named Sondolmok, perhaps 3 miles above the mouth. On its western bank, Sondolmok was guarded by a well-positioned work, Yongdudondae, which the Americans called Elbow Fort. There, *Monocacy* ran onto a rock that holed her. Not long thereafter, the lead-line being used to take soundings on *Benicia*'s steam launch, commanded by Master Seaton Schroeder, fouled the propeller, making the vessel nearly unmanageable in the swift current. Crew members aboard the sidewheel warship effected temporary repairs, those on the steam launch untangled the line from the screw, and the naval force turned about and withdrew southward, firing on the defenses as they did so. On this leg, no fire was returned from the forts, which had been silenced from the large guns aboard the two warships. Two American sailors suffered non-life-threatening wounds.

Rodgers sought a seaward anchorage for his ships while he considered his next move. Finally, he sent word that he would wait ten days for an official apology from the Korean government. There was more, however, to this generous interval than American largesse: By then the period of spring tides would have been succeeded by that of neap tides, with their lesser rise and fall, and thus a higher water level during the ebb to aid navigation. Both Admiral Rodgers and Minister Low agreed that a lesson must be administered lest the Koreans, not to mention other Asian nationalities, believed they could insult the United States flag with impunity. The ranking naval and diplomatic officers also were in accord that punishment should be limited to the offending forts and batteries along the strait (or in their minds, the river).

Although correspondence remained unbroken between the Koreans and the American naval flotilla, with minor concessions offered by both sides, the former refused to budge in either tendering an apology or opening negotiations of any substance. In the staunch refusal by the Koreans to sign a treaty with the Americans,

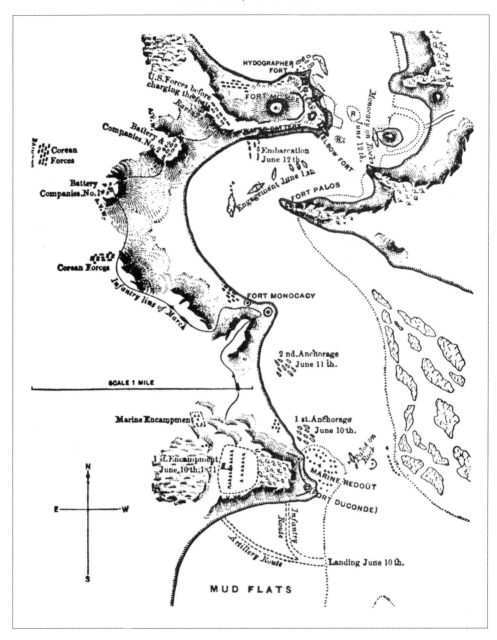

Ganghwa Island on the left of the map shows the progress of the American bluejackets and marines as they pushed north from their landing on June 10, 1871, capturing and destroying the Korean defensive works. The culmination of their mission took place on the 11th with their successful attack on the high citadel, later named Fort McKee for Lt. Hugh McKee, mortally wounded there. The treacherous Yeomha Strait between the island and the mainland on the right was obstructed with unmarked rocks, and subject to troublesome currents and tides, making USS *Monocacy*'s task to provide supporting fire a tricky one. Reproduced in Carolyn Tyson, *Marine Amphibious Landing in Korea, 1871*, 1966.

A council of war took place aboard the screw frigate and flagship *Colorado* not long before the landing force went ashore. Rear Adm. John Rodgers leans over the map table, while from left to right sit Cmdr. Edward McCrea, commanding *Monocacy*; Capt. George Cooper, commanding *Colorado*; and Capt. Edward Nichols, chief of staff to Admiral Rodgers. Standing are Master John Pillsbury, *Colorado*; Cmdr. Lewis Kimberly, commanding *Benicia*; Cmdr. Homer Blake, commanding *Alaska*; Lt. Cmdr. William Wheeler, flag secretary to the admiral; and Lt. Charles Rockwell, commanding *Palos*. Note the cutlass rack on the right. Photograph by Felice Beato. (*Naval History and Heritage Command*)

aside from their long-existing xenophobia, lay the belief, apparently originating with the Chinese, that the United States consisted of little more than an assemblage of rude settlements, and thus remained unworthy of much consideration. However rude they may have been, aboard the ships of the flotilla, the American sailors and marines continued to exercise and practice their martial skills.

The tenth day arrived without a resolution of the issue. Peaceful intentions having been swept aside, battle was now joined.

The Assault

At ten o'clock in the morning on June 10, the American landing force set forth up the strait. Two steam launches led the way, followed by *Monocacy*, Cmdr. Edward McCrea commanding. Her armament consisted of four 8-inch smoothbore muzzleloaders and two 60-pounder Parrott rifles, augmented by two 9-inch Dahlgren smoothbores offloaded from *Colorado*. Her primary task was that of shelling the defensive works located along the passage. Astern of *Monocacy* steamed *Palos*, commanded by Lt. Charles Rockwell, armed with four 24-pounder smoothbore howitzers and two 4-inch rifled howitzers, and towing twenty-two boats that carried the landing force and its field artillery. In overall operational command, Cmdr. Homer Blake (the commanding officer of *Alaska*) would stay afloat on *Palos*. Blake had been captain of the sidewheel gunboat USS *Hatteras* during the war, before being sunk in January 1863 during a one-on-one encounter with the raider CSS *Alabama*. Astern of *Palos*, two steam launches acting in a safety role completed the formation. Each steam launch carried a Dahlgren boat howitzer, either a light 12-pounder smoothbore or a 3.4-inch rifle. During the actual landing phase, these launches, under the command of Lt. Cmdr. Henry F. Picking, would provide close covering fire.

Ashore, Cmdr. Lewis Kimberly, the commanding officer of *Benicia*, had charge of the entire landing force of 651 men. Lt. Cmdr. Silas Casey, off *Colorado*, commanded the infantry battalion of ten companies, A through K (omitting J), and had immediate command of the landing party's right wing, with Lt. Cmdr. William Wheeler, the admiral's flag secretary, as his second-in-command and commander of the left wing. Marines led by Capt. McLane Tilton made up the last two of those lettered companies, I and K, totaling 105 enlisted men and four officers. *Alaska*'s Lt. Cmdr. Douglas Cassel commanded the two artillery batteries of seven Dahlgren howitzers. The right or first battery consisted of four light 12-pounders—two from *Benicia* and one each from *Alaska* and *Colorado*—in charge of Lt. Albert Snow, also off *Alaska*; the left or second battery was composed of one each heavy and light 12-pounders, and a single 3.4-inch rifle, all from *Colorado*, as was the battery commander, Lt. William Mead. Each SB howitzer had upon landing thirty rounds of shrapnel, ten rounds of shell, and ten rounds of canister, with more ammunition on board the two supporting warships.

Lt. Cmdr. Winfield Scott Schley, who would gain fame in the Spanish-American War, served as the adjutant of the expedition, in function the officer responsible for its landing organization and logistics (stopping short of strictly the title of beachmaster, because Schley would move inland with the landing force). A platoon of thirty-six pioneers—in action reports usually called sappers and miners, alluding to their other essential purpose—were detailed from the three largest warships and placed under the command of Mate Quinn (his first name is lost to history) to undertake the heavy work, specifically in assisting the gunners in getting the howitzers off the beach and up and down steep slopes, demolishing the Korean fortifications, and cutting trails where necessary. Chief Engineer Alexander Henderson on *Colorado* asked to accompany the expedition as a volunteer, and his presence and professional experience would prove fortuitous in the rough country. A signal section commanded by Ens. Nelson Houston and a fifteen-man medical party under Passed Assistant Surgeon Charles Wells completed the roster.

The bluejackets and marines prepared to go ashore in light marching order, a concession to both the heat and the distance to be covered. Each man carried sixty rounds of ammunition for rifle, carbine, or pistol, and packed sufficient cooked rations to sustain him for two days.

During the approach of the landing force at the mouth of the Yeomha, a junk appeared, waving a white flag. The steam launch carrying Secretary Drew, the expedition's interpreter, approached the craft and received an official letter addressed to Admiral Rodgers. Drew quickly read the text, and as it contained nothing new, the landing continued as planned.

Picking had selected what appeared to be a broad beach, ideal for putting the infantry and artillery ashore and nicely flanking the lower fort. But appearances proved deceiving, and instead, the expedition landed on a perilous mud flat some 200 yards in depth from high water mark to hard ground. To make the situation worse, the flat was cut by numerous sluices having even deeper and more tenacious mud, deceptively hidden until almost onto them. Tilton later recounted his tallest marines sinking into the muck to above their knees. Men lost their gaiters, shoes, socks, and even parts of their trousers. The landing howitzers fared even worse, plunging at once into the quagmire, such that it oozed up over their axletrees. Extracting the mired guns became impossible for their normal pulling crews, and necessitated the assignment of extra men—a lot of extra men. Kimberly ordered Casey to detach four sailor companies to assist in pulling the guns out of the soft mud. It would need seventy-five men at a minimum to extricate each light howitzer and perhaps ten or twelve more to get out each of the two heavier pieces.

The job required hours to accomplish, and during that long interval, the artillery remained unavailable to the field force. The steam launches stayed close inshore, however, to provide fire support as needed. *Palos* continued up the strait to join *Monocacy* in exchanging fire with the forts there, but struck an uncharted rock in the middle of the narrow passage. She was now not only out of the fight, but in grievous trouble, despite her steam pumps operating at full capacity. Blake ordered *Benicia*'s steam launch to come to her assistance, reducing even further the vessels available for close gunnery support. Having done what he could for *Palos*, Blake transferred to *Monocacy*. Despite the serious difficulties encountered by the Americans in the initial phase, the Koreans unaccountably failed to contest the landing.

Before they cleared the mud flat, the marines had deployed in skirmish order, thereafter passing by inundated rice fields and through a village, both pleasing to the eye. Soon they entered the first defensive work, named Chojijin and by then unoccupied. In the distance, its fleeing garrison loosed a few shots in the marines' direction. The fort's 12-foot curtain wall had a hewn granite foundation topped by a split-rock, mud, and mortar rampart. The work extended north from the lower fort as a water battery mounting a great many guns of various sizes, the largest being two 32-pounders, and ending at a second armed strongpoint, Chojidondae.

Upon being relieved by the main body of sailor infantry, the marines pressed on ahead as skirmishers, now accompanied by the light 12-pounder from Mead's battery. The seamen began the destruction of what became known as the Marine Redoubt, pulling down its walls and casting its artillery into the Yeomha, except for the pair of 32-pounders, which they spiked. Given the delay in getting the artillery off the mud flat, the day was becoming late. Casey decided to bivouac his landing force west of the fort, on flat ground straddled by marsh and rice paddies. The marine encampment lay in an advanced position northwest, about half a mile from the main body. Both camps posted sentries and threw out pickets, with the marine contingent keeping one-third of its force on rotational watch throughout the night.

Sometime between 11 p.m. and midnight, a great racket—wild firing and howling—wakened the camps. Kimberly ordered the long roll beaten and fell in those companies

Bluejackets pose for the camera inside the second fort, Chojidondae, after its capture by the marines, and subsequent renaming Marine Redoubt. Several seamen are exhibiting their almost toylike Remington Model 1867 rolling block carbines, caliber .50-45. The officer on the left wearing the white havelock is Lt. Hugh McKee, who would be mortally wounded in the assault upon the high citadel. Photograph by Felice Beato. (*National Archives and Records Administration*)

Marines and bluejackets exult atop the curtain wall of the third defensive work, on a prominent bluff above the strait, named Deokjindondae by the Koreans and Fort Monocacy by the Americans. The Korean garrison abandoned the work upon the advance of the US naval landing force. Photograph by Felice Beato. (*National Archives and Records Administration*)

not in reserve. The Korean attack proved more noise than substance, however, and a few howitzer rounds fired in the general direction of the din beyond the marine camp drove off the troublemakers.

Reveille beat at 4:00 the next morning, and after eating breakfast, the landing force completed the destruction of the fort, including the burning of military buildings, clothing, and provisions. An hour later, the expedition proceeded to the next fort, on a commanding bluff overlooking the strait. Named Deokjindondae by the Koreans, it became Fort Monocacy to the Americans, because of the great damage sustained from that ship's guns. It and a supporting work had also been abandoned.

Tilton's marines were once more in the advance, supported by two howitzers (the other five guns were farther back, with the bluejackets). In his report, the captain describes the construction as a crenelated wall of chipped granite and earthen fill, covered with a thin veneer of mortar, deceptive in its lack of real strength. The leathernecks began the destruction of the place in a like manner to the first fortification until the sailor infantry came up. Combined, they destroyed thirty-three ancient bronze breechloaders of small bore and four more modern 32-pounders, and pulled down the fort's walls. Unfortunately, the fire set with the purpose of destroying legitimate military buildings and matériel remained unchecked, and afterward consumed the village lying immediately above the fort.

As the landing force advanced north of the Deokjin forts, it found the terrain more challenging, with steep wooded hills separated by even steeper ravines and gullies. In particular, the effort to move the landing howitzers up and down the hills became increasingly difficult. Quinn's brawny pioneers widened narrow trails, bridged gullies with brush and earth fill, and assisted with ropes in lowering the guns into the wider ravines and raising them up on the other side. It proved to be hot and heavy work.

The sidewheel gunboat USS *Monocacy* appears in an Asian port two decades after her close-support role in the action at Ganghwa Island, Korea, in June 1871. She saw service from 1866 to 1903, all but exclusively with the Asiatic Squadron. Variously armed through the decades, the 265-foot-long, 1,370-ton warship was last involved in armed conflict during her participation in the Peking Relief Expedition in 1900. Photograph by Edward H. Hart, Detroit Photographic Co. (*Library of Congress*)

Out in the strait, *Monocacy* had shifted her fire and was now pounding the main citadel, Sondolmokdondae, on a 150-foot eminence on the western shore, and its redoubts lower down, which included Yongdudondae (Elbow Fort) facing the strait. From time to time, she fired on the earthwork occupying much of the sharply outthrusting point on the eastern shore of Sondolmok. The Koreans called this fortification Deokpojin. Although repeatedly silenced, the fort did not remain quiet long, and required the periodic attention of *Monocacy* and eventually the howitzers ashore to suppress its fire. *Monocacy*'s gunnery much impressed Blake, now aboard the sidewheel gunboat, and led him to exult in his action report that he had never seen such accuracy of fire exceeded.

The day became hotter and the arduous march more taxing; not a few men fell out with symptoms of heat exhaustion and even sunstroke. Moreover, the Koreans were gathering in ever-increasing strength atop a high ridge on the landing force's left flank, that is, to its west. When the line of the expedition's advance turned east, toward the high citadel overlooking the Yeomha, that move would put the large hostile force directly in the expedition's rear, thus posing a tactical dilemma. As the opposing forces became more heavily engaged, Kimberly detached Wheeler, the infantry battalion second-in-command, with three companies of infantry—A (Lt. Cmdr. Oscar Heyerman), B (Master Franklin Drake), and F (Master John Pillsbury)—supported by five howitzers in order to keep the hostile force in their rear at bay.

The seagoing officers hardly amounted to ducks out of water. Senior commanders Kimberly, Casey, and Wheeler shared a sound appreciation of infantry tactics and maneuvered their different echelons skillfully. As the demands of the situation changed, Cassel and his battery commanders Snow and Mead moved their guns about like chess pieces, twice to the tops of commanding hills, requiring of their gun crews enormous and repeated exertion. In that arduous task, they often needed the assistance of the pioneers and infantry.

Once in those hilltop positions, the howitzers fired with great effect on the Koreans to their left and rear. Casey thought between four and five thousand hostile soldiers opposed them, but Wheeler, more directly involved in that aspect of the action, estimated between two and three thousand. Whatever the actual number, the rear guard repelled more than one assault, with the credit going mainly to the howitzers. Kimberly's report reads, "It is to the artillery, their precise and rapid firing, that we owe our immunity from attack of the large body of Coreans on our left flank." That of Wheeler testifies, "Great credit is due Lieutenants Mead and Snow, who commanded the artillery, for without the artillery the position which I held would have been perfectly at the mercy of the overwhelming numbers of the enemy; it was only by the aid of the artillery that I was enabled to hold the enemy in check." The reason, Wheeler keenly observed, was that the Koreans very much feared the nasty little howitzers.

With its rear now amply protected, the main body of seven companies—C (Lt. George Totten), D (Lt. Hugh McKee), E (Lt. Bloomfield McIlvaine), G (Master Thomas McLean), and H (Master Robert Brown), with the two marine companies, I (Capt. McLane Tilton) and K (1st Lt. James Breese) in advance as skirmishers—continued to move east toward the high citadel. Two light 12-pounders commanded by Master

Seaton Schroeder provided close support for the advance of the marines, who remained sharply engaged. Lt. Cmdr. Cassel went forward with this howitzer section. To their rear, Mead shifted his two longer-ranged pieces, the 3.4-inch rifle and heavy 12-pounder, to fire over the heads of the main body on the citadel and its supporting redoubts at lower elevations. By now, the artillery ammunition had become badly depleted and required replenishment from the supply aboard *Monocacy*.

The senior officers called a rest halt before the main body reached the crest of the hill from which the attack on the citadel above would be launched. The interval also allowed Mister Houston to signal *Monocacy* to cease firing. On the crest, the assault line formed from right to left: Companies I, K, H, C, D, and G. In front of them lay a precipitous ravine 80 feet deep, which would have to be negotiated before they began to ascend the equally steep slope of the 150-foot hill capped by the citadel.

In the fort above, whose walls had been breached by *Monocacy*'s shells, waited some of the toughest and bravest troops in Korea, the self-styled Tiger Hunters, and regular soldiers commanded by General Eo Jae-yeon. As the assault line moved forward at Casey's command, the Koreans opened a heavy fire with their ancient matchlocks and gingals, inflicting the first death on the Americans, Pvt. Dennis Hanrahan, a member of *Benicia*'s marine guard. The charge gathered momentum, cascading down into the ravine and just as relentlessly and impetuously ascending the opposite slope, with both officers and men racing one another to get to the top and come to grips with the enemy. As the Americans closed and the Koreans found they could not maintain their rate of fire, they resorted to throwing rocks and even dirt. Nothing would halt

Slain Koreans lie on the ground inside the high citadel, named by the Koreans Sondolmokdondae and by the Americans Fort McKee. The work served both as an observation and command post and a keep for the last stand by the Korean defenders. Rumor persisted that photographer Beato had rearranged the corpses somewhat for an image of greater drama. Photograph by Felice Beato. (*National Archives and Records Administration*)

the American advance, however, and the attacking wave rolled up over the walls and through the gun ports.

The first man inside the curtain, Lt. Hugh McKee, commanding officer of D Company, was immediately shot in the groin and then speared in the side. Coming up behind the mortally wounded officer, Lt. Cmdr. Winfield Schley grappled with the spearman, until an enlisted sailor killed the Korean with his carbine. Another American bluejacket, Landsman Seth Allen, lost his life while scaling the wall. Thereafter, the Americans readily won the day, inflicting terrible casualties on the Korean garrison, whose members harbored no thoughts of surrender. Indeed, without exception, in writing about the battle afterward, the senior US officers paid tribute to the great courage shown by their foemen.

Four marines particularly distinguished themselves in the fighting: Pvt. James Dougherty singled out and killed General Eo Jae-yeon, while Pvt. Hugh Purvis and Cpl. Charles Brown captured the general's large yellow *sujagi* or battle standard. Pvt. John Coleman in hand-to-hand combat saved the life of Boatswain's Mate Alexander McKenzie, who in turn had acted to protect the fallen Lieutenant McKee. All five enlisted men would win the Medal of Honor.

With their leader down and at last appreciating the hopelessness of their valor, the Koreans began to withdraw from the fort. The American fire, however, continued unabated. Master Thomas McLean and thirty men from G Company barred the main escape route, while Schroeder's two howitzers punished the fleeing Koreans with canister and shrapnel, and Tilton's marines took them under flank fire.

At the end of the battle, the Americans counted 243 Korean bodies in and around the high fort, with estimates of between 100 and 200 more killed in the lower works and in the wooded areas fired on by Wheeler's howitzers and bluejackets. Those losing their lives included not only the commanding general, but his younger brother Eo Jae-sun and other high-ranking officers. The cost to the invaders amounted to three killed (Lieutenant McKee died subsequently aboard *Monocacy*) and ten wounded, among the more severely, Passed Assistant Surgeon Wells, whose duties anticipated exposure to fire. To this tally must be added the two seamen wounded on June 1 and the handful or two of men felled by heat exhaustion and sunstroke. Twenty Korean prisoners, all wounded to one degree or another, were brought on board and treated, and the nineteen who survived their wounds were soon released. They pleaded not to be, as they expected to be executed for being taken prisoner.

In the late afternoon and again in the evening, *Monocacy* exchanged fire with Deokpojin, the earthen fort on the opposite shore of Sondolmok. After a night of watchful waiting, the landing force embarked unmolested and without a hitch the following morning, the howitzers taken aboard first and the marines last. The Americans renamed the high citadel Fort McKee, in honor of the fallen lieutenant. Although they had not crossed the strait to assail the stronghold on the eastern point, in tribute to its unflinching defiance, it too got a new name, Fort Palos. Such names, of course, were little more than cognomens, in force for only as long as the US Navy lingered in the area.

Aftermath

Minister Low attempted to reopen the negotiations with the Korean emissaries, but he must have known full well that the die had been cast. Indeed, a treaty and diplomatic relations between the two governments would await the year 1882. Despite the immediate failure of its mission, the naval brigade put ashore on the west coast of Korea had been capably done. The howitzers accompanying the force had played a major role in its tactical success, firing more than 200 rounds of shell, shrapnel, and canister, specifically:

Lt. Albert Snow's battery of four light 12-pounders: thirty-one shell, twenty-four shrapnel, and eleven canister;
Lt. William Mead's composite battery: 3.4-inch rifle, twenty-eight shell with percussion fuse, twenty-eight shell with time fuse, two canister; heavy 12-pounder, seventeen shell with time fuse, thirty-four shrapnel, nine canister; light 12-pounder, five shell with time fuse, seventeen shrapnel.

Snow attested to the accuracy of his pieces, with a comparative few of the fused rounds failing to explode, and those perhaps a function of age. The reliability of the copper cartridges for the Remington carbines differed substantially, however, failing to detonate in far too many instances. The indefatigable Tilton carried out an investigation soon after the action and found the source of the problem: destructive oxidation around and in the primer cup, which occurred only when the cartridges had been packed in paper boxes, but never when they had been placed in wooden ones. He made his strong recommendation accordingly. The navy ceased issuing copper-cased cartridges in the spring of 1873, at the same time it ordered the Remington carbines withdrawn from service.

Within months of the expedition, the US Navy awarded fourteen Medals of Honor, eight to seamen and six to marines, all of whom had behaved gallantly. The rather large number of such medals conferred must be attributed, in part, to the still evolving standards to be met for the bestowal of this highest military honor. Nonetheless, under the tightened standards, in December 1915 the navy awarded a fifteenth Medal of Honor for the Korea expedition, to former Chief Quartermaster Patrick Grace, who had died nearly twenty years before.

References Consulted

Bartlett, M. L., and Sweetman, J., "River Raid on Korea," *Naval History*, December 2001, 15(6): 43–47.

Bauer, K. J., "The Korean Expedition of 1871," US Naval Institute (hereafter USNI), *Proceedings*, 1948, 74(2): 197–203.

Bechtol, B. E., Jr., "Avenging the *General Sherman*: The 1871 Battle of Kang Hwa Do," master's thesis in military studies, Command and Staff College, US Marine Corps, 2001.

Chang, G. H., "Whose 'Barbarism'? Whose 'Treachery'? Race and Civilization in the Unknown United States–Korea War of 1871," *J. Amer. Hist.*, 2003, 89(4): 1331–65.

Dictionary of American Naval Fighting Ships, Naval History and Heritage Command (online). Consulted for each US naval warship and auxiliary mentioned in this chapter.

Duvernay, T., "The *Shinmiyangyo*," Royal Asiatic Soc., Korea Branch, *Transactions*, 2014, 89: 1–49.

——————, *Sinmiyangyo: The 1871 Conflict Between the United States and Korea* (Duvernay, 2021), pp. 7–8, 29–54, 92–94, 199.

Ellsworth, H. A., *One Hundred Eighty Landings of United States Marines 1800–1934*, History and Museums Division, HQ, US Marine Corps (Washington, DC: Govt. Print. Office, 1974), pp. 57–59.

Howell, G., "Our Brief Clash with Korea," USNI, *Proceedings*, 1935, 61(11): 1624–36.

McAulay, J. D., *Rifles of the United States Navy & Marine Corps 1866–1917* (Woonsocket, RI: Mowbray Publishing, 2017), pp. 10–11, 18–21, 24–29, 32–34.

Provan, J., *Wild East: The British in Japan 1854–1868* (Stroud: Fonthill Media, 2020), pp. 130–43.

Rankin, R. H., *Small Arms of the Sea Services: A History of the Firearms and Edged Weapons of the U.S. Navy, Marine Corps, and Coast Guard from the Revolution to the Present* (Fort Lauderdale: N. Flayderman & Co., 1972), pp. 84–85, 103, 118–23.

Schley, W. S., *Forty-five Years under the Flag* (New York: Appleton, 1904), pp. 73–96.

Schroeder, S., *A Half Century of Naval Service* (New York: Appleton, 1922), pp. 29–30; 41–43; 47–55.

Sterner, C. D., *Shinmiyangyo: The Other Korean War*, 2002 (online).

Tyson, C. A., comp., *Marine Amphibious Landing in Korea, 1871*, Historical Branch, HQ, US Marine Corps (Washington, DC: Naval Historical Foundation, 1966), pp. 2–12, 14–16, 18–21.

US Navy Dept., *Annual Report of the Secretary of Navy*, 1871, pp. 12–13; Appendix No. 18, "Expedition to Corea," pp. 275–313.

——————, *Ordnance Instructions for the United States Navy*, 4th edit. (Washington, DC: Govt. Print. Office, 1866).

4

The 1870s–1880s and a New Landing Howitzer

Origin

The Dahlgren boat and landing howitzers had served the US Navy reliably and well through a long war and in numerous prewar and postwar encounters, whether showing the flag or keeping the peace. As the nineteenth century approached the three-quarter mark, however, it had become clear that the navy needed to design and adopt a more up-to-date piece to support naval landings. In keeping with ordnance available in both Europe and the United States, the navy decided upon a breechloading rifled gun for the task. In conformance with their Dahlgren predecessors, period correspondence often describes the new pieces as howitzers. The inscription atop their barrels, however, attests only to a 3-inch rifle. The original intention of the navy was to combine the rifled BL howitzer with fixed metallic cartridges, such as in use by guns made by both Hotchkiss and Krupp abroad. In the end, however, because of limited funds and perhaps a loss of nerve, the navy settled on bagged propellant for service use.

Wartime experience had badly soured most naval and military officers on the continued use of ordnance made of iron, whether cast iron (too brittle) or wrought iron (insufficiently hard). Given the need for greater range and accuracy, and thus the requirement for a rifled gun, the choice of the reliable, hardened bronze also offered a less than ideal solution. Rifling in bronze wore poorly, and after wartime service, not a few Dahlgren 3.4-inch rifled howitzers were bored out, converting them to heavy 12-pounder smoothbores. The answer lay in mild carbon steel, which provided variance in tensile strength and elasticity. While American metal manufacturers had made cast and crucible steel on a relatively small scale, the state of the country's steel companies lagged far behind that of the industrially advanced and technologically savvy nations of Western Europe. The European arms makers could manufacture steel of high tensile strength, coupled with the necessary elasticity and elongation, in massive ingots and in great quantities.

Therefore in 1873, Commo. William Jeffers, the newly appointed chief of the Bureau of Ordnance—he had briefly commanded USS *Monitor* during the war—initiated a series of steps that led to the new steel 3-inch boat and landing howitzer, a piece he could claim as his brainchild. What must be appreciated in the lengthy and difficult struggle that ensued is that Jeffers had a rather bigger fish to fry. His real intention lay in assisting in the creation of a domestic steel industry that could be relied upon to manufacture large steel guns and heavy armor plate to arm and protect the warships of the fleet. It would be the New Navy arising from the ashes of the Old during the 1880s, after his own tenure in office had passed. Indeed, following a prolonged and desultory production, only forty steel light boat and landing howitzers would augment the older Dahlgren howitzers (whose manufacture Jeffers ordered discontinued in April 1873). In addition to these standard landing howitzers, the navy acquired a handful of heavy boat howitzers made of steel, two heavy pieces composed of hardened bronze, and thirteen bronze light howitzers, a dozen of which would see use as training pieces at the naval academy, with the remaining one assigned to first-line service.

Commodore Jeffers, who by any standard must be considered among the best chiefs of ordnance that fortune cast the navy's way, acted decisively and at once to improve the service's guns. Although for his entire eight years in office, he bucked a persistent refusal by congress to meet the funding needs of the navy (or the army)—indeed a legislature indifferent to the reality that a warship belonging to a third-rate naval power armed with first-rate European guns could at long range reduce any American cruiser to blazing wreckage—there were things he could accomplish of a highly professional nature. To foster a greater interest in and knowledge of naval ordnance in general, Jeffers directed his deputy, Cmdr. Montgomery Sicard, to issue a copy of the current *Ordnance Instructions for the United States Navy* (Sicard was the junior author of its 5th edition) to each line officer who duly applied. In another move, he sent officers to gather intelligence in both Europe and Russia, in order to keep an eye on the fast-developing changes there. Whether abroad or back home, the officers assigned had fluency in the language of the country in which the ordnance originated, as in the case of the Reffye gun.

The naval attaché in London, Cmdr. Francis Ramsay, had gotten wind of the French competitive trials of the Vavasseur and Reffye breechloading guns. Jeffers was determined that the US Navy should not be forced into building yet another generation of muzzleloading ordnance, and he urged Ramsay to keep his ears open to the winds blowing from across la Manche. Ramsay did better than that: he got his hands on a French military pamphlet describing the gun designed by Gen. Jean-Baptiste Verchère de Reffye, which he sent to Jeffers. The ordnance chief passed the publication on to Cmdr. Joseph Marvin, in charge of the Naval Experimental Battery at Annapolis. Marvin in turn gave it to his assistant, Lt. Cmdr. William M. Folger, instructing him to translate it. The information convinced Jeffers that the Reffye gun, and particularly its breech mechanism, should serve as the basis for the US Navy's new landing howitzer. When the translated pamphlet was issued to the fleet in November 1873 for general information, it credited the Reffye gun as being the origin of the new steel howitzer.

In February 1874, the Washington Navy Yard duly received two breechloading test howitzers fabricated of smooth-turned chrome steel that the service had ordered from the Chrome Steel Company, Brooklyn, New York. After finishing the pieces, the yard sent one of them to the experimental battery during the summer of 1875 to begin test firing. In these trials, 7-pound projectiles were propelled by charges contained in both

The Reffye interrupted screw breech mechanism that equipped at least two types of French field guns, including the 75mm Canon de campagne de 5 de Reffye modèle 1873 shown here, considerably piqued the interest of the US Navy's Bureau of Ordnance when devising its 1870s 3-inch breechloading howitzers. The French equipment originated with Gen. Jean-Baptiste Verchère de Reffye, who superintended the ordnance works at Meudon. *Matériel de l'artillerie et des équipages militaires*, 1877. (*Bibliothèque nationale de France*)

serge bags and copper-clad cartridges. Elevations varied between 1 and 10 degrees, with muzzle velocities, times of flight, and ranges measured.

A paper written by Commo. Foxhall A. Parker, Jr., and published in Volume 1 of the US Naval Institute's periodical in 1874 (indeed, Parker was one of the institute's founders), likely hastened the transition to the new ordnance. He had been present as chief of staff in January of that year, when the navy's combined fleet had carried out maneuvers in Florida Bay, lying between the southern tip of mainland Florida and its southwesterly extending Keys. The fleet had assembled there to prepare for a possible war with Spain, owing to the *Virginius* crisis during the previous fall. One of the navy's most forward-looking thinkers and the recently appointed superintendent of the naval academy, Parker carried authority with his pen. Although he had good things to say about the beach landings and mock battle on Key West carried out by the 1,700 bluejackets and marines, he wrote a great deal less kindly about the boat howitzers providing the supporting fire and their deployment once the guns had been landed ashore, describing them as an utter failure. Such performance must have been particularly irksome for the officer who had authored both *The Naval Howitzer Ashore* and *The Naval Howitzer Afloat* less than a decade before.

That same year, Commander Sicard, inspector of ordnance at the Washington Navy Yard and the officer most closely associated with Jeffers in the development of the steel howitzer, ordered the casting of five experimental bronze howitzers. Three had the Reffye interrupted-screw breech mechanism, and two the Hotchkiss sliding wedge block, resembling the type produced by the Krupp armaments firm. All would fire metallic cartridges of Hotchkiss design, which offered a renewable gas check that prevented excessive wear from the high-pressure gas generated upon the ignition of the propellant, otherwise worse with bagged powder ignited centrally (axially) through the breechblock. The bureau inclined toward the Reffye mechanism, because it more readily accommodated either metallic cartridges or bagged powder. Going with the wedge type, the navy rationalized, would also have extra metal required on the breech end.

Meanwhile, until high quality steel became readily available in the United States, the ordnance chief decided upon hardened bronze as an acceptable transitional gun metal. Another factor making bronze less objectionable lay in its affordable cost, because the navy had maintained a plentiful supply of that renewable metal. Again a reflection of their Dahlgren predecessors, the new guns would be built in two weight classes, using bronze as the basis of those weights: a light piece of 350 pounds and a heavy one of about 500 pounds. The foundry at the Washington Navy Yard cast thirteen 350-pound and three 500-pound bronze 3-inch rifles (in actuality, the latter averaged 524 pounds).

Parallel with this production, the navy converted thirty-six Civil War-era Dahlgren bronze 4-inch (or light 20-pounder) rifles to breechloading pieces by cutting off their breech ends and installing Reffye-pattern breech mechanisms. Unlike the 12-pounder howitzers, these larger and heavier guns remained as shipboard ordnance, each typically mounted on a bronze Marsilly carriage, and saw use in both the US Navy and the Treasury Department's Revenue Cutter Service.

When subsequently completed, the steel 3-inch howitzers had the same dimensions as their bronze counterparts, but were of lesser weight by a considerable margin. The steel light howitzers had a mean weight of 320 pounds, while twenty-three of the heavy pieces averaged 462 pounds. Ten of the heavy howitzers fabricated by the Midvale Steel Company in 1885–1886, however, differed in carrying a circumferential band to bear the gun trunnions and averaged 481 pounds. Thus in the instance of the finished steel rifles, the weights of 350 and 500 pounds inscribed—atop the barrel of the light type and most of the heavy type, and atop the trunnion band of the ten even heavier exceptions—represented class designations only.

In October 1875, while the building of the bronze pieces was underway, Jeffers informed Sicard that the navy would henceforth proceed with constructing only the new BL howitzers made of steel. Moreover, just the steel pieces would be put into service—or at least service in the US Navy's first-line warships—and the need had become so immediate that no competitive tests would be done on bronze versus steel ordnance. Jeffers declared that test firing could continue, however, with the bronze howitzers using existing lead-coated shells, but with the steel howitzers, once available, firing the newly designed projectiles bearing a copper rotating band.

The ordnance chief based his decision to proceed quickly to steel howitzers on sample ingots from two of the four firms that had bid on this preliminary fabrication: Nashua Iron and Steel Company, Nashua, New Hampshire, and Midvale Steel Company, Nicetown, Pennsylvania (the others were the Atlantic Steel Works, New York City, and Black Diamond Steel Works, Pittsburgh, Pennsylvania). These firms had contracted in February, March, and August 1875 to submit a single experimental steel ingot each to the bureau.

The Reffye interrupted screw breech mechanism was that component of the 1870s French field guns most needed by the US Navy's Bureau of Ordnance to solve the difficult problem of obturation (sealing the breech) in its contemporary breechloading landing howitzers. The rendering shows three stages in opening the Reffye breechblock, with the heavy collar acting as the carrier for the breechblock readily seen. *Matériel de l'artillerie et des équipages militaries*, 1877. (*Bibliothèque nationale de France*)

Because of its size and weight, the navy restricted the Dahlgren bronze 4-inch (light 20-pounder) rifle to shipboard use. Thirty-six of them were converted from muzzleloaders to breechloaders during the 1870s, receiving Reffye-inspired breech mechanisms and modified rifling. Such guns must be regarded as an important step in the introduction of breechloading to US Navy ordnance. This example rests on a concrete plinth in Garden City, KS. (*Nelson H. Lawry*)

Nashua Iron and Steel, which had been capable of open-hearth steelmaking since the immediate postwar years, delivered its ingot promptly, within five weeks of the contract being let. The tensile strength and elasticity were within the guidelines for the 350-pound howitzer, but needed improvement for larger guns, including the 500-pound howitzer. Accordingly, Nashua obtained an order for five more ingots in order to better its first attempt. Although it delivered them before Midvale Steel even shipped its first ingot to the navy, the New Hampshire company would thereafter receive no more howitzer orders. After the Nashua and Midvale steel ingots had been forged into howitzers at the Washington Navy Yard ordnance shops in 1876, one howitzer originating from each firm joined the Naval Experimental Battery at Annapolis, firing projectiles having metal cartridge cases.

During the mid-1870s, while the steel companies in the United States groped uncertainly toward the making of an acceptable product, all too often it became difficult to ascertain where experimentation ended and unschooled floundering began. Outright failure was not uncommon. The ingot provided by the Park Brothers' Black Diamond Steel Works was condemned, having met none of the specifications. Even worse, the Atlantic Steel Works' efforts came to nothing and the company subsequently withdrew from the manufacture of steel ordnance. A few years later, the navy cancelled its contract with the Otis Iron and Steel Company, after the Cleveland, Ohio, company had failed to deliver any of the eight forgings for 500-pound BL howitzers.

Despite the nascent state of the American steel-forging industry, three (perhaps four) firms agreed to supply the steel ingots for the production howitzers. At first, they simply provided the shaped ingots to the ordnance shops at the Washington Navy Yard for the gun making. By 1882, two of the new steel manufactories had undertaken forging of the guns, using a heavy steam hammer to more closely shape the red-hot steel ingot. In all cases, the Washington yard did the final finishing work on these guns. Time, experience, and increasing sophistication showed that quenching with oil and heat tempering during the steel making, and thorough hammering during the forging process, much enhanced the quality of the steel, in turn leading to better, more reliable, and longer-lasting guns.

The most noteworthy of the new steel manufactories was Midvale Steel Company, situated in the industrial part of Philadelphia, thus near the anthracite coal fields of the Keystone State and the transportation system carrying such coal. This firm would develop in the coming decades into a well-respected steel maker, noted for its metallurgical research and innovation. During the post-Civil War period bearing the name of the William Butcher Steel Works, however, it had struggled. Its setbacks included the failure to provide steel members of sufficient strength for the Eads Bridge spanning the Mississippi River at St. Louis. Following the ousting of its senior founder, the production ability of the reorganized Midvale Steel Company improved considerably. Beginning in March 1875, it contracted with the navy to provide twenty-three ingots for 350-pound howitzers and sixteen ingots for 500-pound howitzers. Midvale Steel began forging its ingots for the navy's breechloading howitzers in 1882. The job initially required the company to borrow experienced hammerman George Roxborough from the army's Frankford Arsenal near Philadelphia. Between 1885 and 1888, Midvale made and forged the final fourteen ingots for 500-pound howitzers. As will be related, in the late 1880s it furnished steel forgings for a special order of two additional 350-pound pieces.

Naylor & Company of Boston, known otherwise as the Norway Steel & Iron Works, contracted for a dozen ingots to be fashioned into 350-pound 3-inch howitzers and another dozen into 500-pound 3-inch howitzers. The company delivered its final seven light units in 1882–1883 as forgings. In a subsequent and ultimately unsuccessful bid in March 1883 to provide eight ingots for 500-pound pieces, Naylor tendered the option for either unforged or forged pieces, the latter being far more expensive.

Three things of significance happened with the 3-inch boat and landing howitzers in 1876. Early in the year, Jeffers sent twelve of the thirteen hardened bronze howitzers to the US Naval Academy for evaluation, and subsequent training and drill by the midshipmen. The thirteenth piece, specified as the first service howitzer, with a breech mechanism differing from its dozen near-sisters, saw sea duty aboard the second-rate screw frigate *Trenton*. In December 1881, the gun became part of dispatch vessel *Despatch*'s armament, and went to the bottom nearly a decade later, when that vessel foundered off Assateague Light, Virginia.

Uncertain of the ability of the emergent American steel plants to produce ingots sufficiently large and satisfactory for breechloading guns, Commo. William Jeffers, chief of the Bureau of Ordnance, ordered the first examples for both sizes of 3-inch BL rifles to be made of hardened bronze. Twelve of the thirteen bronze light landing howitzers soon went to the US Naval Academy for evaluation and training. A pair of ten-round ammunition chests straddle this bronze piece in Prophetstown, IL. (*Courtesy of Glen M. Williford*)

Also in 1876, at Jeffers's behest, Commander Sicard, in charge of the howitzer production at the Washington Navy Yard, authored *Description of Naval 3-IN. B.L. Howitzers, with Instructions for Their Use and Care*. The bureau reprinted Sicard's pamphlet in that year's report of the secretary of the navy. While the two experimental steel howitzers underwent testing and evaluation, however, the production of these first steel pieces lost priority to larger steel gun projects undertaken by the navy. As a result of limited funding, continued problems by the steelmakers, and the diminished importance of these small guns in the naval ordnance hierarchy, their distribution to the fleet stretched out over several years.

Six light 3-inch howitzers saw completion in 1877, five made from Nashua steel and one made from Midvale steel, followed by eleven in 1878 and ten in 1879, all built from Midvale steel. The Pennsylvania company provided its last ingot for a light service piece in 1880, the same year that Naylor delivered its first five light ingots. At that point a hiatus ensued. With the resumption of light howitzer production in 1882, Naylor provided seven steel forgings. The navy thus received forty steel light howitzers for boat and landing use, all but seven of which had been shaped ingots that the gun shops at the Washington Navy Yard had to forge and finish.

Midvale produced the first two heavy ingots in 1879, and Naylor provided a dozen heavy ingots in 1881–1882. Five forgings completed by Ames in 1882 and 1883 became heavy 3-inch howitzers. Because of an inexplicable paucity of correspondence regarding them, including any contract or remuneration data, little has been learned about the precise identity of Ames. The thin, indirect evidence available reveals that while the venerable Ames Manufacturing Company of Chicopee Falls, Massachusetts, possessed a hammer capable of forging these light artillery pieces, it had no capacity to make sufficiently large steel ingots. Who or what Ames was during the early 1880s, whether a company, an individual, a forging process, or a unique type of steel, thus remains a mystery. Whatever, four of the 500-pound howitzers made from Ames forgings—the Bureau of Ordnance rejected at least two other such forgings—were purchased by the army in June 1884. Finally, Midvale Steel made and forged the final fourteen steel heavy howitzers from 1885 to 1888.

The Treasury Department purchased most of the thirty-three steel heavy 3-inch howitzers so constructed for its revenue steamers, and the army acquired four (Nos. 16–19), in 1884, perhaps with an altered breech thread, and the last four of standard design (Nos. 30–33) in 1887 and 1888. The navy took seven or eight for itself (Nos. 1, 2, 7, 8, 15, 28, and 29 for certain) and used them in shipboard service. Some or most of the army's heavy 3-inch landing howitzers went to the New Jersey National Guard (until March 1869, its state militia), which preferred light artillery pieces without the need for horses, as required for the army's standard field guns. In late August 1888, during the national guard's summer drill and live firing at Sea Girt, New Jersey, two men died because of the failure to completely close the breechblock on one of the howitzers.

A final two steel light 3-inch howitzers were built in 1887 at the behest of Samuel Colt's twenty-nine-year-old scion, Caldwell Colt, who put them aboard his yacht, the schooner *Dauntless*, a contender in the second America's Cup race seventeen years before. Although contrary to later practice, the Naval Gun Factory—built at the Washington Navy Yard in 1886—numbered them 41 and 42, a continuation from where the service production had ended. Colt paid for the labor and material to construct these last pieces, as well as a supply of ammunition.

Description of the External and Internal Design Details

Whether built of bronze or steel, the external aspect and dimensions of the gun in each weight class remained very much the same. In the 350-pound piece, or standard landing howitzer, its barrel length, muzzle to breech, measured 45 inches. Two short cylinders comprised the reinforce toward the breech and a truncated cone made up the chase. In all such pieces, the pair of trunnions projected from filleted rimbases located two-thirds the distance breechward. The steel light howitzer averaged 30 pounds less than its bronze counterpart. The heavy 500-pound piece measured 55.75 inches muzzle to breech, but there was only one cylinder for the reinforce, with the chase carried farther breechward. In twenty-three of the heavy howitzers, their trunnions projected from filleted rimbases located three-fifths of the distance breechward; ten heavy pieces bore them on a trunnion band screwed onto the gun at about the same distance breechward. The trunnioned steel heavy howitzer averaged 62 pounds less weight than its bronze counterpart; the heavy steel howitzer mounting the trunnion band averaged 43 pounds less than the bronze.

The cylinders making up the reinforce were connected by fillets, namely concave strips of metal rounding off the interior angle between two surfaces having different elevations. A flat cheek lay on either side of the end of the breech cylinder. The right cheek carried the collar hinge for the breechblock and the left received the nose of the collar latch, as well as the holder for the tangent sight, which was angled left—initially 2 degrees but later 3 degrees, 11 minutes—to offset the drift of the projectiles rotating from the right-hand twist of rifling. The underside of that cylinder carried the elevating lug.

Sixteen each lands and grooves comprised the rifling, with the lands 0.04 inch high and in the shape of a truncated wedge; the bore diameter between the lands measured 2.92 inches. Whereas in the bronze howitzers, the twist in the rifling increased muzzleward from 0 to 6 degrees, that in their steel counterparts assumed a constant 6-degree twist. Behind the rifled bore lay the shot chamber, with its compression slope that took the belt or rotating band of the projectile; it was followed by the powder chamber and its centering slope. At the rear of the powder chamber in the original design lay the ring recess containing the copper Broadwell gas check ring. The screw box fitting the breechblock followed rearward from the gas check.

The screw box or breechblock recess had three threaded surfaces and three blank surfaces, equidistantly spaced around its circumference, to match the interrupted screw breechblock—or in naval terminology, the breech plug—which was made of bronze or soft steel. Thus the closure of the breech, after the plug was thrust home, required a clockwise turn of the operating lever amounting to 60 degrees, or one-sixth

PLATE II (top)

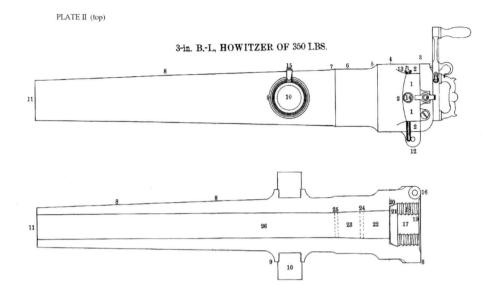

The sectional drawings of a light 3-inch breechloading howitzer show the two reinforce cylinders connecting gracefully by fillets, the pair of trunnions attaching to the gun by rimbases, and the operating lever and other details of the breechblock. The lower drawing reveals the alternation of the threaded and blank sectors in the screw box. When inserted, the breechblock (plug) was turned 60 degrees clockwise so that the threads meshed in order to withstand the detonation of the propellant. Montgomery Sicard, *Description of Naval 3-in. B.L. Howitzers*, 1876. (*Courtesy of James Schoenung*)

of the circumference of the cylinder. Both ends of the breech plug were recessed, the forward end to fit the boss of the nose plate, and the rear end simply to make the block lighter.

At the rear of the screw box, a shallow guide recess accepted a projection on the heavy collar within which the breechblock moved, and aided the proper closure of the block. The breechblock was hinged on the right and latched on the left, with the heavy collar supporting the breechblock while the latter was withdrawn from the screw box. The inside of the collar bore an interrupted screw and engaged the threads of the breech plug, and not only constituted the means whereby the plug passed through the collar, but also prevented the edges of the threads on the plug and screw box catching on each other. Three equidistant guide bolts around the collar periphery engaged a like number of guide slots on the rear of the breech plug, each configured as a reverse L, which stopped the travel of the plug through the collar and helped to stabilize the plug while within the breech.

The collar also bore a locking catch that prevented the withdrawal of the breechblock until intentionally unlocked. After throwing the operating lever, the plug man opened and closed the breechblock by means of the handgrip on its rear.

The Firing Mechanism

In the guns made before 1878 or so, a passage 1 inch in diameter ran through the axis of the breechblock, in turn holding the stem through which the flame of the primer traveled, and thus constituting the breech vent. The forward end of the breech plug abutted against the boss on the rear of the steel nose plate, from which arose the stem that ran rearward to the breech. The forward face of the nose plate bore against the Broadwell gas check ring after the locking of the breechblock. The vent passed forward through the nose plate, which was recessed on its forward face. Within the recess lay the vent check, a disk of mild steel secured by four equidistant steel screws permitting a play in the disk of about five hundredths of an inch. The design permitted the flame from the primer on the breechblock, traveling through the stem, to push the disk muzzleward in order to reach the propellant, normally contained in a serge bag in the powder chamber. Upon the detonation of the powder charge, the pressure forced the disk in the opposite direction up against the nose plate, and thus it served as a gas check.

On the opposite or breech end, the top of the flange on the breechblock bore a nipple into which the gunner inserted a friction primer. The stem descending from the nipple reached an elbow connection within the external breech hollow, and there made a 90-degree turn inward through the axis of the block (plug). The customary tug on the lanyard ignited the primer, whose flame detonated the propellant charge through the axial route described, or did in most cases.

The "plumbing" involved in the elbow configuration for priming proved not always reliable, with back pressure now and then preventing the flame from reaching the powder chamber. Sometimes blowbacks occurred, and not infrequently oil occluded the vent. On occasion, a tiny piece of the serge powder bag would lodge in the vent. Given those difficulties, the Bureau of Ordnance ceased installing axial venting in the higher serial numbers in the productions runs, and changed to the centuries-old, tried-and-true method of radial priming, by drilling a hole through the reinforce directly into the powder chamber. The primer placed in this copper-bouched hole acted directly upon the propellant and no longer needed the longer route. In the 1880s, some guns of both weight classes and both metal compositions were retroconverted, some not, likely according to their locations around the globe.

The Broadwell ring, commonly used by Krupp in Germany, soon proved less than ideal as a gas check device. In 1878, after considerable testing, the Bureau of Ordnance began replacing it with the cup gas check, as devised and improved by the Elswick Ordnance Company in Britain. This gas check, a flanged ring joined to the front face of the nose plate and fitting into the ring recess, offered a smaller surface area to the force of the detonating propellant, and thus put less strain on the breech mechanism. But again, replacement of the original gas checks in the guns of the fleet proved fragmentary.

On those occasions when the gun was altered to fire metallic fixed rounds, which served as their own gas check, gunners undertaking such firing substituted a nose plate without a vent check and removed the Broadwell ring. Extensive experimental firing took place using such fixed rounds. Had the navy opted for those fixed metallic

Axial priming is displayed in this steel light 3-inch BL howitzer. The primer nipple (soldered shut) appears at the top center on the flange circle, whereupon the vertical stem descends to the elbow to connect with the axial vent extending through the breech plug to the powder chamber. The center nubbin of the handgrip is pierced with a hole, which allowed for the insertion of a priming wire to clean out the vent. The nubbin is solid in guns with radial venting. (*Nelson H. Lawry*)

The outer part of the breechblock is missing in this steel heavy 3-inch BL howitzer on display in Chester, NH, and the axial vent through the block has been sealed. The new radial priming vent (arrow) was drilled through the reinforce just forward of the breech, and thereafter bouched with copper. Unlike most of the 3-inch BL boat and landing howitzers, a distinct trunnion band bears the trunnions on this piece. Note the rear and forward sights, the latter on the left trunnion. The wheeled carriage is not an authentic reproduction. (*Nelson H. Lawry*)

rounds in fleet service, as was intended from the inception of the new boat and landing howitzers in the mid-1870s, instead of the serge powder bags subsequently decided upon, it would have been spared the not insubstantial headache of achieving reliable gas checks and foolproof primer venting, problematic in the use of bagged propellant.

Boat and Field Carriages

Both the light and heavy 3-inch BL howitzers served as boat guns and were provided with a single type boat carriage, although the heavy piece needed an adapter to be fitted to the carriage. Service as a landing gun was for the most part restricted to the light (or standard) howitzer mounted on a field carriage.

In the boat carriage, the howitzer's trunnions were placed in bearings (or seats) in ribbed bronze transoms of the top carriage. These transoms took the form of four legs, the front pair being nearly vertical and connected by a transverse transom, and the back pair extending well rearward; the entirety constituted the sliding part of the top carriage. This part slid on flat wrought-iron rails welded and riveted to a wrought-iron front transom, in turn riveted to a pivot plate. The bronze transom connecting the rails at the rear also held India rubber bumpers. The rails were riveted to cheeks of quarter-inch steel 6 inches high, with the rivet heads countersunk on the upper side. Tie bolts and stay bolts sensibly placed stiffened the structure and prevented the cheeks from spreading under the stress of firing.

The boat carriage operated through a frictional compressor having a flat rectangular plate, made of bronze and designed to follow the movements of the carriage. The underside of the compressor plate incorporated a heavy rib, which expanded into a boss, in turn drilled to receive the compressor screw on the top carriage. The recessed upper surface of the compressor plate held lignum vitae that provided the frictional surface acting on the underside of the rails of the slide. A large nut was turned on the upper end of the compressor screw until it bore against a boss on the upper side of the carriage transom. A lever shipped onto the nut was then pushed, forcing the carriage down and the compressor plate up to bind against the underside of the slide. The action of the nut thus squeezed the slide rails between the two surfaces. The action of the compressor plate also acted as a clip to prevent the forward end of the carriage from rising off the slide during the recoil from firing. Such recoil was 10 to 15 inches for the light howitzer and somewhat more for the heavier piece.

It remained essential that the elevating screw be allowed to oscillate while firing the 3-inch howitzer from the boat carriage, so the gun crew attached it by a nut hung on trunnions near the rear of the carriage. As in the field carriage, it worked by means of a hand-operated disk.

At Jeffers's direction, in late 1879 the Washington Navy Yard equipped at least one boat carriage with a pneumatic recoil cylinder. The yard proof-tested it beginning in June 1880.

The backbone of the field carriage, of overall riveted construction, consisted of two vertical cheeks and two transoms of quarter-inch sheet steel. At their forward end, the cheeks elevated the gun 12 inches above the center of the 3-inch diameter tubular wrought-iron axle for the wheels. The breast transom, set at an angle, connected the front ends of the cheeks, and the vertical after transom did so just rearward of the elevating mechanism. The cheeks converged rearward to form the trail, whose upper edges had angle steel welded on to stiffen these members; a stay-bolt connected the two trail cheeks to provide additional strength.

The paired trunnions of the gun rested in bronze trunnion seats high on the carriage cheeks, with hinged capsquares locking them in place. The trunnion seats were riveted in place, angling downward from the rear, and having the hinge at the rear and the eye and forelock on the forward end.

Measuring 3 feet in diameter, each wheel consisted of oaken felloes and spokes (fourteen of the latter), a wrought-iron tire, and a bronze hub. Bronze washers and spring linchpins held the wheels in place; bearings secured the axle for the wheels. Both threaded ends of the axle coupled with a threaded wheel brake, a hollow bronze cylinder attached to a short lever. When thrown, the lever caused a coned member inside the brake cylinder to bind to the hub, and thus restricted the gun's recoil upon firing to about 6 feet. Their action was automatic in the sense that once the brakes had been set, they tightened upon recoil and slacked off when the gun was again run up.

A bronze elevating mechanism connected the bottom of the breech with the carriage cheeks. Its handwheel with a knurled edge allowed the breech to be raised and lowered—and thus the muzzle moved in the opposite direction—in relation to the horizontal axis. The maximum effective angle was around 10 degrees.

The two trail cheeks ended with a bronze shoe, flattened and curved on the bottom to provide a suitable surface for recoil after the gun crew unshipped the trail wheel. To the rear of the shoe, the trail pieces curved upward to fit the 12-inch-diameter bronze trail wheel. A strap and pin arrangement permitted the wheel to be shipped for travel and unshipped for firing. A wrought-iron eye affixed to the rear of the trail wheel permitted the trail of one howitzer to be hitched to the pintle hook beneath the center of the axle housing of a second howitzer, so that the two pieces could be hauled together. The field carriage for the light howitzer had a mean weight of 455 pounds; the mean weight of the field carriage for the heavy howitzer stood at 536 pounds.

A pair of bronze rests for ammunition chests straddled the howitzer, inboard of the wheels. The rest consisted of a longitudinal strip parallel to the axis of the gun, with a cross piece at each end. The forward cross piece bore hooked lugs, to act as stops for the ammunition box shoved in from the rear. Rear lugs prevented the ammunition boxes from shifting laterally. In addition, the inside rear lug had an extension that connected to the carriage cheek and prevented the chest from turning with the axle. Other parts and devices lent stability to the ammunition box in all three dimensions, but when properly operated, permitted the removal of the box when the need arose. The chest itself was of pine and held ten rounds. Bronze handles, straps, and clinch hooks adorned the boxes, the latter providing a secure fastening when the gun moved over rough terrain. A key locked and unlocked the ammunition chests as circumstances determined.

Propellant, Projectiles, Primers, and Fuses

Twelve ounces and 16 ounces of large-grain (or coarse) cannon powder, respectively, constituted the standard loads of the 350-pound and 500-pound 3-inch BL howitzers. They were contained in cylindrical serge bags closed by a thin wooden sabot, with a single groove on its circumference. These loads generated a muzzle velocity of somewhat less than 1,100 feet per second and somewhat greater than 1,100 feet per second, respectively, in the small and large boat howitzers when firing shells. At 5 degrees elevation, the light howitzer achieved a range of somewhat greater than 1,900 yards when firing shell; at 10 degrees elevation, a thousand yards more. The heavy howitzer hit well in excess of 3,000 yards.

When loaded and fused, the shell weighed 7 pounds and in its original form, the shrapnel weighed 8 pounds. Both projectiles measured 2.5 calibers in length, with the ellipsoidal head of about 1 caliber and the cylindrical base of 1.5 calibers. The wall of the shell was 0.4 inch thick and that of the shrapnel 0.3 inch thick. The shrapnel consisted of fifty-five lead balls of .52-caliber circumference, set off by a 2-ounce bursting charge held in an axial tube running through the payload. In August 1879, the shrapnel round was redesigned to reduce its length consistent with a weight reduction to 7 pounds, equal to that of the shell. Fourteen months later, the powder chamber in the shrapnel round had to be enlarged in order to better accommodate the Boxer fuse, described farther on.

Each projectile had a copper rotating band 2 inches in length, very slightly tapered toward the front end and grooved on the rear half. A steadying ring one quarter inch broad circled the projectile 0.3 inch from its base. Upon the round being fired and driven forward, the bore squeezed the steadying ring and so prevented the rear end of the projectile from battering the rifling, particularly as bore wear progressed.

A combination of lanyard and friction primer remained the method of firing the propellant. As already described, radial ignition using a copper-bouched primer hole drilled through the gun's reinforce into the powder chamber replaced the original axial ignition through the breechblock. The original method relied upon the primer inserted into a nipple on the flange of the breechblock, and its flame traveling down the vertical secondary vent, and then a right angle turn into the main breech vent and into the powder chamber. The higher serial numbers in the building sequence of the steel howitzers received radial priming from the outset, with lower numbers converted as the opportunity arose. Such radial priming holes were also drilled into some of the bronze light howitzers. With the axial priming in place, the nubbin in the middle of the breechblock handgrip had a centrally located hole to permit the insertion of a priming wire or boring bit to clear the vent; with radial priming, there was no need to clear a horizontal vent, so the handgrip carried a solid central nubbin.

Although the Naval Experimental Battery at Annapolis tested a variety of fuses in the new landing howitzer, the Bureau of Ordnance issued the Boxer or wooden stock time fuse for fleet service. The device was the invention of Col. Edward Boxer of the Royal Artillery, in the shape of a truncated cone, with beech being the wood of choice for the case. The case enclosed the fuse itself, wrapped in paper. The chief of piece or the quarter-gunner set the fuse and inserted it into the nose of the projectile. The small or lower end went in first, with the large or upper end on the outside. Brass wire wound around both the upper and lower ends of the cone strengthened it for insertion and made it snugger within the nose of the projectile.

Inside the beechwood cone ran the more steeply conical composition powder channel, offset from the central axis of the wooden cone by 0.1 inch. At the head of this channel lay an empty cavity, from which ran four lateral passages that served as gas vents. They angled slightly upward and opened above the point of the projectile. Atop the cavity sat the igniter, a bronze cylinder containing a percussion cap affixed to a nipple. A brittle wire composed of copper and lead inside the igniter restrained the firing plunger, and the cylinder had a flange on top, to prevent its being driven rearward upon the discharge of the gun. Shellacked paper or linen sealed the top of the Boxer fuse.

Two other channels, of rather smaller diameter, extended upward from the bottom of the fuse. They loosely resembled partial ladders cut closely parallel to the side rails, leaving the stubs of the rungs in place. In actuality, they constituted five side-directed time holes, spaced equidistantly along each channel but staggered from the five holes in the other channel; all connected with the outside of the wooden cone. Only the bottommost hole of each side channel ran from the exterior all the way through to the near-central composition channel. Mealed gunpowder filled the entirety of these side channels and time holes. The staggered nature of the paper-covered holes in the two narrow side channels permitted the fuse setter to select the precise time determined to be correct for the range from the target. The quicker ignition times were thus closer to the nose and graded longer toward the base.

When the fuse setter pierced the selected time hole, it now communicated all the way to the composition channel. When the gunner fired the projectile, the flimsy copper-lead retaining wire broke, allowing the plunger to strike the percussion cap, igniting the mixture within the larger channel. That powder burned down to the pierced hole and flashed into the side channel to ignite the faster burning mealed powder within, to continue to the base of the fuse and set off the bursting charge of the round at the selected time. If for any reason the powder in the time hole failed to ignite, the composition mixture would still burn to the longest or default time of the fuse and detonate the round. If between the selected and default times, the projectile hit the target, the fuse would shatter and set off the bursting charge as a concussion fuse, or at least that remained the hope.

As with its Dahlgren predecessor, the 3-inch howitzer did not have a fixed size in its gun crew, but rather that number remained contingent to immediate need. Nonetheless, those required for the new piece were generally fewer because of the greater ease in serving a breechloader. In many cases, whether the chief of piece or the quarter gunner set the fuses depended upon how hot and heavy the action got.

Service

Likely because of the much diminished size of the postwar navy, the number of the new 3-inch boat and landing howitzers available for service at sea numbered around fifty. Forty steel light howitzers, the standard pieces, were augmented by a single bronze light howitzer, two bronze heavy howitzers, and seven steel heavy howitzers. An additional dozen bronze light howitzers, sent originally to the naval academy for evaluation early in 1876, had soon been relegated to landing drill, field exercises, and blank firing. These bronze pieces were followed by the receipt of a single steel light howitzer for actual firing by the midshipmen.

Given this finite number of the new howitzers, the continued use of the bronze Dahlgren howitzers within the fleet through the decade of the 1880s remained essential, a reality that hardly displeased many serving officers in the navy. Although these men tended to be older conservative officers of middle rank, who remained either unfamiliar with or mistrustful of breechloading guns in general, more than a few younger junior officers rather familiar with the new BL howitzers felt the same. The predominant objections with the new howitzers were twofold: they could not fire canister and their firing mechanism represented an Achilles heel. Although inferior in range and accuracy to the BL rifled howitzers, the older Dahlgren muzzleloading 12-pounder with its larger bore delivered lethal canister at close range, ideal for both mob control and street fighting, and for cutting swaths through the ranks of "savages" armed with primitive weapons who would ostensibly seek close-in fighting. Also, the threaded breechblock of the newer landing howitzer, in their opinion, could not stand up to salt water and beach sand, and would easily jam.

A good example of this mixed ordnance is provided by the screw frigate *Trenton*. Upon commissioning in February 1877, she took aboard four boat and landing howitzers: two bronze 12-pounder Dahlgrens and two bronze 3-inch breechloading rifles, one light (No. 13) and one heavy (No. 2). During the following decade, two steel BL howitzers—light No. 31 and heavy No. 15—supplanted the earlier bronze BL howitzers. After a typhoon had wrecked *Trenton* in Apia harbor, Samoa, in March 1889, the navy recovered her boat and landing howitzers, and the breechloading pieces saw further service.

Well thought-out essays and papers by junior officers published by the US Naval Institute between 1879 and 1888 reached opposite conclusions on this issue (although virtually all of the writers tacitly admitted that hostile tribesmen encountered had by then armed themselves with modern weapons). In his 1879 paper on the use of boat guns as light artillery for naval landing parties, Lt. Theodorus Mason, a modernist and soon to be the navy's first chief intelligence officer, strongly advocated the exclusive use of the new 3-inch howitzer and derided those officers who clung to the old smoothbores. During the discussion period on Mason's paper, several officers disagreed with that conclusion, including that navalist of future fame, Cmdr. Alfred T. Mahan. One year later, Lt. John Soley's paper on the deployment of the naval brigade specified both types of howitzer seeing use. In the discussion of the Mason paper, Soley had expressed his preference for the Dahlgren ML howitzer, but in his own paper,

Cadets at the US Naval Academy train on a light 3-inch BL howitzer. Its trail wheel is unshipped and its breechblock is open, while one of the cadets either examines its screw box or peers up the barrel. The time is 1893, and by decade's end these boat and landing guns will be gone from the fleet. (*US Naval Academy Archives*)

Members of the port watch of USS *Trenton*, both two-legged and four-legged, pose in the mid-1880s with the ship's 3-inch BL landing howitzer. *Trenton* not only served as a fleet training vessel, but also tried out various new armaments, including boat and landing guns. She was destroyed in a hurricane in March 1889 at Samoa. Photograph by Edward H. Hart, Photo-Gravure Co. (*Naval History and Heritage Command*)

he may have simply been acknowledging the reality of both types of howitzer being found aboard. Also in 1880, in writing on the navy's new or converted rifled guns, Lt. Henry Lyon extolled the facility in the use of the 3-inch BL howitzers, particularly by then with the improvements in their primer vents and cup gas checks. Clearly, it had become a matter of different strokes for different folks.

Two additional papers discussing the formation and deployment of the naval brigade appeared within a year of each other in the late 1880s. That written by Lt. Charles Hutchins and published in 1887 proved to be that year's prize essay. Significantly, he espoused that only the 3-inch BL howitzer, the Hotchkiss revolving cannon, the Hotchkiss single-barrel piece, and the Gatling gun should be landed. Significantly, Hutchins added, "The smoothbore howitzers have had their day and should *not* be landed." The discussion afterward proved that certain officers continued to disagree with that belief in this late year, still desiring the odd Dahlgren howitzer loaded with shrapnel or canister for street fighting and mob control. Hutchins advised against too many guns being landed, because they then became burdensome to the landing force, but argued that each piece landed should have a large enough gun crew and a limber, the latter because ammunition supply had always been a logistical problem ashore. He bemoaned the continuing paradox of the marines being armed with single-shot rifles, whereas the sailors were sent ashore with modern repeaters. That reality stemmed from the influence of the army on marine equipage and weaponry, and in turn from the reactionary nature of many senior army and navy ordnance officers. Hutchins also stressed the role of the beachmaster in superintending the landing and defending the beach position.

In 1888, Ens. William Rodgers added his thoughts, declaring that the members of the landing brigade must not be underarmed or include raw recruits. He expressed his preference for the Hotchkiss revolving cannon or rapid-fire gun, believing the larger 3-inch BL howitzer lacked power proportional to both its bore and the weight to be moved by the brigade. He warned that the fire of the revolving cannon was difficult to spot in combat (a reality that came to pass in the Boer War at century's end, given the trouble in spotting the fire of Maxim pompom guns of identical caliber, 37mm). Rodgers urged that once ashore, the field guns and machine guns be kept tactically separate. He wrote that volley versus independent fire had to be a function of range, and he recommended the proper positioning of the battery commander and forward observer for the battery to function effectively, as well as the type of projectile ideal to a given target and range.

However lofty the discussion generated in academic halls, numbers dictated the continued use of the Dahlgren bronze howitzers alongside their steel successors. Landings by American armed naval parties took place frequently during the 1870s and 1880s, in response especially to disturbances in Latin America, Africa, and Asia. Those parts of the world were described at worst as half-civilized to non-civilized, and at best as having unstable governments, rife with coups, revolutions, and civil wars. Under those circumstances, naval landing parties sought to protect not only the lives of Americans and the citizens of other "civilized" nations, and their consulates, but also the private property belonging to those citizens. Virtually to a man, the US

Navy's officer corps believed in the expansion of American commercial interests and the exportation of their products abroad. The influence exerted by large companies, more than likely based in New York, upon the American naval establishment could be enormous. Nonetheless, a delicate ballet had to be danced in order to preserve at least the appearance of US neutrality toward the internal affairs of those nations into which the navy sent landing parties to protect American lives and property.

In that light, both ship and squadron commanders grasped every opportunity to practice for the actual event, not only to keep their crews well drilled, but to impress upon troublesome nations that the US Navy would resort to force if necessary. In 1882, while at Chefoo, China, the commanding officer of the screw sloop *Swatara* requested and eventually received permission to carry out a tactical shore exercise of limited duration. The ship landed nine officers and ninety-four bluejackets and marines, along with two field howitzers—a 12-pounder Dahlgren and a light BL howitzer, each having a thirteen-man gun crew. The two-day maneuver saw practice firing by small arms and the BL howitzer, whereas the party did not fire the Dahlgren smoothbore, rather using its field carriage as a provision cart. On the first day, the entire force deployed to defend against an attack by an imaginary foe; on the second day, the marines separated to act as a hostile force, with the sailors required to send out scouts to locate them, and then to carry out an assault on their defensive positions. At the end of the mock battle, the force re-embarked, with the marines acting as the rear guard in line with current doctrine. The exercise showed that the sandy ground, along with wind and rain, had not been injurious to either the Hotchkiss magazine rifle, newly issued to the naval infantry, or to the 3-inch breechloading howitzer, contrary to the fears expressed by some officers.

The Panama Expedition, April 1885

Despite the frequency in which American bluejackets and marines went ashore to maintain order and protect American interests during the two decades following the Civil War, barring the Korean expedition, there occurred few instances to deploy landing guns, much less fire them. A notable exception, in Colombia's Isthmus of Panama in April 1885, did see a variety of landing guns put on shore, with a handful placed aboard an armored train on the transisthmian railroad.

It is both fair and accurate to describe the civil unrest or outright insurrection that disturbed Panama in the second half of the nineteenth century as nearly interminable. In September–October 1860 and March 1865, American naval landing parties went ashore at Panama City on the Pacific side, typically to occupy the US consulate and/or the station of the American-owned Panama Railroad. It became the turn of the Atlantic side in April 1868, when screw gunboat *Penobscot* sent a party into Aspinwall (subsequently renamed Colón) to protect the railroad and its passengers from widespread rioting. Panamanian insurrectionist forces were at it again within five years. On May 7, 1873, Rear Adm. Charles Steedman brought screw steamer *Pensacola* and screw sloop-of-war *Tuscarora* of the South Pacific Squadron to the

Bay of Panama. At the behest of the American consul and other prominent persons in Panama City, Steedman landed about 200 sailors and marines, along with four howitzers—given the year, they were Dahlgrens—in order to protect non-Colombian nationals, the American consulate, and the railroad. Troubled conditions in September brought in screw sloop *Benicia*, followed by *Pensacola* and Admiral Steedman (soon relieved by Rear Adm. John Almy), this time to put 100 men ashore with a pair of howitzers. Reinforcements joined the original landing party over the next several days, until the unrest ended in October. Nonetheless, *Benicia* lingered in the bay until mid-December.

It took another twelve years until the pot began to bubble again in the state of Panama. The episode, involving armed forces of the United States, became significant for three reasons: (1) It represented yet another example of the influence exerted by powerful and well-heeled American private citizens, resulting in military action being taken to protect the investments and property owned by those persons. (2) The military force landed by the United States this time was especially large, tantamount to the naval brigades deployed by the major European powers. (3) The event introduces Bowman H. McCalla, a physically brave, impetuous, and swashbuckling US naval officer, who will appear twice more in the course of this book.

After several preliminary rounds, including the brief landing in January 1885 of a party from the screw gunboat *Alliance*, bigger things began stirring. As always, whenever confronted by revolution in a country in which American citizens had heavily invested in property, the United States attempted to maintain absolute neutrality in protecting those lives and properties. In consequence, both the governments of the United States and Colombia vacillated aplenty. All of that ended in late March, when insurgent forces seized the Pacific Mail steamship *Colón*, loaded with American arms and ammunition. When the Americans attempted to negotiate peacefully, the rebels took captive Consul Robert Wright and his clerk, two officers from USS *Galena* then at Colón, and two senior staff members of the Pacific Mail Company, and threatened to shoot them all if the US Navy retaliated. Cmdr. Theodore Kane, commanding *Galena*, called the insurgents' bluff and all of the Americans eventually escaped unharmed. When beaten back by loyalist forces, the insurrectionists burned much of the city. The conflagration included the deliberate torching of the US consulate by the rebel leader, Pedro Prestan.

Although crew members from *Galena* freed the American merchant vessel, fought the fire, and saved the offices of the Panama Railroad Company, the wharf of the Pacific Mail Steamship Company, and some foreign-owned buildings in the city, they could not save the consulate. Moreover, Prestan threatened to kill Consul Wright when next he laid eyes on him. If that was not sufficient, another rebel band had captured Panama City on the opposite side of the isthmus, and its chieftain, Rafael Aizpuru, promised to kill *all* Americans. Between Prestan and Aizpuru, they had put a halt to transisthmian rail traffic. That factor as much as anything brought the US Navy ashore in strength.

As the situation in Panama steadily worsened, on April 1, Secretary of the Navy William Whitney ordered Rear Adm. James Jouett, commanding the North Atlantic Squadron,

Wearing flat hats and carrying Winchester-Hotchkiss caliber .45-70 rifles, US Navy bluejackets stand by aboard the screw steamer *Galena*, perhaps off Colón in March 1885. Beginning in 1879, the navy purchased nearly 2,500 of these bolt-action rifles for use by its sailors (but not by its marines, who continued to use the single-shot trapdoor Springfields in the same caliber). (*Naval History and Heritage Command*)

to send the screw sloop *Swatara* from New Orleans directly to Colón. Later that day, Whitney received a frantic telegram from Wright, urging an immediate naval landing, "or Americans must abandon the Isthmus." In New Orleans, Jouett's flagship, the screw frigate *Tennessee*, had been made ready for sea and she stood out the next morning. On the day following, the Pacific Mail steamer *City of Para* took aboard 200 marines in New York and also sailed for Colón. Another 500 marines and bluejackets waited to board Pacific Mail's *Acapulco* if the situation required additional troops, although Whitney remained reluctant to invest too large a force. Meanwhile, senior officers of several New York-based firms, which enjoyed a healthy business on the Isthmus of Panama, clamored for an overwhelming American military presence there. The firms included the Panama Railroad (which had actually been sold to Ferdinand de Lesseps's French canal company three years previous), the Pacific Mail Steamship Company, the Atlas Steamship Company, and the Central and South American Telegraph Company. Under this kind of pressure, Whitney caved and ordered Cmdr. Bowman McCalla to put 250 marines and 150 sailors aboard *Acapulco* and steam for Colón.

When the determined marines, mustered from barracks in Philadelphia, New York, and Boston, embarked aboard *City of Para* in New York, J. B. Houston, president of the Pacific Mail Steamship Company, observed that they were the finest body of men he had ever seen. It was a thought echoed by others among the wealthy Americans with an investment in Panama whose bacon the marines were saving. Had it been another five years on, when Rudyard Kipling would publish his *Barracks Room Ballads* collection of poems, some of those marines, perhaps appreciated only when they were engaged in such a safeguarding effort, might have liberally paraphrased, "But it's the thin blue line of heroes when the trooper's on the tide."

With this exceptionally large American force on the way to Panama, Secretary Whitney began to have second thoughts. Accordingly, he cautioned Admiral Jouett to remain circumspect and to continue to play the neutral card. After all, Whitney rationalized, the legitimate government of Colombia had in no way requested American intervention. Within a few days, however, the American navy began putting its men ashore on both sides of the isthmus.

On the west coast, screw sloop *Shenandoah*, commanded by Capt. Charles Norton, arrived at Panama City on April 7, 1885, and on the following day landed her marine guard, two sailor companies, and the artillery unit to defend the Pacific end of the railroad against the rebel forces of Rafael Aizpuru. Lt. Cmdr. Edwin Longnecker commanded the American force, whose landing coincided with the return of *Alliance* to Colón. On the 10th, *Tennessee* arrived at that port, and *City of Para* disembarked the marine battalion commanded by Maj. Charles Heywood, which began guarding the trains and protecting the railroad property across the isthmus. Finally, on the 15th, *Acapulco* landed her 400 men, with supporting artillery, and Commander McCalla took command of the entire Panama expeditionary force ashore. Six days later, he moved to Panama City, in order to establish his new headquarters there.

Lt. William Kimball supervised the installation of boiler plate on two railroad flat cars, which subsequently protected a Gatling gun, a howitzer (likely a 3-inch steel breechloading piece), and a Hotchkiss revolving cannon. These cars were attached to the west- and eastbound consists, and thus made for a primitive but effective armored train. The first such train proceeded from Colón to the west coast on April 11, carrying men to garrison the station at Matachin and to bolster the force at Panama City, in order to prevent Aizpuru from burning the Pacific coast city, as Prestan had done at Colón.

At sea, Admiral Jouett instructed his ship captains to stop, board, and search all vessels, to prevent suspicious persons and arms and ammunition from being landed. The admiral admitted later that the actions he took were no longer neutral, but were clearly detrimental to the insurgents. He felt, however, that he had no choice if he was to restore order to Panama. Jouett's actions very much met the approval of Commander McCalla and his boss back in Washington, Capt. John Walker, chief of the Bureau of Navigation. Both men were strong interventionists, and ardently believed the United States should impose a strong force in Panama, of a lengthy albeit temporary nature. The choice of McCalla as second-in-command to Jouett had not been the admiral's. The Bureau of Navigation remained also the navy's personnel department, and officer assignments originated with it. Thus Captain Walker determined who should go as Jouett's second-in-command, and for that job he tapped his own deputy.

Walker went to the questionable length of instructing McCalla personally to keep the Navy Department informed of the events in Panama, in effect telling him to go around his superior there, Rear Admiral Jouett. Furthermore, Walker wished the compliant McCalla to survey the islands of the Bay of Panama, using the steam launch from *Shenandoah*. Some of these islands had facilities erected on them by the Pacific Mail Steamship Company, and Walker wanted McCalla to look them over carefully, with an eye toward permanent occupation by the United States at some time in the future. With Walker staunchly behind him, McCalla wired Secretary Whitney early in April, strongly urging that his artillery on shore be doubled from one to two batteries. Whitney's displeasure at this notion was an immediate, unequivocal, and indignant "No!"

Down in Panama, Jouett soon came to regret both the presence of McCalla and his large troop reinforcement aboard *Acapulco*, perceiving the former to be a loose cannon and the latter to be overkill. When word of three shiploads of loyalist troops approaching Panama City reached rebel leader Aizpuru, he ordered barricades thrown up across the city. McCalla used the erecting of such an obstruction between his headquarters in the railroad station and the offices of the Central and South American Telegraph Company, which thus impeded his ability to communicate with the outside world, as an excuse to occupy the entire city. Surprisingly, insurgent Aizpuru was easily pacified, allowing McCalla to proclaim the American occupation and the assumption of police responsibilities within the city, a move that horrified Secretary Whitney and Admiral Jouett.

Although Whitney once again found himself forced by powerful New York businessmen with interests in Panama to back down from his intention to reduce substantially the occupying force, Jouett entrained for Panama City to curb his subordinate. By this time, McCalla had issued a joint statement with Aizpuru—the latter imputing to be the president of a sovereign Panama—lending the strong perception that the United States had taken sides. Jouett managed to smooth things over with the loyalists and the revolution ended in late April 1885. American-operated transportation and communications companies in Panama again functioned without restraint, and the US Navy withdrew its expeditionary force from both sides of the isthmus.

The Register of Naval Guns indicates that the navy landed at least three steel light BL howitzers at Aspinwall (Colón) in April 1885, serial numbers 5, 27, and 29. None of them fired a shot as far as is known. The second piece specified would be lost on *Vandalia* four years later.

At decade's end, four warships of the new steel navy, officially designated the Squadron of Evolution but quickly nicknamed the White Squadron because of their common livery, departed for Portugal. Commanded by John Walker, now a rear admiral, the squadron thereafter steamed south and entered the Mediterranean. In February 1890, the ships lay off Corfu in the Ionian Sea, where 700 seamen and marines of the squadron carried out daily landing exercises for two weeks. In so doing, the United States Navy reminded the European naval powers that its operations were no longer restricted to the Western Hemisphere. Closer to home, it promoted the image of the navy at a time of increased naval spending and reminded congress that the nation's money had been well spent, a suggestion that Walker never lost sight of.

The famous ABC protected cruisers of the New Navy, left to right, *Chicago*, *Boston*, and *Atlanta*, as well as the more distantly seen gunboat *Yorktown*, are anchored early in their voyage in 1889. Not only steam-powered, these otherwise-modern warships were provided sail rig to extend their range, though it was eventually discarded as redundant. At Corfu on the Ionian Sea, the squadron carried out extensive landing operations in order to impress the European powers. Photograph by Edward H. Hart, Detroit Photographic Co. (*Library of Congress*)

Service Losses of the 3-inch Breechloading Howitzer

Although deployed ashore rarely, at least four of the breechloading howitzers were lost in service. Ironically, two of them had served together aboard *Trenton* in early 1877.

Bronze light howitzer No. 13, which had seen use on *Trenton*, went aboard the screw dispatch vessel *Despatch* in December 1881, and was lost with that vessel—which had also served as a training ship for the midshipmen (soon redesignated naval cadets) of the naval academy and as the first presidential yacht—when she foundered off Assateague Island, Virginia, during a gale in October 1891. The men at the nearby US lifesaving station assisted in saving *Despatch*'s entire crew.

Bronze heavy howitzer No. 2, also seeing early service aboard *Trenton*, drew assignment to the paddlewheel steamer *Tallapoosa* in July 1883, along with sister bronze heavy howitzer No. 3. *Tallapoosa* served as a dispatch and training vessel, and thirteen years before had carried Admiral David Farragut on the final voyage before his death at the Portsmouth Navy Yard. Howitzer No. 3 was removed in 1884 at the New York Navy Yard, but No. 2 stayed on board, to be lost when *Tallapoosa* collided in August of that year with the schooner *J. S. Lowell* off Vineyard Haven, Massachusetts. *Tallapoosa* herself was salvaged at no little cost, rebuilt, and returned to the fleet, serving until 1892.

Steel light howitzer No. 27, one of the pieces landed at Aspinwall (Colón), Panama, during the April 1885 expedition, went aboard screw sloop *Vandalia* in March 1888. One year later, it was lost with that vessel while anchored in Apia harbor, Samoa, when a ferocious cyclone struck without warning. Although most of *Vandalia*'s crew went safely aboard the adjacent frigate *Trenton*, also wrecked, she lost forty-three men to the typhoon, including her commanding officer, Capt. Cornelius Schoonmaker. The third American warship present, screw gunboat *Nipsic* would be salvaged, rebuilt, and returned to the fleet.

Steel light howitzer No. 38 was assigned to screw sloop *Kearsarge* twice, in July 1883 and September 1888. After the Civil War, when she had gained fame as the victor of the June 19, 1864 duel with CSS *Alabama* off Cherbourg, *Kearsarge* saw various duties. In the Caribbean, steaming to Bluefields, Nicaragua, on February 2, 1894, she struck a reef off Roncador Cay. *Kearsarge* was abandoned over the next day without the loss of any her crew. Interestingly, the loss of BL howitzer No. 38 is given in the naval ordnance register as February 10, suggesting that some salvage effort was made in the days following, but it did not include the howitzer.

Another, very likely steel light 3-inch BL howitzer, serial number uncertain, may have been abandoned during the Peking relief expedition in 1900. The details known so far are provided in the relevant chapter.

Reflections

To the United States Navy of the nineteenth century, the 3-inch breechloading howitzer represented a small but useful guinea pig in the evolution toward modern steel guns. After the early 1880s, boat and landing howitzer construction lost priority and such production dwindled, in large measure because of funding. The manufacturing and testing program remained valuable to the navy, however, leading in particular to increasingly sophisticated and effective designs of its breech and gas check mechanisms. The howitzers themselves continued aboard American warships well into the 1890s, until replaced by the Fletcher generation of guns just prior to the Spanish-American War. Some of the steel BL howitzers could still be found during that brief conflict, aboard older warships, auxiliaries, and lesser vessels. Given the limited number of these howitzers, a variety of other pieces served as well in the boat and landing role, most commonly Gatling and Hotchkiss guns, both types found in differing calibers as the nineteenth century came to an end.

References Consulted

Allard, D. C., "The Influence of the United States Navy upon the American Steel Industry, 1880–1890," master's thesis, Georgetown University, 1959, pp. 23–25, 32–34, 38–39, 41–42.

Author not specified, "The Midvale Steel Company, Fiftieth Anniversary 1867–1917" (Philadelphia, 1917), pp. 20–21.

Cooke, A. P., *A Text-Book of Naval Ordnance and Gunnery, Prepared for the Use of the Cadet Midshipmen at the United States Naval Academy*, 2nd ed. (New York: John Wiley, 1880), pp. 393–97, 631–32.

Dictionary of American Naval Fighting Ships, Naval History and Heritage Command (online). Consulted for each US naval warship and auxiliary mentioned in this chapter.

Ellsworth, H. A., *One Hundred Eighty Landings by United States Marines 1800–1934*, History and Museums Division, HQ, US Marine Corps (Washington, DC: Govt. Print. Office, 1974 [originally printed 1934]), pp. 46–51.

Glasow, R. D., "Prelude to a Naval Renaissance: Ordnance Innovation in the United States Navy during the 1870s," doctoral dissertation, Department of History, University of Delaware, 1978, pp. 120–22, 124, 128–30, 132, 174–78, 181–86, 192–96, 201, 243–44, 254–56, 310–12.

Hagan, K. J., *American Gunboat Diplomacy and the Old Navy, 1877–1889*, Contributions in Military History No. 4 (Westport, CT: Greenwood Press, 1973), pp. 5–6, 53–56, 160–87.

House, K., "Caldwell Hart Colt: The Man, The Mystery," *The Texas Gun Collector*, Fall 2012, p. 29.

Hutchins, C. T., "The Naval Brigade, Its Organization, Equipment, and Tactics," US Naval Institute (hereafter USNI), *Proceedings* 1887, 13(3): 303–41. Discussion following, 13(4): 511–46.

Ingersoll, R. R., "The Present Course in Ordnance and Gunnery at the Naval Academy," USNI, *Proceedings*, 1886, 12(2): 91–103.

Jaques, W. H., "The Establishment of Steel Gun Factories in the United States," USNI, *Proceedings*, 1884, 10(4): 531–909.

Lyon, H. W., "Our Rifled Ordnance," USNI, *Proceedings*, 1880, 6(1): 1–13. Discussion following, 6(1): 14–15.

Mason, T. B. M., "On the Employment of Boat Guns as Light Artillery for Landing Parties," USNI, *Record*, 1879, 5(3): 207–23. Discussion following, 5(3): 224–30.

McAulay, J. D., *Rifles of the United States Navy & Marine Corps 1866–1917* (Woonsocket, RI: Mowbray Publishing, 2017), pp. 53, 59–62, 74–77, 95–96, 113–15.

Meade, R. W., and Sicard, M., *Ordnance Instructions for the United States Navy*, 5th ed. (Washington, DC: Govt. Print. Office, 1880), pp. 148–49, 191–92, 195–201, 210, 227–65.

Misa, T. J., *A Nation of Steel: The Making of Modern America 1865–1925* (Baltimore: Johns Hopkins Univ. Press, 1995), pp. 96–97.

Parker, F. A., *The Naval Howitzer Ashore* (New York: Van Nostrand, 1865).

———, *The Naval Howitzer Afloat* (New York: Van Nostrand, 1866).

———, "Our Fleet Manœuvres in the Bay of Florida, and the Navy of the Future," USNI, *Record*, 1874, 1(8): 163–76.

Rodgers, W. L., "Notes on the Naval Brigade," USNI, *Proceedings*, 1888, 14(1): 57–96.

Sicard, M., *Description of Naval 3-in. B.L. Howitzers, with Instructions for Their Use and Care*, Bureau of Ordnance, US Navy, 1876.

Smith, J. E., *Small Arms of the World*, 10th rev. edit. (Harrisburg: Stackpole, 1973), p. 129.

Soley, J. C., "The Naval Brigade," USNI, *Proceedings*, 1880, 6(3): 271–90. Discussion following, 6(3): 291–94.

Stone, C. A., "A General Description of the Ordnance and Torpedo Outfit of the U.S.S. *Trenton* (2nd Rate)," USNI, *Record*, 1877, 3(6): 89–93.

US National Archives, Record Group 74, Records of the Bureau of Ordnance, Washington, DC, and Waltham, MA.

US Navy, "Landing Exercise, U.S.S. *Swatara* 1882," Bureau of Ordnance Circular No. 13, *in* USNI, *Proceedings*, 1882, 8(3): 499–502.

US Navy Dept., *Annual Report of the Secretary of the Navy, 1875*, p. 19.

———, *Annual Report*, Bureau of Ordnance, *1873*, p. 103; *1874*, p. 83; *1876*, pp. 111–12, 201–28; *1878*, p. 65; *1880*, p. 80; *1881*, pp. 184–85; *1882*, Vol. 3, pp. 5–6.

Wicks, D. L., "The First Cruise of the Squadron of Evolution," *Military Affairs*, April 1980, 44(2): 64–69.

Wingfield, T. C., and Meyen, J. E., (eds.), "Lillich on the Forcible Protection of Nationals Abroad," Appendix I. "A Chronological List of Cases Involving the Landing of United States Forces to Protect the Lives and Property of Nationals Abroad Prior to World War II," *International Law Studies* 77: 131–32.

5
Smaller Caliber Additions for Landing Parties

Interlude

In the 1870s and 1880s, the US Navy purchased a sizable number of shoulder arms, machine guns, and small cannon, both to enhance its capabilities afloat and ashore, and to act as experimental pieces for the modern age so clearly presaged by the developments in Europe and Russia. These acquisitions came hand in glove with two important developments for the naval service during that decade: the evolution of the New Navy and the building of the Naval Gun Factory. The navy contracted for its first modern steel ships during the 1880s, to be armed with large guns built at its gun factory from forged steel provided by commercial firms. In 1883, after surveying the national steel companies, the joint navy-army Gun Foundry Board, with Rear Adm. Edward Simpson as president, visited the armament facilities of England, France, and Russia—Germany, as represented by Krupp, declined the request—and one year later issued its report. That report recommended that the army gun factory be built at Watervliet Arsenal near Albany, New York, and the naval facility at the Washington Navy Yard. The Naval Gun Factory was formally established at that yard in 1886.

Shoulder Arms and Revolvers

Soon after the end of the Civil War, the services decided to adopt small arms firing metallic cartridges, with the shoulder weapons loading from the breech. That determination led the navy to purchase a succession of Remington rolling block types, first the Model 1867 carbine firing the .50-45 cartridge, and then two successive versions of the Model 1870 rifle firing the .50-70 Government cartridge and differing essentially in the location of the rear sight. A curious arrangement allowed the navy to sell back most of the first variant purchased in order to acquire a greater number of the second. All of those shoulder weapons fired single rounds, however, and contrary to the lukewarm interest by the army, the navy realized the need for magazine-fed, bolt-action repeating rifles to arm its sailors before decade's end. Considerably assisting the task, by mutual accord during the 1870s, the services reduced the infantry round from .50-caliber to .45-caliber, with 70 grains of black powder propelling a 405-grain (and later 500-grain) bullet. The navy's decision for adoption thereupon hinged on what type of magazine would feed the breech action.

In 1879, in conformity with army purchases, the navy took delivery of the first of nearly 2,500 Hotchkiss-designed bolt-action rifles, possessing an early type bolt that operated through a single rear-locking lug. The rifle was fed from a tubular magazine in the buttstock holding five .45-70 Government rounds, loaded from the receiver rearward and then pushed forward by a coil spring. This rifle would be acquired in two different variants, assembled by Springfield Armory from Winchester-made parts, the second variant being provided in the early 1880s. The difference between the two variants lay in the nature of the magazine cutoff, that is, the device that forced the operation as a single-shot weapon, with the rounds in the magazine then held strictly in reserve. In the first model Winchester-Hotchkiss, the circular cutoff device was large and located only on the right side of the wooden stock between the receiver and the trigger guard; its size often contributed to the splitting of the single-piece wooden stock. In the second model, the stock remained a single piece, but the smaller and higher cutoff lever functioned from both sides of the receiver. The third model, brought out in 1883, which the navy failed to adopt but the army did acquire for additional field testing, had a two-piece wooden stock separated by a solid metal receiver. This part held the much-modified cutoff lever, at last solving the problem of the splitting stocks. The Winchester-Hotchkiss rifle, of either model adopted by the navy, saw service aboard screw sloops *Alaska*, *Galena*, *Lancaster*, *Shenandoah*, *Swatara*, and *Wachusett*, as well as screw gunboat *Yantic*, sidewheel gunboat *Monocacy*, screw tug *Palos*, and sail training ship *Monongahela*, to specify only several ships.

Under the watchful eyes of a commissioned officer and a senior petty officer, young apprentices, many of them teenagers, undergo drill with their Winchester-Hotchkiss bolt-action rifles, caliber .45-70. Their ship is the converted sailing sloop *Monongahela* and the date is June 1891. It was the ship's first tour as a training vessel. A century later such sailors-in-training are called seamen recruits. Photograph by Frank H. Child. (*Naval History and Heritage Command*)

To continue the testing of the other magazine types, in the early 1880s the navy purchased 250 Remington-Keene rifles, the result of the collaboration between Remington Arms and inventor John Keene of New Jersey. The piece held nine rounds in a tubular magazine beneath the barrel, with the weapon having an offset cleaning rod to accommodate the under-barrel feed. The bolt incorporated a single locking lug and a sliding magazine cutoff on the left side allowed single-shot fire. The belief existed that the men found a bolt action initially awkward, so this possibility may explain the otherwise odd coupling of such action with a familiar external hammer—actually a hammer-shaped cocking device on the end of the bolt, with the rounds fired by a striker within the bolt. If so, more the irony, because the rifle was never popular. The Remington-Keene saw service aboard the steam frigate *Trenton* (eleven 8-inch MLR), which frequently acted as a guinea pig for new ordnance before her loss at decade's end, and on the Great Lakes sidewheel gunboat *Michigan*, as well as aboard the older steam frigates *Powhatan* and *Tennessee* (various armaments). Most of these rifles had been removed from service by the mid-1880s.

During their years of use, extending into the 1890s, the navy found certain aspects of both the Winchester-Hotchkiss and Remington-Keene rifles to be disadvantageous. The major objection in each case remained the tubular magazine, with the contact between the fore end of one bullet and the center primer of the round in front of it. When the rifleman jostled his weapon in negotiating rough and broken ground, there existed the inherent danger of discharging that primer, and thus its round, and then the other rounds in a chain reaction injurious to that individual. Although exaggerated, that concern persisted.

The navy considered the most successful of the new shoulder arms to be the Lee (or Remington-Lee) bolt-action rifle, with a detachable box magazine positioned just forward of the trigger guard. That magazine held five .45-70 rounds and did not possess a cutoff device. However, the rifle could be fired as a single-shot weapon by not inserting the box magazine. The navy contracted initially for 300 Lee rifles, Model 1879, manufactured by both Sharps and Remington, followed by 700 more, manufactured solely by Remington after Sharps had failed in 1881. These rifles served aboard the first warships of the New Navy, the so-called ABCD ships, which included the protected cruisers *Atlanta* and *Boston* (both two 8-inch BL guns).

Several improvements incorporated from an interim trials rifle made for the army appeared in the Model 1885 Remington-Lee, caliber .45-70. Apart from the somewhat longer barrel, most noticeable externally were the operating handle moved to the rear of the bolt, and the guide grooves stamped onto the box magazine. The navy ordered a total of around 3,400 of this model, the first of which went to the protected cruiser *Philadelphia* (twelve 6-inch guns). This model rifle eventually filled the arms racks of a wide variety of ships in the late 1880s and early 1890s, including several additional protected cruisers, the unprotected cruiser *Detroit* (two 6-inch guns), the large-gun monitor *Monterey* (one each twin 12-inch and twin 10-inch gun turret), the gunboats *Concord* (six 6-inch guns) and *Castine* (eight 4-inch guns), the pneumatic dynamite gun ship *Vesuvius* (three 15-inch pneumatic guns), and the ex-whaler/Arctic rescue vessel *Thetis* (one 47mm or 53mm revolving cannon), to name only several. In the early 1890s, the navy also provided the Remington-Lee to various state naval militias, including those of California and Massachusetts. The Remington-Lee served to the satisfaction of the navy until the reduction in the caliber of the infantry rifle during the mid-1890s.

The members of the US Marine Corps, tightly tied to the dictates and fashions of the army, continued to use obsolete or obsolescent rifles throughout this period. During the 1871 landing in Korea, while the sailors had landed with Remington rolling block carbines firing metallic cartridges, the marines had gone ashore lamentably armed with muzzleloading Springfield and Plymouth muskets. Throughout the following decades, the marines continued the unfortunate trend, first with upgraded single-shot Springfield .50-caliber rifles having the Allin conversion, and later with much the same rifle in .45-caliber using the improved trapdoor action. In his prize essay presented before the US Naval Institute in 1887, Lt. Charles Hutchins deplored the prolonged equipping of "this well trained corps" with arms from "bygone days." It was only in the mid-1890s that seamen and marines were issued the same model repeating rifle.

In the late 1870s through the mid-1880s, the US Navy purchased and tested three bolt-action repeating rifles in caliber .45-70, adopting the first and third in sizable numbers. The Winchester-Hotchkiss (top) had a tubular magazine in the buttstock, accepting five rounds. The Remington-Keene (middle) bore a tubular magazine beneath the barrel, loading nine rounds. To the rear of the bolt, the hammer-like device did not fire the bullet, but cocked the firing pin. Both of these rifles incorporated a magazine cutoff, providing a single-shot option. The Remington-Lee (bottom) offered the most advanced design, with a five-round detachable magazine under the receiver. The one shown is the later variant. (*From the author's collection*)

Carefully observed by their officer and his field music (bugler), marine skirmishers from the battleship *Maine*, seeking cover behind a railroad embankment, reflect the end of an era. They still wear chasseur-style caps (and white gloves for reason unknown) and carry .45-70 single-shot trapdoor Springfield rifles, both items decades old. Two or three years distant, the mid-1890s will bring transformative change to the corps, both in its uniform items and shoulder arm. The distant cows remain indifferent to it all. Photograph by Edward H. Hart, Detroit Photographic Co. (*Library of Congress*)

In 1873, the US Navy made the decision to adopt revolvers firing metallic cartridges, but some confusion remains about the details of the Colt revolvers it obtained. It appears that the service's single-action Models 1871 .38-caliber revolvers were conversions of the Colt Models 1851 and 1861 Navy .36-caliber cap-and-ball percussion pieces, largely using spare parts. Its single-action Model 1872 .44-caliber revolver was based on the design of the Colt Model 1860 Army .44-caliber percussion revolver, but manufactured *de novo* by Colt's in the early 1870s. For these modernized pistols, a loading gate was added on the right rear of the cylinder and a spring ejector rod under the barrel, replacing the rammer assembly, to expel the spent cases. The lack of a top strap above the cylinder persisted as a significant weakness from the original percussion revolver design, indeed leading the army to reject the design. The navy made a similar arrangement with the Remington Arms Company in 1875 to convert its solid-frame .36-caliber New Model Navy cap-and-ball revolver to fire .38-caliber centerfire rounds. During the following years, the navy obtained a number of Colt and Smith & Wesson revolvers on an experimental basis, but the next large investment occurred with the change to the Colt New Navy double-action .38-caliber revolver in 1889.

Machine Guns

The Gatling hand-cranked machine gun, manufactured by the Colt's Patent Fire Arms Company, remained in standard use by both the army and navy during the last few decades of the nineteenth century. The namesake of inventor Richard Jordan Gatling, M.D., then living in Indiana, the gun consisted of five, six, or ten barrels in cylindrical array that when cranked, revolved around a central axis. Because most of these machine guns mounted their crank on the side, perpendicular to the barrels, a gear reduction was necessary to turn the barrel assembly. Each barrel had its own bolt (or lock), through which the firing pin assembly ran. In the various 1870s models, mildly differing arrangements of lugs and cams operated the firing mechanism as each barrel cycled, pulling its firing pin back and compressing a coiled spring. After a cartridge had been fed into the breech, a cam released the firing pin, allowing the spring to propel the pin forward against the cartridge primer, firing it. When that barrel reached the bottom of the array, the mechanism expelled the spent case through the ejection port. A cartridge that had failed to discharge did not affect the firing cycle, and the intact cartridge ejected in the same manner as did a spent case. When these usually reliable guns jammed, the blame more often than not lay with faulty ammunition feed or improperly designed and/or manufactured cartridge cases. Importantly, cases made of drawn brass worked best.

When in his poem, "Vitaï Lampada" (1892), Henry Newbolt presumably alluded to the battle of Abu Klea in the Sudan, January 1885, "The Gatling's jammed and the Colonel dead, And the regiment blind with dust and smoke," the poet exercised a bit of license. The malfunctioning machine gun in British hands had actually been a Gardner. Of interest, however, the small naval contingent present manned that machine gun in the Khartoum relief force consisting mostly of army units. On an experimental basis, the US Navy also purchased the Gardner machine gun, as it did the Lowell battery gun. Similar to the Gatling, both the Gardner and Lowell machine guns were hand-cranked.

The navy adopted two basic Gatling models during the mid-1870s, the long and the short, each of ten barrels. The long Gatling gun had 32-inch exposed barrels, measured 49 inches overall, and weighed 200 pounds; the early short Gatling had 18-inch barrels, measured 35.5 inches overall, and weighed 135 pounds. A later short naval Gatling possessed 25- or 26-inch barrels, with proportionately larger dimensions than its predecessor. Both short models generally had their barrels encased in a bronze jacket, as did the Gatling Bulldog, also acquired in a small number by the navy. The Bulldog was essentially available off the shelf in five, six, and ten 18-inch barrels. This diminutive Gatling gun mounted its firing crank through the cascabel plate on the rear of the breech, obviating the need for reduction gears to turn the barrels and permitting a faster rate of fire, as high as 1,000 rounds per minute. The Bulldog found useful deployment as a small and handy boat gun.

Although the bronze shroud protecting the rotating barrels offered suitable protection for service at sea, this addition proved an expensive one. After the decade between 1879 and 1889, during which all government-purchased Gatlings possessed

This early 1890s photograph shows naval cadets in formation at the US Naval Academy preparing for a tactical exercise. The infantry carry Winchester-Hotchkiss .45-70 rifles, with the artillery present consisting of long Gatling guns of that caliber, and beyond, 6-pounder breechloading landing guns that shortly before had replaced the 3-inch BL howitzers of the previous generation. (*US Naval Academy Archives*)

such a jacket, naval models thereafter returned to the exposed-barrels configuration. A subsequent long version borrowed another Bulldog innovation. In the Model 1883, the operating crank could be moved from the standard location on the right side of the breech casing to the rear of the cascabel plate to achieve a faster firing rate.

The feed mechanisms for the Gatling gun evolved over the course of four decades. In the 1860s version, a simple hopper, offset from the axis of the barrels, served for the powder and ball rounds held in ejectable steel cases requiring percussion caps. With the advent of fixed metallic cartridges during the 1870s, the Gatling gun came into its own. By mid-decade, the previous .50-caliber guns had been supplanted by those firing .45-70 Government rounds. Such conversion was readily accomplished, because many of the operating parts from the larger bore gun served well enough in the smaller bore successor, and did not require changing out, at least in the near term.

In the period just before the caliber change, the company introduced the Broadwell drum feed, devised by Gatling employee Lewis Broadwell. The first variant held 400 cartridges, apportioned in twenty stacks of twenty rounds each that emptied by gravity, assisted by a weighted follower on top of each stack. After one stack had been fired, a gunner turned the drum manually against its base plate (much as did the magazine on the twentieth-century Kodak Carousel slide projector) to line up the next stack. This configuration proved to be too heavy, so the company redesigned the drum to hold 240 rounds, in sixteen stacks of fifteen rounds each. To load, the gun crew placed the drum onto a central spindle atop the gun, fitting two ribs on its base plate into two notches on the breech casing that diverged muzzleward from near the spindle. Thus mounted, the drum was held in place horizontally by the rib-and-notch arrangement and vertically by gravity. Too great an elevation or depression of the gun, however, could cause problems with both the ammunition feed and the retention of the drum atop the gun.

From the early 1880s, the navy found a much better and more foolproof loading arrangement with the Bruce feed, designed by Gatling employee Lucien Bruce. This feed consisted of a tall, vertical bronze frame mounting a steel bar with two T-shaped slots milled into its face to accept the rim of the cartridge. The bar swung from side to side, so the two slots alternately aligned with the lower central channel leading to the gun's feed hopper. To load, a cardboard box—or wooden box made by the ship's carpenter—holding twenty rounds with their noses inward was placed against the upper extension of the bronze plate and brought downward so that the rims of the two rows of cartridges in the loading box entered the channels. The box was then pulled forward, away from the cartridges now loaded into the tracks, and

The Model 1877 Gatling Bulldog, in this instance having ten .45-caliber barrels, mounted its operating crank on the rear cascabel plate, obviating the need for gear reduction and thus permitting a faster rate of fire. For loading rounds, the Bruce gravity feed shown here most often saw use, finding preference in US service until the Gatling gun became obsolete. The two ammunition chests transported on the landing carriage could be tilted forward in order to achieve a wider arc of fire for the gun. Boxes for loading the Bruce feed lie behind the right wheel. (*Armas de las Islas Filipinas*, online)

which fed by gravity into the hopper. After one track had emptied, the weight of the cartridges in the other track caused it to shift over to the central channel without gun crew intervention. This operation permitted uninterrupted fire as long as the crew continued to insert rounds efficiently from the ammunition boxes into the feed. In the Model 1879 Gatling gun, the Bruce feed required an adapter base to clamp onto the hopper designed for the earlier tin feed, but in the Model 1881, the breech casing included a dedicated hopper for the Bruce feed.

The Accles positive feed drum, invented by James Accles, yet another Gatling employee, became available for fleet use during the mid-1880s. It was a donut-shaped device, with the hole of the donut aligned perpendicular to the long axis of the gun (the gunner could peer through the hole toward the muzzle end of his gun). This feed worked on a mechanical basis rather than by gravity. On the plus side, this feed permitted extreme angles of elevation and depression in firing the piece. Although enjoying a period of popularity, the feed proved overly heavy and complicated, and troublesome to load, even with new chargers designed independently by Accles and Bruce. The Accles feed failed to stand up to the rigors of military and naval use, and the services retired it in the late 1890s. During that decade, a modified positive feed apparatus of the Bruce pattern supplanted the Accles feed on the Gatling guns previously acquired. That ammunition feed continued as the device of choice until the Gatling gun, by then chambered to .30-caliber, went out of service during the early twentieth century.

In the last quarter of the nineteenth century, there arose numerous occasions in which the US Navy landed Gatling guns to support shore parties. A number of different field carriages mounted these machine guns, one early model nearly identical to the iron carriage for the Dahlgren howitzer, and a sturdier one resembling the army field carriage, although lighter in construction. It required the ammunition chests straddling the gun to be tilted forward at a considerable angle, in order to provide a clear line of sight for firing. An interesting early feature of the gun lay in its automatic traverse, accomplished by an oscillator device that worked off a multi-tracked cam at the end of the crankshaft, and caused the gun to swing back and forth on either shipboard mount or field carriage.

More often than not, the short Gatling was brought ashore and mounted on a tripod having rugged wooden legs, atop which a yoke and turntable sat, to mount and swivel the gun. The Model 1877 long Gatling did away with the cam-operated oscillator, utilizing instead the turntable and yoke introduced in the tripod, and that feature enabled a traverse of 80 degrees and a wide dispersion pattern.

Although the Gatling machine gun firing the large .45-70 Government cartridge proved to be a deadly combination, that reality was not always fully appreciated by naval officers more attentive to long-range fire by big guns. In December 1879, Lt. Cmdr. William Folger, newly commanding the screw sloop *Swatara*—and to be chief of the Bureau of Ordnance in the early 1890s—complained that shipboard detachments were investing too little time training on their Gatling guns, and were thus often deficient in the guns' proper use. Because both naval academy cadets and apprentice seamen in navy boot camps were receiving adequate instruction on Gatling guns, the deficiency likely had to lie either with the ships' commanding officers or in the Navy Department's standing instructions.

The final decade of the nineteenth century would inaugurate the era of the fully automatic machine gun, embodied in the wicked effects of the arm invented by Maine Yankee Hiram Maxim, and covered in the next chapter.

Small-Caliber Cannon

Between the late 1870s and the 1890s, the navy adopted guns whose bores measured between the infantry caliber rifles and machine guns, and the standard 3-inch boat and landing howitzer. The attractive feature of this ordnance lay in its dual nature: it served aboard ship on suitable deck mounts, but could be quickly transferred to boat and field carriages.

Among its first acquisitions, around 1878, the navy obtained a Hotchkiss revolving (or revolver) cannon by purchase or transfer from the army. Benjamin Hotchkiss, born in Connecticut and thoroughly trained at Colt's Hartford plant, began ordnance work in Paris in 1867. One year later, in the Congress of St. Petersburg, the European military powers voted to prohibit the use of explosive infantry rounds whose projectiles weighed less than 400 grams (0.88 pound). Following the Franco-Prussian War of 1870–1871, so disastrous for France, Hotchkiss & Cie began the manufacture of their signature revolving cannon, commonly available in three allowable calibers: 37mm (1-pounder), 47mm (3-pounder), and 53mm (4-pounder). The nations either purchasing these guns outright or building them under license included France, Germany, Russia, the Netherlands, Argentina, Brazil, and China. In the United States, Pratt & Whitney of Hartford, Connecticut, began a contractual relationship with Hotchkiss in the 1880s.

When in naval service, this shoulder-pointed gun was intended primarily to defend warships from the fast and agile torpedo boats appearing ever more numerously across the globe, particularly in the navies of those countries unable to afford large capital ships. Such a gun mounted in the bow of a warship's steam launch (e.g. that of *Hartford*) or steam cutter (e.g. that of *Philadelphia*) yielded a small gunboat for investigating coves and mouths of creeks or for supporting landings. The navy obtained seventy-seven Hotchkiss 37mm revolving cannon as its standard weapon—the first one from the army, the subsequent thirty-two by the mid-1880s from Hotchkiss in Paris, and the final forty-four between 1889 and 1891 from Pratt & Whitney—along with twenty-six 47mm guns and a handful of 53mm pieces during the 1880s.

The Hotchkiss revolving cannon resembled externally the Gatling machine gun, with the clockwise movement of its barrels accomplished by turning a hand crank. It differed greatly, however, in its firing action. Whereas the Gatling possessed a firing bolt for every barrel in the gun, with some phase of loading, firing, and ejecting being done at a given moment by a different barrel in the cycle, the Hotchkiss possessed only one fixed firing lock, with the five barrels alone rotating. The action was thus intermittent, with three of the five barrels in play, one near the top being loaded (though not completely), one at the bottom firing, and one having its spent cartridge case extracted, before the array advanced one barrel. The remaining two barrels "in the queue" held live rounds being gradually and gently pushed home.

Reduced to the simplest terms, an irregular and continuously rotating worm gear carried on the crankshaft accomplished these ends. The worm was partly helical and partly circular, with the former portion causing the barrel assembly to turn 72 degrees and the latter portion thereupon to lock the assembly in place. During that turning, a combination of a spiral cam, lug, and leaf spring cocked and fired the 37mm cartridge (the two larger caliber guns were fired by a trigger on a pistol grip under the breech end). Gears connected the loading piston and extractor, such that they worked reciprocally to load a new cartridge and extract the spent case.

Generally, the 37mm revolving cannon, 20 calibers in length and weighing 462 pounds, was the only Hotchkiss revolving cannon brought ashore. The 47mm/25-caliber piece at 1,265 pounds and the 53mm/27-caliber piece at 2,200 pounds were too heavy and ungainly for use as landing guns. A field carriage became available for the 47mm piece, although not adopted or used by the US Navy. The light Hotchkiss field carriage weighed 638 pounds and the limber nearly 1,000 pounds when carrying its 300 rounds of ammunition, so together even the light gun, field carriage, and

In 1896, six company-grade marine officers, wearing undress blouses (also called patrol jackets), pose beside two landing guns assigned to the Washington Marine Barracks. On the left is an encased short Gatling gun in caliber .45-70, and beside it, a Hotchkiss 37mm revolving cannon. Both the weight and the gunners' seats positioned atop the larger field carriage indicate animal haulage to be necessary ashore. From left to right: Capt. Paul Murphy, Capt. Thomas Wood, 2nd Lt. John H. Russell, Jr. (he became the sixteenth commandant of the US Marine Corps), 2nd Lt. Louis Magill, 2nd Lt. John Myers, and 1st Lt. Joseph Pendleton (the last three officers won fame in marine actions during the next few decades). (*US Marine Corps Archives*)

loaded limber weighed more than a ton and presented a serious challenge to sailors ashore. Jury-mounting on other existing carriages may have been done, but more often than not when brought to shore, the light Hotchkiss revolving cannon was placed in a fixed position or mounted aboard a protected railroad car. On the field carriage, the gun had an elevation of 12 degrees and a depression of 5 degrees, as contrasted with an elevation of 15 degrees and a depression of 35 degrees on the shipboard mount. The gun could hit at 3,000 yards, with much more reliable accuracy achieved at a thousand yards less.

One man of the three-man gun crew fed ammunition in clips into an angled, offset hopper on the receiver holding ten rounds, although the normal practice thereafter was to replace each round as fired. The gun captain both fired the gun and adjusted the elevation and traverse, and the third man brought ammunition to the loader. The projectiles available for the 37mm revolving cannon were shot, explosive steel shell, and cast-iron common shell, as well as canister (inexactly termed case). Although the navy experimented with the base-fused steel shell, meant for deep penetration into hostile naval vessels, the standard shipboard round remained the nose-fused common shell. Its 1-pound projectile held a bursting charge of a bit more than three-quarters of an ounce and bore a soft brass rotating band in order to impart spin. The canister round contained twenty-eight 7-ounce lead balls packed in sawdust, with a payload weight of 1.25 pounds. Having 2.8 ounces of propellant, the common shell complete weighed 1.4 pounds and the canister round 1.6 pounds. The cartridge cases were

Two sailors man a Hotchkiss 37mm revolving cannon mounted on the ship's rail, with the man on the right loading rounds into the angled feed on the left side of the receiver. These guns could be removed from their shipboard mountings and placed upon field carriages. Two hand weapons hang from the men's belts: an Ames cutlass and a Colt Navy model revolver, by this decade likely converted to fire metal cartridges. On deck mounts, these light cannon were sometimes provided shields against small arms fire. Drawn by Thure de Thulstrup for *Harper's Weekly*, May 1890.

made initially of coiled brass, but after a period necessary to perfect them, drawn brass cases replaced the original type around 1885. This evolution assumed particular importance in the canister rounds, for which the brass case needed to be quite thin.

Ashore, both common shell and canister saw use. Because of the intention for the shell to explode upon impact, detonation was by means of a Hotchkiss percussion fuse. Most armies and navies set great store by this fuse, not only for its reliability and safety, but because it could be used for both direct and grazing fire.

The fuse's outer brass cylindrical body fitted into the nose of the shell, having an upper, ogival-shaped shoulder to secure it in place. The main body held a cylindrical brass plunger (or slider) weighted with lead. The plunger's central chamber or magazine was filled with gunpowder, above which sat a percussion cap containing fulminate of mercury. A small brass wire was fashioned into a U-shaped loop within the plunger, such that its two ends projected into the base of the plunger. Finally, a lead safety plug fitted tightly into the hole in the bottom of the main body, forcing against the brass wire ends and holding the plunger firmly in place. Upon firing, the projectile's momentum thrust the plunger rearward, expelling the safety plug. The wire ends still steadied the plunger against the rotation of the shell and the descending arc of the trajectory. Upon impact, the plunger shot forward against the conical firing pin in the nose of the fuse and the percussion cap exploded the magazine, which set off the bursting charge of the projectile.

Neither time nor tide waits for any man or machine, however, and the Hotchkiss revolving cannon all too soon passed into obsolescence, although the odd survivor would soldier on into World War I. Torpedo boats and other potential target vessels had gotten bigger and tougher, with an increasing thickness in their steel plate. The introduction of guns having much the same bore as the Hotchkiss revolving cannon surpassed the earlier pieces in the size of their rounds, muzzle velocities, range, and general performance. The new 37mm (1-pounder), 47mm (3-pounder), and 57mm (6-pounder) guns were single-barrel, single-shot pieces having a vertical sliding wedge breechblock and a shoulder piece for pointing. These pieces lacked the sheer mass of metal and thus awkwardness of the revolving guns. Some of them possessed semiautomatic breech mechanisms, in which the combination of recoil and counter-recoil manipulated the breechblock and ejected the empty cartridge case without intercession by the gun crew, making them quite fast in action. In various types and sizes, they would see use as shipboard guns during the final decade of the nineteenth century and well into the twentieth century, easily shifted from their deck mounts to boat and landing carriages. During the 1890s, too, the base fuse on the projectile gained preference over the nose fuse.

From the mid-1880s, the navy purchased from either Hotchkiss or Pratt & Whitney twenty-four of the light, short 1-pounder guns, firing rounds with metal cartridge cases 94 mm long, identical to those chambered in the revolving cannon. The navy soon found them inadequate in power, given their muzzle velocity of about 1,320 feet per second, although coupled with a violent recoil when mounted on a field carriage. By the early 1890s, the more powerful heavy, long 1-pounders became available, using rounds with cases 136 mm long and generating a muzzle velocity of 1,800 feet

During their first annual training exercise on Fishers Island in summer 1891, members of the New York Naval Militia deploy a Hotchkiss short 1-pounder landing gun. The shoulder arms issued to this militia, organized two years earlier, are clearly a mixed lot. The kneeling sailor on the left cocks the hammer on the rolling block of his Remington Navy Model 1870, firing .50-70 rounds, while the prone man to the right of the landing gun reaches for a .45-70 cartridge for his Winchester-Hotchkiss second model bolt-action rifle, both weapons being navy hand-me-downs. (*New York National Guard*)

Apprentice seamen from the Great Lakes Naval Training Station carry out a landing exercise on Lake Michigan in 1918, with the gunner about to discharge the Hotchkiss long 1-pounder landing gun. Firing the longer 37mm round, the gun generated an appreciably higher muzzle velocity than its short predecessor. Such light artillery remained in useful naval service well into the twentieth century. The national ensign shows a union of forty-eight stars, a 1916 change from thirteen on US Navy small boat ensigns in use for decades. Photograph by Chief Yeoman William Sato, USN. (*Naval History and Heritage Command*)

per second. Mounted on better designed carriages, their recoil proved more easily controllable. Over the years, these trunnioned, then trunnionless guns were built by Pratt & Whitney, American Ordnance Company, Driggs Ordnance Company, and the Naval Gun Factory. By the time such purchases ceased, the navy had such amassed these guns in the several hundreds. Not only were the long 1-pounders popular in the regular naval service, but such pieces on specially designed field carriages with thirty-six-round—shell or canister—ammunition boxes went to numerous naval militia units and provided fine training. Long 1-pounders mounted on small craft would see action during the early phase of the Veracruz landings in April 1914, and mounted on field carriages would perform harbor defense duty in World War I a few years later.

References Consulted

Canfield, B. N., "19th Century Military Winchesters," *American Rifleman*, March 2001, 149(3): 36–41, 77.

———, "A What? The Remington-Keene U.S. Navy Rifle," *American Rifleman*, April 2009, 157(4): 56–57, 74–76.

Cooke, A. P., *A Text-Book of Naval Ordnance and Gunnery, Prepared for the Use of the Cadet Midshipmen at the United States Naval Academy*, 2nd ed. (New York: John Wiley, 1880), pp. 397–404.

Dictionary of American Naval Fighting Ships, Naval History and Heritage Command (online). Consulted for each US naval warship and auxiliary mentioned in this chapter.

Hutchins, C. T., "The Naval Brigade: Its Organization, Equipment, and Tactics," US Naval Institute (hereafter USNI), *Proceedings*, 1887, 13(3): 303–41.

Kimball, W. W., "Magazine Small Arms," USNI, *Proceedings*, 1881, 7(3): 231–53.

———, "Machine Guns," USNI, *Proceedings*, 1881, 7(4): 405–35.

Koerner, A., *The Hotchkiss Revolving Cannon: A Description of the System, Its Employment in the Field, in Fortifications, etc., and for Naval Service* (Paris: privately printed, 1879), pp. 20–21, 40–42, 44–48, 56–59.

Lee, J. P., "The Lee System for Small Arms," USNI, *Proceedings*, 1881, 7(3): 325–27. Discussion following, 7(3): 328–29.

McAulay, J. D., *Rifles of the United States Navy & Marine Corps 1866–1917* (Woonsocket, RI: Mowbray Publishing, 2017), pp. 24–29, 32–34, 40–44, 47, 49–57, 59, 62, 65, 70–74, 76–77, 79–91, 107–11, 129.

Mellichamp, R. A., *A Gun for All Nations: The 37 mm Gun & Ammunition*. Vol. I. *1870–1913* (Houston: Mellichamp, 2010), pp. 1–2, 11–14, 21–23, 25–26, 40–42, 47, 53, 57–62, 71–75, 145–48, 151–56, 263, 301–06, 457–61, 505–11.

Myszkowski, E., *The Remington-Lee Rifle* (Tucson: Excalibur Publications, 1994).

Rankin, R. H., *Small Arms of the Sea Services: A History of the Firearms and Edged Weapons of the U.S. Navy, Marine Corps, and Coast Guard from the Revolution to the Present* (Fort Lauderdale: N. Flayderman & Co., 1972), pp. 124–26, 179.

Smith, J. E., *Small Arms of the World*, 10th rev. edit. (Harrisburg: Stackpole, 1973), pp. 61–63, 100–03, 129, 166.

Stone, C. A., "A General Description of the Ordnance and Torpedo Outfit of the U.S.S. *Trenton* (2nd Rate)," USNI, *Record*, 1877, 3(6): 89–93.

United States, *Report of the Gun Foundry Board as Organized by the President in Accordance with the Act of Congress Approved March 3, 1883*, Rear Adm. Edward Simpson, president of the board (Washington, DC: Govt. Print. Office, 1884), pp. 39–51.

US National Archives, Record Group 74, Records of the Bureau of Ordnance, Washington, DC, and Waltham, MA. See especially Entry 112, Ordnance Registers, 1842–1900.

US Navy Dept., *Annual Report of the Secretary of the Navy, 1875*, p. 19; *1880*, p. 10.

———, *Annual Report*, Bureau of Ordnance, *1873*, pp. 102–03; *1874*, p. 83; *1876*, pp. 111–13; *1878*, pp. 65–66; *1879*, pp. 62–63; *1880*, pp. 80–81; *1881*, pp. 183–84; *1882*, Vol. 3, pp. 6–7; *1883*, Vol. 1, pp. 413–14; *1884*, Vol. 1, pp. 413–14; *1886*, p. 249; *1887*, pp. 233–34; *1888*, pp. 187–88; *1889*, Part 2, pp. 440–41; *1890*, pp. 240, 246; *1891*, p. 209; *1892*, pp. 234–41.

US War Dept., *Annual Report of the Secretary of War, 1877*, Chief of Ordnance, Appendix O: "Trial of the Hotchkiss Revolving Cannon, Caliber One and a Half Inches," pp. 609–26; Plates I–IV.

Very, E. W., *The Hotchkiss Revolving Cannon: Descriptions and Illustrations of the Systems as Designed for Naval Service, Field Artillery and Flank Defense* (London: Waterlow and Sons, 1885).

Wahl, P., and Toppel, D. R., *The Gatling Gun* (New York: Arco Publ. Co., 1965), pp. 49, 52, 57, 61–62, 77, 80, 95, 97–99, 114, 117, 119, 125.

6
Into the New Age

The United States Strives to Stay Abreast

As the world's military and naval powers moved into the modern era of steel breechloading rifles firing smokeless powder, the ordnance departments of most of the world's great powers recognized that an appreciable downsizing of the bores of troop-level ordnance had become possible. The British Enfield and US Springfield rifled muskets, used respectively in the Crimean War of the 1850s and the American Civil War of the 1860s, fired .58-caliber Minié bullets. Following a brief period that saw a reduction to .50-caliber rounds, shoulder arms and comparable machine guns issued during the 1870s and 1880s typically were of .44- or .45-caliber, still designed for black powder use. By the 1890s, across the globe the new weapons firing nitro-based propellants underwent a substantial reduction in bore to around .30-caliber. In a like manner, medium fieldpieces in most armies were now standardized between 75 and 77 mm in bore, those diameters closely spanning 3 inches.

Most of the difficulties in obtaining trustworthy breech loading and obturation had been solved and a multitude of breech mechanism designs had made their appearance in this final decade of the nineteenth century. Small and medium artillery pieces and naval guns used fixed rounds, with the projectile inserted into a cartridge case made of drawn brass or some other suitable metal, the case handily ejected after firing. Such guns had become a safe, reliable, and rapid means of delivering fire in support of infantry, cavalry, and naval landing forces. Carbon steel was now not only being produced in dependably large quantities, but experimentation on alloy steels, particularly that incorporating nickel, promised even stronger, more durable, and less corrodible metals for both guns and armor plate.

Those and other major ordnance developments began to make themselves known from the great armament firms of Euro-Asia and America: Whitworth, Armstrong, and Vickers-Maxim in Britain; Krupp and Gruson in Germany; Schneider-Creusot, St. Chamond, and Hotchkiss in France; Hontoria in Spain; Skoda in Bohemia; Aboukhoff (Obukhov) in Russia; and Driggs in the United States. Most of these nations also had government arsenals busily engaged in ordnance design and production.

Against this backdrop of ferment and rapid change, the US Navy recognized the limitations of its 1870s–1880s generation of small ordnance, and like all other nations, pondered which of it needed immediate change and which could stand as it was for a bit longer. A thorough look at the ordnance adopted by the navy for its landing parties during the final decade of the nineteenth century reveals the uncertainty and confusion that confronted the service and in the end led it to make some unfortunate choices. As always, money was the root of all preferences.

Arming the Naval Militia

In approaching the final decade of the nineteenth century, the United States Navy remained acutely aware that, unlike the major European powers and the US Army, it possessed no armed and trained reserve contingent to flesh out its manpower in the event of war. Indeed, a decade previous, William Jeffers, chief of the Bureau of Ordnance, advised that the resupplying of the sloop *Swatara* with ammunition be held off because her putting to sea had been delayed, and pointed out that the navy then had more ships than men (although this last may have been apocryphal). The navy had ample precedent at home and abroad of how to set up such an auxiliary force. Although the United States continued to eschew a federally administered reserve of men who had previously seen regular service, the army had relied upon militia organized and maintained by the constituent states since the inception of the nation.

Naval officers remained mixed in their opinions whether their service similarly needed backup by a partially trained militia, equipped with obsolescent arms and equipment. That doubt, however, flew in the face of another consideration. The chances were greater for a more sizable congressional appropriation to arm the fleet with entirely new rifles, machine guns, and landing guns, if the navy could dispose of its older but still usable infantry ordnance to naval militia units, newly organized in the home states of some of the more influential members of the House and Senate.

John Soley, who as a navy lieutenant had written persuasively in 1880 for the *Proceedings* of the US Naval Institute about the need for the naval brigade, had since left the service. He remained highly interested in naval affairs, however, and had been active in the creation of a sizable naval militia in Massachusetts, the first state to muster such a volunteer armed force. The membership of what became the state's Naval Battalion of Voluntary Militia emerged from the Dorchester Yacht Club, soon to be renamed the Massachusetts Yacht Club. The battalion, consisting of four companies of infantry and artillery, was in place by 1890, quartered in Boston's South Armory and commanded by a lieutenant commander, with lieutenants as his company commanders. The unit personified the major objection by professional naval officers: that naval militia, attracting as it did university students and graduates, and wealthy weekend yachtsmen, would be more prone to thinking, not fighting.

The US Navy nonetheless supplied that militia, and those whose establishment quickly followed—of Rhode Island, Connecticut, New York, California, and Michigan—with ordnance either contemporary or only somewhat obsolescent. Those arms included .45-caliber infantry rifles (typically the Remington-Lee), like-caliber Gatling machine guns, Hotchkiss long 1-pounder rapid-fire guns, and both 3-inch breechloading and 12-pounder muzzleloading howitzers. The navy equipped the crew-served guns with boat and field carriages in order to train the volunteer gunners for service on the water and on shore. Also sent to the various militias were the necessary ammunition (including separate charges and cartridge cases), primers, and fuses. These ordnance stores were either loaned or bestowed, in the latter case paid for by the congressional Appropriation for Arming and Equipping the Naval Militia. Soley wrote again for the *Proceedings* in 1891, chronicling the genesis of the naval militia.

The navy found assistance in arming the naval militia of the various states, coincidentally, by the recall of all bronze ordnance from the fleet or shore stations. In December 1889, for example, Ordnance Bureau chief Montgomery Sicard directed the Boston Navy Yard to send 111 bronze 24-pounder howitzers to the Washington Navy Yard. These heavy pieces, unsuitable for boat and landing service, were not provided to the naval militias, so one must assume that some other fate awaited them, perhaps being melted down. In June 1892, the bureau issued a general directive to the eastern navy yards to send all bronze ordnance, except that used for saluting and like purposes, to the same navy yard. The yard thereafter sent the smaller 12-pounder

Seemingly oblivious to the steam drifting by, three immaculate junior officers of the New York Naval Militia serving aboard monitor *Nahant* pose with a Driggs-Schroeder long 1-pounder rapid-fire gun on a cone mount. During the Spanish-American War, the navy recommissioned a number of these old ironclad monitors laid up since the Civil War at League Island, Philadelphia Navy Yard, and sent them to eastern port cities. Such old warships would have proved of dubious value in defense. Detroit Photographic Co. (*Library of Congress*)

howitzers, light and heavy, as both training guns to naval militia units and monument pieces to Grand Army of the Republic (GAR) posts.

The 3-inch breechloading howitzers not surprisingly proved of greater value than the muzzleloaders to the volunteer gunners of the 1890s. Initially, the dozen bronze howitzers that had seen training in the hands of the cadets at the naval academy became available; later in the decade the navy distributed some of their steel counterparts. Particularly for the bronze pieces, the navy expressed concern about their safety, and so the barrel of each one was thoroughly inspected for pitting and cracking. If the piece passed the test, its right trunnion (typically) was stamped "1893" to show it was now safe to fire. The last bronze howitzer available went to the Connecticut Naval Militia in May 1895.

Not only were the various naval militias the beneficiaries of the navy's largesse. In August 1892, at the behest of Governor Robert Pattison of Pennsylvania, Col. Thomas Hudson, chief of artillery of the state national guard, wrote to Commo. William Folger, chief of the Bureau of Ordnance. Hudson inquired about the possibility of obtaining 3-inch naval pieces on field carriages, given their more suitable size relative to that of the standard army light field gun. The bureau responded with an exhaustive list of components and prices—total cost $2,144.03 per 350-pound howitzer—as ordered from the Naval Gun Factory. It is certain that at the very least, New Jersey National Guard units did acquire such howitzers, as shown by photographs taken at various summer encampments. Whether Pennsylvania did is rather less definite. In October of the same year, the navy extended permission to the Massachusetts Nautical Training School, located on a Boston pier (now the Massachusetts Maritime Academy on Buzzards Bay), to retain the 32-pounder deck guns and the 3-inch BL howitzers aboard the screw sloop *Enterprise*, loaned for training purposes.

A pair of highly polished naval 3-inch BL howitzers in army militia hands, straddling an army Model 1885 3.2-inch field gun, are deployed at Orchard Beach, NY. If this photograph is a wartime one, these troops may be of the New Jersey National Guard, known to have been armed with the 1870s 3-inch naval howitzers. Near this place some 120 years before, British troops landed in an attempt to outflank and envelop Gen. George Washington's army. (*Courtesy of James Schoenung*)

A New Shoulder Arm

The bore reduction undertaken by the American navy in its infantry-caliber rifle turned out to be more radical than that made by virtually anyone else, including the US Army. The navy's new Model 1895 Winchester-Lee "straight-pull" bolt-action rifle fired a 6mm or .236-caliber round of 135 grains (later reduced to 112 grains), contained in a semi-rimmed cartridge case. As with its Remington-Lee predecessor, the rifle held five rounds in its external magazine in front of the trigger guard, but now permanently fixed in place, which the sailor—and following a delay in adoption, the marine—recharged by an en bloc clip closely resembling a stripper clip. A great advantage in the use of the light round lay in the sheer number of them that the rifleman could carry, particularly in rough or mountainous country. In the numerous firefights during the Boxer Rebellion, the riflemen of other nations too frequently exhausted their .30-caliber or similar-sized rounds, whereas the American marines and sailors continued to have ample ammunition. The rifle fixed an 8-inch knife bayonet, a design considerably ahead of its time, in contrast to the long unwieldy bayonets of the period that proved so disadvantageous in hand-to-hand trench combat during World War I.

Winchester Arms manufactured approximately 15,000 of these rifles during the mid-to-late 1890s. Though nicely accurate, the Lee proved awkward to work in

A bluejacket aboard the armored cruiser *New York* in the late 1890s is shown in light marching order, holding a Winchester-Lee Model 1895 6mm rifle with its short bayonet fixed. The cognomen for its bolt action, "straight pull," was decidedly misplaced, and its small caliber round lacked stopping power. Behind the seaman is a recently issued Mark I 3-inch field gun (landing gun), with its trail wheel removed. Photograph by Edward H. Hart, Detroit Photographic Co. (*Library of Congress*)

combat—its straight-pull cognomen decidedly a misnomer—and certain components of its cam-operated bolt action, including the ammunition feed and floating extractor, were nothing if not fragile under field conditions (as well, the extractor could fall out and be lost). The biggest complaint, however, concerned the round's diminutive size and limited man-stopping power, despite its deep penetration. Criticism arose, too, about the erosion of the Metford rifling, which the reduction in the weight of the bullet attempted to address, but of course making the round even less of a man-stopper. In December 1898, a board that included officers from all three services strongly recommended a common infantry-caliber arm, and two years later, American marines would begin to turn in their Winchester-Lees for the army's harder hitting Krag-Jørgensen bolt-action caliber .30-40 rifles, either Model 1896 or 1898. Many sailors, however, would still have their Lees going into 1903.

A New Machine Gun

The navy's choice in its machine guns proved as unfortunate as that of its shoulder arm. Despite the trend elsewhere toward adopting the fully automatic Maxim gun—the devilish invention of Mainer Hiram Maxim, who had moved to England to sell his wares worldwide—and the vehement protests from the Young Turks of the American officer corps, the Maxim machine gun was largely eschewed in the inventor's country of birth. Both the navy and the army continued to use the hand-cranked Gatling gun, even going so far as to reduce its caliber from the .45-70 black-powder round to .30-caliber smokeless. After a brief flirtation with the Accles drum, the services went back to the tried-and-true Bruce feed. Gatlings could still be found in use by all three services into the first decade of the twentieth century.

Finally listening to the prescient counsel that the Gatling gun had had its day, offered during the previous decade by then Lt. Charles Hutchins, the navy decided it needed a fully automatic machine gun, and so tested a number of them. Although the service conceded the numerous advantages to be had with the Maxim machine gun, it decided upon the domestic Colt-Browning, and moreover chambered it for the 6mm cartridge used in the Winchester-Lee infantry rifle, with the automatic gun also designated the Model 1895. It was gas-operated, belt-fed, and air-cooled. At some point the Colt-Browning machine gun earned the sobriquet of "potato digger," a not-so-nice reflection of its peculiar hinged operating lever underneath, similar to the manipulation of a lever-action rifle. That awkward lever required several inches of clearance when firing the gun. The automatic gun had to be unloaded immediately after extensive firing, because while its barrel remained hot, it had the dangerous predilection of cooking off one to several rounds. The option existed to mount the Colt-Browning either on a tripod for greater flexibility or on a light field carriage for more sustained operation. The latter carried two ammunition boxes, each holding 2,000 rounds.

Although the Colt-Browning automatic gun saw some improvement in .30-caliber during the years of the following century, in the 1890s it remained an unreliable machine gun prone to jamming or outright breaking at the most inopportune times.

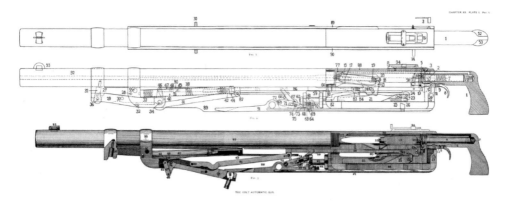

Construction of the Colt-Browning Model 1895 machine gun, caliber 6mm, is shown in these sectional drawings. The gun's ungainly hinged operating lever that affected its ground clearance and prompted the derisive tag "potato digger" is evident in the bottom drawing. This automatic gun had the awful habit of jamming at critical moments, more than once costing lives. William F. Fullam and Thomas C. Hart, *Text-Book of Ordnance and Gunnery*, 1903. (*Eugene L. Slover, online*)

The Mark I 3-inch Field Gun

The American navy hardly needed reminding that its 3-inch breechloading boat and landing howitzers, devised twenty years before, had become rather long in the tooth. The time had come to design and manufacture a modern gun and issue it promptly to the fleet. Its closest friend and fiercest rival, the Royal Navy, despite the continued use well into the last decade of the century of muzzleloading landing guns, had made the same conclusion. In 1894, that navy introduced a quick-firing (fast breechloading) 12-pounder 8-hundredweight landing gun to its fleet. For a brief period, a British-built American warship would carry a pair of those QF 12-pounders.

Other than the decision to retain bagged powder, the US Navy had taken a bold and innovative stride in the design and adoption of its 3-inch breechloading howitzers during the 1870s, and it wished to repeat that success. On this occasion, however, what eventually became the Mark I 3-inch field gun—though in purpose a landing gun—would find itself lagging behind nearly every good idea as the 1890s progressed. Indeed, the desired general issuance of the new landing guns to the fleet was badly set back, and the steel 3-inch landing howitzers of the previous generation could still be found aboard some ships into the first decade of the twentieth century.

In its quest for a replacement, the navy considered the merits of a 6-pounder (57mm) landing gun for fleet use, and during the early 1890s it began comparative testing of such guns firing fixed metallic rounds and equipped with either the Hotchkiss or Driggs-Schroeder breech mechanism. Also, the Naval Gun Factory at the Washington Navy Yard undertook the construction of five such guns of its own design. The navy favored the Driggs-Schroeder mechanism, but the tests soon revealed the 6-pounder to be inadequate in power for landing purposes. Another significant limitation lay in its round being too small to carry shrapnel. The army came to the same conclusion in the early twentieth century with its own 6-pounder, and estimated the smallest projectile that could hold shrapnel at 7½ pounds—and then only of marginal usefulness—which translated to a 2.38-inch or 60mm gun. The army built a single small field gun of this model for the support of cavalry and fixed coast artillery batteries, before shelving the project. For its part, beginning in 1892 the navy consigned the 6-pounder field guns it had acquired to the US Naval Academy as sufficient for the training purposes of the cadets, just as it had done with the dozen bronze 3-inch BL howitzers in 1876. Thereafter the service returned to the 3-inch gun as best suited for its landing force needs.

As part of its design, the navy wished to build the new gun of a more durable alloy steel, and in 1892 contracted with the nearly always reliable Midvale Steel Company to produce a steel forging containing 25 percent nickel. On this occasion, however, Midvale failed to deliver, more than likely because of the very high percentage of nickel in this first-ever attempt at such alloy steel in American gunmaking. This forging proved to be unsatisfactory in quality and was condemned in 1893. The navy immediately ordered as its replacement a forging of ordinary gun steel.

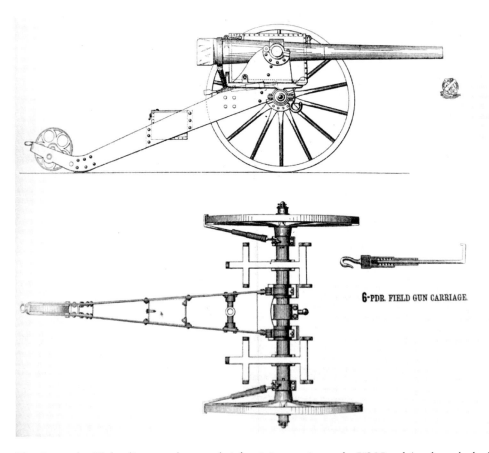

The 6-pounder BL landing gun that saw brief training service at the US Naval Academy lacked an on-carriage recoil apparatus, and was thus a modern breechloading gun firing metallic cartridges, mounted on an old-fashioned field carriage. The spring device shown on the right attached to both the wheel and the axle, one on each side, and helped restrain the recoil of the piece over the ground. *Annual Report of the Secretary of the Navy, 1890.*

In June 1893, Pratt & Whitney forwarded to the Bureau of Ordnance a set of drawings for a 3-inch/21-caliber naval landing gun of Lt. Frank F. Fletcher's design. The company declared the piece would fire a 10-pound projectile at 1,150 feet per second, and described its breech mechanism as having simplicity and other advantages. The company requested an order to manufacture a pilot gun, at a cost of $1,350. The bureau requisitioned that gun in October, and Pratt & Whitney completed it for only $2.24 more than the estimate. One year after submitting the drawings, the company shipped the piece to the Naval Proving Ground at Indian Head, Maryland, where it was placed on a test carriage and fired 174 times over the next twenty-one months.

The Bureau of Ordnance appointed a board of three junior officers in late November 1894, in order to appraise the three guns in competitive trials to be the navy's next boat and landing gun. Lieuts. Newton Mason and Templin Potts, and Ens. Friend Jenkins were appointed for that task. In addition to the Pratt & Whitney piece having the Fletcher slotted screw breech mechanism, a Driggs-Schroeder 3-inch gun submitted by the Driggs Ordnance Company and a Hotchkiss 10-pounder completed the trio to be tested. The latter two guns had been thoroughly tested by early December, but a delay ensued with the Fletcher gun. They were able to test that piece in early January 1895, with representatives from the Driggs and Hotchkiss companies in attendance.

The board reported its findings on January 19. In one letter, the two lieutenants ranked the Hotchkiss and Driggs-Schroeder guns a close first and second, giving the nod to the Hotchkiss 10-pounder because of the simplicity of its breech mechanism and the ease in dismounting and breaking down the assembly to change the firing pin or broken extractor. In his minority opinion, Ensign Jenkins considered the Driggs-Schroeder 3-inch gun superior because of its rapidity of fire, but conceded the Hotchkiss to be a close second. All of the officers were in solid accord that the Fletcher gun, with its awkward breech mechanism, lay in third place, and with its reliability questioned, a poor third at that.

Four days later, Commo. William Sampson, chief of bureau, requested that Ensign Jenkins, the junior member of the board, be directed to prepare a further report, in which he would suggest improvements to the breech mechanisms of all three guns in order to make them more efficient in the roles intended. That choice at first glance may be considered unusual. Lieutenant Mason, the senior member, was currently serving his second assignment with the Bureau of Ordnance, this time in charge of the Naval Proving Ground, and had the greater experience. Following Jenkins's new report, Pratt & Whitney rebuilt the Fletcher gun, apparently incorporating the ensign's suggested changes.

On May 6, 1895, the bureau directed the board to reconvene and retest only the Fletcher gun, and in the absence of company representatives. This time the board's conclusion was unanimous and unequivocal: the Fletcher gun, even as modified, was still unsatisfactory and in no way superior or even equal to its Driggs and Hotchkiss competitors. Perhaps to everyone's great surprise, however, the bureau decided that the Fletcher gun was most suitable and deserving of adoption. When Commodore Sampson announced that the Fletcher pattern gun had received the bureau's approval, and 100 such pieces would be built at the Washington Navy Yard, the Driggs Ordnance Company reacted vehemently.

Learning with interest that half of those guns would be built at once and the other half made at some time yet to be determined, the company filed a formal protest with Secretary of the Navy Hilary Herbert and then took its complaint to the influential *New York Times*. The paper printed the story on October 19, 1895, describing the gun as "vastly inferior" in the opinion of the trial board and reporting that despite the later tinkerings, the piece had never worked satisfactorily. Most damning was the disclosure that Driggs had underbid Hotchkiss—the latter because of its association with Pratt & Whitney, which controlled the Fletcher patents—and still had been rebuffed for the contract to build 100 landing guns. The *Times* observed that the decision did not appear to be in accordance with wise public policy. Moreover, it would discourage private industrial plants, likely needed to support the government in future national emergencies, from remaining open and available. The upshot was that the kerfuffle led to the formation by 1896 of an industrial conglomerate involving both the Hotchkiss and Driggs ordnance companies. It was located in Bridgeport, Connecticut, and named the American Ordnance Company.

The determination in favor of the Fletcher gun cannot simply be dismissed as favoritism for fellow naval officer Frank Fletcher. Both William Driggs and Seaton Schroeder at the time also served as naval officers (Schroeder would in fact rise to flag rank during the twentieth century). Benjamin Hotchkiss had been born and had learned his trade in Connecticut, and his reliable armaments had won worldwide renown. Nonetheless, months later the decision remained inexplicable. It was an incredible and unfortunate one that would cost the navy dearly in treasure and wasted effort for more than a decade to come. A sad additional note was added upon the death of the newly promoted Lt. (j.g.) Friend Jenkins, who had done the follow-up report, when the capital ship *Maine* exploded and sank in Havana harbor on February 15, 1898.

The Mark I 3-inch field gun, as officially designated, in its production configuration was a single-forged, rapid-fire weapon 21 calibers in length, weighing a bit less than 400 pounds complete, with a muzzle ring on one end and the newly adopted Fletcher breech mechanism on the other. It may be fairly stated that the Fletcher interrupted-screw breech mechanism would soon become obsolescent. As with most other rapid-fire breech mechanisms, the single-motion breech lever generated three operations of the plug or breechblock—rotation, translation, and swinging-out. A carrier ring, cut on its inner circumference into raised and blank sections and engaging automatically by a locking bolt, supported the plug when withdrawn from the screw box. The carrier ring thus acted as both a collar and a side-swing plug tray. The plug and screw box possessed four each threaded and blank sections, which did not occupy the entire length of the plug, because its rear portion remained blank for the carrier ring. The extractor, cut on its inner end to fit under the rim of the spent cartridge case, pivoted in a slot through the jacket of the gun, rather than being carried on the plug. A hinge plate was attached to the rear face of the jacket, furnishing the bearing for the hinge pin on which both the carrier ring and the operating lever rotated.

The operating lever, bearing a handle on its left end and bent down to clear the firing lock, pivoted on the hinge pin located on the right side of the breech. With the breechblock closed and locked, the inner end of the locking bolt rested in the guide slot of the plug and the outer end engaged the recess of the hinge plate, locking the carrier ring to the breech.

The lever engaged the circumferential rack on the plug, with the extractor claw under the rim of the cartridge case. When the lever was thrown to the right, its teeth engaged with those of the plug rack and rotated the plug 45 degrees counterclockwise, so disengaging its threads from those of the screw box and thus permitting it to be withdrawn (or translated). At that juncture, the locking bolt reached the end of the circumferential part of the guide slot in the plug and prevented the plug from rotating farther.

As the operating lever continued to turn, the first of the withdrawing teeth came opposite the threaded portion of the plug, and thereafter the teeth engaged the plug threads and caused the plug to move to the rear through the carrier ring. The locking bolt then moved in the dogleg part of the guide slot parallel to the plug axis, and the persisting movement of the operating lever completed the withdrawal of the breechblock. When the end of the locking bolt reached the depression at the front end of the guide slot, the spring forced the opposite end of the locking lever up, thus pushing the locking bolt down and disengaging it from the recess in the hinge plate. With the final rotation of the operating lever, the plug thereupon locked to the carrier ring in its withdrawn position and swung open with the carrier. As the plug and carrier cleared the breech, the rear face of the lever acted as a cam on the outer end of the extractor, causing it to draw out the spent cartridge case, at first slowly and forcefully, and thereafter more quickly.

The bluejacket aboard armored cruiser *New York* in the late 1890s, carrying the extra gear for heavy marching order, has partially opened the breech plug of the Mark I 3-inch field gun. The awkward location of the elevating shaft and wheel under the breech interfered with the independent loading and elevating of the piece by the gun crew. Two pairs of ammunition chests are in place on the field carriage, each chest carrying eight rounds. Photograph by Edward H. Hart, Detroit Photographic Co. (*Library of Congress*)

The Badger firing lock that ran through the axis of the breechblock accomplished a number of things almost as one. The gunner pulling the lanyard smartly and smoothly to the rear brought the firing pin to the rear by the cocking pawl and compressed the firing spring, until the pawl was lifted by the trip toe on the rear of the firing mechanism case. Thereupon the firing pin flew forward and struck the plunger, which in turn struck the cartridge primer and fired the gun. Thus the firing point was not the firing pin itself, but rather part of the plunger struck by the firing pin. The spring surrounding the plunger acted against the firing spring, ensuring that the latter had almost no compression when the cocking lever remained uncocked, such that the closure of the plug caused no blow to the primer. Despite the precautions built into the design, it was still possible to fire the gun before the breech plug was turned and locked; for example, if the lanyard caught on something and became taut. A few years into the twentieth century, Lt. Joseph Graeme designed a safety device that blocked the firing pin from striking before the plug man closed and locked the breechblock.

The fixed round weighed 15 pounds, which when its 14 ounces (400 grams) of smokeless powder detonated, generated a muzzle velocity of 1,150 feet per second. The gun trials that began in 1894 lasted until the following year, the entirety on the proving ground test carriage. Halfway through the necessary trial protocol, the navy made the decision to accept the gun for landing and boat use, and the service let a contract with Bethlehem Iron Company for 100 steel forgings. In 1896, the navy determined to build half of the 100 new field guns at its gun factory and a bit later to contract the other half, perhaps with little coincidence, to the American Ordnance Company. The Bureau of Ordnance also resumed testing of the type gun, this time on one of the newly built field carriages. The gun and carriage together weighed more than 1,360 pounds, and with a full ammunition load added—four removable boxes, each carrying eight rounds—about 1,860 pounds.

Properly defined as part of the carriage because that component must absorb part or all of the recoil of the gun, the recoil mechanism was a peculiar one, even for that era. It consisted of a hydraulic recoil sleeve engirdling the barrel, with a collar attached and keyed to the barrel; the collar acted as the piston and the barrel the piston rod. The sleeve bore on the gun just at the ends, which had stuffing boxes fitted, and recoil fluid composed of glycerin and water filled the intervening space. A filling hole allowed the addition of fluid. Inside the cylinder, rifling fitted recesses on the edge of the collar-piston and prevented the rotation of the gun upon firing. The counter-recoil spring was secured in slight tension between the collar-piston and the rear cylinder head. Externally, the sleeve bore a pair of trunnions that fitted into seats (pockets) on the trail and were locked in place by capsquares (trunnion caps).

The trail mounted a drag wheel, which when unshipped permitted the spade to dig into the ground in order to absorb part of the recoil. As configured, the recoil mechanism restricted the gun to a short recoil, with the remainder of the recoil action absorbed by a run-back over the ground. The carriage lacked a traversing gear and had to be shifted by its trail stave. The elevating worm under the gun permitted the muzzle to be lifted 10 degrees, providing an effective range of 4,400 yards. Oddly, the elevating shaft and hand wheel thrust rearward just under the breech, which proved awkward for the gun crew in action. Trouble occurred during the proving ground trials, and adjustments had to be made by enlarging the size of the piston ports and replacing the return spring with a stronger one. As will be seen in the following chapter, the original design of the recoil mechanism was fundamentally flawed, and the sleeve suffered, often badly, from the too short and violent recoil.

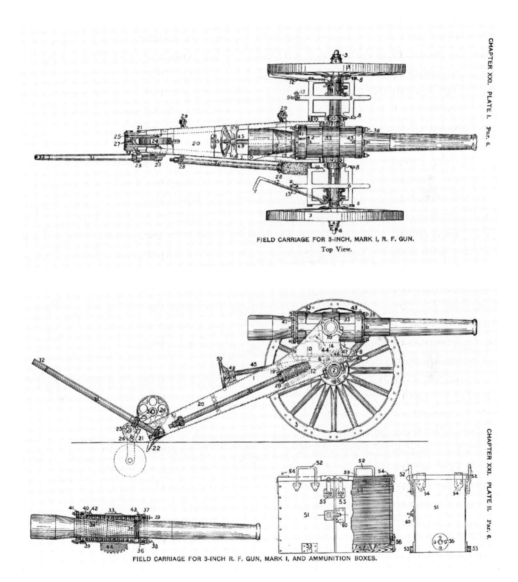

The recoil sleeve or cylinder engirdles the barrel just forward of the breech of the Fletcher Mark I 3-inch RF gun on its field carriage. The collar within the outer cylinder acted as a piston and the recoiling barrel the piston rod. Given the violent jump of short-recoil guns, this configuration simply did not work, but the navy futilely spent a great deal of time, effort, and treasure attempting to fix an unfixable design. The ammunition boxes are shown on the lower right. Fullam and Hart, *Text-Book of Ordnance and Gunnery*, 1903. (*Eugene L. Slover, online*)

At the same time that the adjustments described were being made, experiments were done to ascertain the gun's pull upon recoiling, as would be exerted on the fastenings in a boat. Numerous firings determined a force of about 14 tons, thus requiring a heavily or specially built craft when the piece served in the manner of a boat gun, a role that was steadily becoming less important with the passage of the last decade of the nineteenth century.

Bethlehem Iron delivered the 100 forgings in 1896, and construction of fifty field guns began at once at the Naval Gun Factory. They were completed, along with their wheeled carriages, by September 1897, and the allocation to large warships began. At that time, American Ordnance Company had delivered none of its fifty field guns and fifty field carriages, or its 10,000 rounds of 3-inch fused shrapnel. Some time later, likely before the brief Spanish-American War began, the company completed its contract and thereupon the US Navy possessed, in addition to the type gun, 100 3-inch field guns of the new design, with ample ammunition. The Mark I field carriage mounted the Mark I 3-inch gun, and the registration or serial numbers of both guns and carriages ran from 1 to 100.

The year the first Mark I guns and carriages were completed, 1897, is more notable as that when the French 75 or *Soixante-quinze*, or more formally, the Canon de 75mm modèle 1897, was introduced to French army service. The truly innovative part of the gun lay in its carriage, more specifically the hydropneumatic recuperator, which permitted both a long recoil and subsequent counter-recoil with a compressed air component—casually known by field artillerymen as an "air spring"—in lieu of a steel spring. During recoil and counter-recoil, the carriage absorbed the entirety of the kinetic energy generated and did not move out of place in battery, permitting a very high rate of fire. But even when the navy's Mark II field carriage superseded the less-than-satisfactory original and modified Mark I carriages early in the twentieth century, it remained the short-recoil type and was clearly obsolete upon arrival.

The British QF 12-pounder 8-hundredweight Landing Gun in US Naval Service

As the nineteenth century reached its final decade, two types of steel, rifled muzzleloading (RML) landing guns, a 7-pounder and a 9-pounder 6-hundredweight, placed in service two decades before, remained aboard ships of the Royal Navy. The former piece was a diminutive but rugged one, readily disassembled and reassembled. It continued to see regular use as both a naval landing gun and an army mountain gun to the end of the 1890s, and by colonial establishments into the twentieth century. Aware that it needed a more powerful landing gun, the Royal Navy contracted with the Elswick Ordnance Company (EOC) owned by Sir William G. Armstrong, to develop a suitable quick-firing piece for that role. The resulting 12-pounder 8-hundredweight gun, equipped with a single-motion Elswick QF breech mechanism and firing separate-loading metallic ammunition, came into British fleet service in 1894. Such guns saw action in both the Boer War and World War I, most notably during the campaigns in Gallipoli and colonial Africa. All such British landing guns measured 3 inches in bore, as did their American counterparts.

Although technologically the world's leading naval service, the Royal Navy had hitherto given less attention to its landing guns than had the US Navy. Aboard pre-dreadnought *Camperdown* in 1895, the gun crew exercises its steel 7-pounder rifled ML landing gun, about the same size as the American navy's contemporary 3-inch BL howitzer. Behind the gun may be seen the towing harness and ammunition limber. Two years before, *Camperdown* had collided with the more modern battleship HMS *Victoria*, and the latter had sunk with substantial loss of life. *Navy and Army Illustrated*, December 20, 1895.

The United States and Spain went to war in April 1898 and remained so for the next four months. Aware that Spain had an interest in two protected cruisers being built at the Elswick works for Brazil, *Amazonas* and *Almirante Abreu,* the United States moved preemptively and purchased them in March, one month before war began. Renamed *New Orleans* and *Albany*, respectively, the two cruisers became the sole units of the American navy's New Orleans class. *New Orleans* commissioned in March 1898 and would be the only one of the pair to serve in the upcoming war. After a transatlantic passage, she called briefly at New York and Norfolk, and continued on to join Commo. Winfield Schley's Flying Squadron off Santiago de Cuba. On three occasions in May and June, she joined with other American warships in exchanging fire with Spanish ships and harbor defense batteries at Santiago. In late June, she steamed to Key West to take on coal and thus missed the sea battle off the Cuban port on July 3. Upon her return, *New Orleans* patrolled the war zone in the Caribbean, and in mid-month, she shelled and burned the Spanish merchantman *Antonio López*, previously grounded near Dorado, Puerto Rico.

Because *New Orleans* was purchased as equipped by the Elswick works, she had aboard two Armstrong-Whitworth QF 12-pounder 8-hundredweight landing guns. She also carried other armament non-standard in the US Navy: six QF 6-inch and four QF 4.7-inch guns in her main and secondary batteries, and three 5-barrel Maxim-Nordenfelt 7mm volley guns, which could be shifted as boat and landing guns (two field carriages were carried on the fantail). Also for her landing party, the protected cruiser carried 110 Mauser 7mm rifles, in addition to thirty-six Winchester-Lee 6mm rifles. Other than her large deck guns, it remains doubtful if any of the smaller guns suitable for landing operations ever saw use. During her refit in the spring of 1899, the New York Navy Yard divested the cruiser of her Maxim-Nordenfelt and Mauser ordnance; near year's end, it took the 12-pounders, too. When *New Orleans* sailed again, she had aboard five Hotchkiss 1-pounder boat guns and 174 Winchester-Lee infantry rifles, both types standard in American naval service.

Albany did not become available until May 1900, nearly two years after the war had ended. She did not take aboard the QF 12-pounder 8-hundredweight landing guns slated for her—they had been duly purchased and arrived in the United States aboard the merchant vessel *Baltimore*—but she did mount smaller ordnance intended to support landing operations. They included eight Hotchkiss 1-pounders, two of which could be shifted to her sailing launch and first cutter, and two Colt-Browning automatic guns, either one movable to the ship's steam cutter. Both a field carriage and a tripod were available for landing the machine guns. As with her sister, *Albany* carried 174 Winchester-Lee 6mm infantry rifles.

In late September 1904, the Bureau of Ordnance drafted plans for the complete rearming of the New Orleans-class cruisers, including ten 5-inch/50-caliber guns for their main battery and one 3-inch field gun for landing purposes. The navy sold three of the 12-pounder landing guns in that same year to Norwich University, Vermont's military college, and just past decade's end, it donated the fourth piece as a monument gun to a town in Pennsylvania. The navy had found no particular fault with them, but they required a completely different family of ammunition—a substantial expense for so few guns.

During the East Africa campaign in May 1916, a British QF 12-pounder 8-hundredweight naval landing gun fires from a native village. These guns lacked an up-to-date field carriage to absorb any part of the recoil generated. With the purchase of the two New Orleans-class protected cruisers from Elswick Ordnance Company just prior to the onset of the Spanish-American War, the US Navy also acquired their four 12-pounder 8-hundredweight landing guns. The pair issued to *New Orleans* remained aboard during her brief wartime service off Cuba and Puerto Rico. *The Illustrated London News*, May 1916.

The US Navy Would Use What It Had

Whatever its faults, the new Mark I naval field gun arrived in time to participate in the first and last of the three major imperial adventures the United States engaged in during the final three years of the nineteenth century: the Spanish-American War, April–August 1898; the intervention in Samoa, March–April 1899; and the attempted relief of the besieged legations in Peking, China, June–July 1900, a part of the more extended Boxer Rebellion. Its equally flawed consort, the Colt-Browning gas-operated 6mm machine gun, played a role in all three of these conflicts.

References Consulted

Beyer, H. G., "Observations on the Effects Produced by the 6-mm Rifle and Projectile—An Experimental Study," *J. Boston Soc. Med. Sci.*, 1899, 3(5): 117–36.

Byline not specified, "Gunmakers Have a Grievance: Protest Against the Manufacture of Guns at Washington Navy Yard," *New York Times*, October 20, 1895.

Dictionary of American Naval Fighting Ships, Naval History and Heritage Command (online). Consulted for each US naval warship mentioned in this chapter.

Fullam, W. F., and Hart, T. C., *Text-Book of Ordnance and Gunnery* (Annapolis: US Naval Institute, 1903), pp. 246–70.

Goldsmith, D. L., *The Browning Machine Gun*. Vol. I. *Rifle Caliber Brownings in U.S. Service* (Cobourg, ON: Collector Grade Publications, 2005).

Hanson, J., "The 6mm USN: Ahead of Its Time," *Rifle Magazine*, 1977, 9(1): 38–41.

Hutchins, C. T., "The Naval Brigade: Its Organization, Equipment, and Tactics," US Naval Institute (hereafter USNI), *Proceedings*, 1887, 13(3): 303–40. Discussion following, 13(4): 533.

———, "Infantry Fire Tactics, Fire Discipline, and Musketry Instruction and Practice with Rapid-Fire Cannon," USNI, *Proceedings*, 1887, 13(4): 547–60.

Lawry, N. H., "Difficulties Encountered with the Early, Rapid-fire Seacoast Ordnance. Part I. Guns in Search of Users: Light Pieces for Battery Defense, 1898–1908," *Coast Defense Study Group J.*, 1999, 4(1): 4–13.

McAulay, J. D., *Rifles of the United States Navy & Marine Corps 1866–1917* (Woonsocket, RI: Mowbray Publishing, 2017), pp. 129, 140–149.

Mellichamp, R. A., *A Gun for All Nations: The 37 mm Gun & Ammunition*. Vol. I. *1870–1913* (Houston: Mellichamp 2010), pp. 40–41, 47, 57–61, 71–75, 145–48, 151–54.

Myszkowki, E., *The Winchester-Lee Rifle* (Tucson: Excalibur Publications, 1999).

Niblack, A. P., "The Naval Militia Movement," USNI, *Proceedings*, 1895, 21(4): 781–87.

Professional Note. "Caliber of the New Navy Rifle," USNI, *Proceedings*, 1893, 19(3): 305–09.

———, "The Fletcher Rapid-Fire Breech Mechanism," USNI, *Proceedings*, 1894, 20(3): 605.

Rankin, R. H., *Small Arms of the Sea Services: A History of the Firearms and Edged Weapons of the U.S. Navy, Marine Corps, and Coast Guard from the Revolution to the Present* (New Milford, CT: N. Flayderman & Co., 1972), pp. 126–27, 179–80.

Smith, J. E. *Small Arms of the World*, 10th rev. edit. (Harrisburg: Stackpole, 1973), pp. 63, 116–17.

Soley, J. C., "The Naval Brigade," USNI, *Proceedings*, 1880, 6(3): 271–90.

———, "Naval Reserve and Naval Militia," USNI, *Proceedings*, 1891, 17(3): 469–96.

US National Archives, Record Group 24, Records of the Bureau of Navigation, Entry 118, Ships' Log Books: deck logs of *New Orleans* (Mar. 1898–Sept. 1900) and *Albany* (May 1900–Apr. 1902), Washington, DC.

———, Record Group 74, Records of the Bureau of Ordnance, Washington, DC.

US Navy Dept., *Annual Report, 1892*, Board of Visitors of the Naval Academy, p. 89.

———, *Annual Report, 1895*, Board on Small Arms, pp. 301–10.

———, *Annual Report, 1898*, Bureau of Navigation, Appendix, p. 11.

———, *Annual Report*, Bureau of Ordnance, *1890*, pp. 240, 246, 253; *1891*, pp. 209, 216, 250, 255; *1892*, pp. 205, 214–15, 269–70; *1893*, pp. 228, 235, 298; *1894*, pp. 234, 249–50, 266, 303, 318; *1895*, pp. 201, 215–18, 271–73; *1896*, pp. 25–26, 30, 276, 284, 288, 294–95, 325–26, 337; *1897*, pp. 20, 277, 279, 281, 283, 301–02, 315, 320–23, 330–31; *1898*, pp. 483, 485, 490; *1899*, pp. 516, 557; *1902*, p. 502.

———, *Annual Report*, Marine Corps, *1897*, pp. 558–59, 566, 572–73; *1900*, pp. 1096–97.

US War Dept., *Annual Report, 1895*, Chief of Ordnance, "Driggs-Schroeder 12-pounder Gun on Minimum Recoil Carriage," p. 53.

Various officers, "Instructions for Infantry and Artillery, United States Navy," Bureau of Navigation, in USNI, *Proceedings*, 1891, 17(4): 569–751.

Wahl, P., and Toppel, D. R., *The Gatling Gun* (New York: Arco Publishing Co., 1965), pp. 98–99, 114, 119, 125, 140.

Wieand, H. T., "The History of the Development of the United States Naval Reserve, 1889–1941," doctoral dissertation, Princeton University, 1952, pp. 30–31, 35–39, 45–46, 52–56.

7

American Imperial Adventures

The United States Acquires a Territory with the Help of the Cruiser *Boston*

A number of rebellious acts had disturbed the Hawaiian Islands in the 1880s, culminating in the overthrow of the monarchy during the following decade. Those agitating for control of the islands were mostly rich and powerful sugar planters born in the United States or Great Britain. In 1892, a mostly white secret society that advocated the deposing of the monarch, Queen Lili'uokalani, evolved into the Committee of Safety, and allied itself with a revolutionary military arm, the Honolulu Rifles. The nearly bloodless revolution began on January 16, 1893, abetted by John Stevens, the US minister to the Hawaiian Kingdom.

For some years, the US Navy had leased Ford Island in Pearl Harbor, and the protected cruiser *Boston* (two 8-inch guns) had been on station in the islands for five months. At the behest of Minister Stevens, her commanding officer, Capt. Gilbert Wiltse, sent ashore a large landing force led by Lt. Cmdr. William Swinburne and comprised of two sailor companies and a marine company. Ostensibly they were there to protect the US legation and consulate. To accentuate the point of their presence, however, one of those officers saw fit that a Gatling gun and a Hotchkiss 37mm revolving cannon mounted on field carriages would accompany the landing party. *Boston*'s bluejackets and marines did not fire a shot or enter onto the palace grounds, but their mere presence acted to coerce Lili'uokalani and her supporters.

The queen was deposed by proclamation the next day, which angered President-elect Grover Cleveland, who wanted no part in what he considered a shameful coup. He attempted to overturn her deposing, but the congress did not support him. On May 4, 1898—perhaps with little coincidence, three days following the US naval victory at Manila Bay—the islands became a US territory, and sixty-one years later, the nation's fiftieth state.

Commanded by Lt. Lucien Young (1852–1912) on the right, a part of protected cruiser *Boston*'s landing force appears in formation in front of the Arlington Hotel in Honolulu, following the January 1893 overthrow of Queen Lili'uokalani (1838–1917). The bluejackets are armed with Remington-Lee bolt-action rifles and the marines beyond with trapdoor Springfield rifles, bayonets fixed, both loading .45-70 Government rounds. The end of the trail of a landing gun may be seen on the left. The landing party did not otherwise participate in the armed rebellion by moneyed sugar planters. (*Hawaiian State Archives*)

US Marines Assault Guantánamo Bay during the Spanish-American War

The reasons for war between the United States and Spain, declared on April 25, 1898, included the bloody revolution in Cuba, the revelation in the press of a letter written by Enrique Dupuy de Lôme, the Spanish minister to the United States, insulting to President William McKinley, and the sinking in mid-February 1898 of the American battleship *Maine* (two twin 10-inch gun turrets) in Havana harbor, from an unknown cause at the time. Nine days before, on April 16, Lt. Col. Robert Huntington, the officer commanding the Marine Barracks, New York Navy Yard, had received orders to form two battalions of marines for possible action in Cuba. It soon became obvious that two such units could not be put together with the men readily at hand, so the commandant instructed Huntington to organize a single battalion, but reinforced with two additional companies. The men were drawn from several east coast navy yards and receiving ships. The 650-man First Marine Battalion consisted of five rifle companies, A through E, and a single artillery battery, F Company, commanded by Capt. Francis Harrington. His unit was armed with four of the new Fletcher 3-inch field guns issued by the yard in Brooklyn. The battalion also had some Model 1895

Colt-Browning machine guns firing 6mm rounds, the same as for the Model 1895 Winchester-Lee bolt-action rifles with which the marine infantry units were equipped.

From the Brooklyn docks on April 22, the men boarded USS *Panther*, an auxiliary cruiser and troop transport converted from the former banana boat, SS *Venezuela*, and armed with six 5-inch guns as her main armament. En route to Florida, after a stopover at Hampton Roads, Huntington ordered that instruction be carried out on the Lee rifle, following which each enlisted man fired ten rounds from his own weapon. As well, the field gun crews each fired one shrapnel round. The battalion disembarked at Key West one week later, ordered off by the ship's captain, Cmdr. George Reiter. He kept half of the marines' 6mm and 3-inch ammunition for ballast, because his ship had been assigned as a tow vessel. After protest, the marines managed to regain their rifle ammunition, but the artillery rounds stayed on board. In Key West, the battalion's officers and men carried out battle exercises, made long marches, and underwent general toughening up, while they waited for more suitable transport to Cuba. Whereas all of the field grade officers and some of the company grade officers were overage, many of the men were raw recruits.

Fortune did not smile on the battalion, however, for Secretary of the Navy John D. Long ordered the contingent to reboard the hot and cramped *Panther* and proceed immediately to Cuba. The overly officious Commander Reiter remained at constant odds with Colonel Huntington and his officers in regard to the command and conduct of the marines while aboard his ship. This relationship would only get worse when the ship reached Guantánamo Bay, and indeed would threaten to put the landing force in jeopardy.

Guantánamo Bay became vital after the arrival of the Spanish squadron commanded by Admiral Pascual Cervera y Topete at the port of Santiago de Cuba, having outfoxed both Rear Adm. William Sampson's North Atlantic Squadron and Commo. Winfield Schley's blockading Flying Squadron. The Spanish warships' presence now demanded of the US Navy a constant watch on the port, but the frequently rough water off the southern coast of Cuba made coaling of the blockading warships difficult. A sheltered coaling station 40 miles east at Guantánamo Bay would be a godsend to the American naval force if it could be captured from the Spanish. Once it had realized the excellent potential of the bay, however, the US Navy began casting covetous eyes on it for the longer term.

Transport *Panther* with the First Marine Battalion arrived off the entrance to the bay on June 10, joining other American warships present, the largest being the unprotected cruiser *Marblehead* (nine 5-inch guns). The cruiser's captain, Cmdr. Bowman McCalla, was both the senior officer present afloat and the officer commanding the landing operations. He had been a central figure in the American landings on the Panama isthmus thirteen years before and remained a proponent of such operations, thus he possessed considerable experience along those lines. Because the Spanish harbor defenses had already been shelled and most of them put out of action, the marine landing commenced shortly following *Panther*'s arrival, with the laden whaleboats and cutters towed into shore by the steam launches from the various warships present.

On April 22, 1898, the First Marine Battalion marches down Navy Street in downtown Brooklyn, before returning to the New York Navy Yard to take ship to Key West, FL, and then Guantánamo Bay, Cuba. The battalion's commanding officer, Lt. Col. Robert Huntington (1840–1917), may be seen mounted at the head of the column, with his Mameluke sword drawn. This blade was standard for marine officers since 1826, commemorating the actions fought in the First Barbary War, 1805. (*USMC History Division*)

After disembarking from the transport *Panther* at Key West, Huntington's battalion underwent hard training and long marches, and as seen here, close order drill, despite the late April heat. At Guantánamo Bay in June, those marines not issued light tropical uniforms often shed their blue blouses to engage in combat, but all of the men continued to wear their blue 1897 Pattern forage caps. (*USMC History Division*)

Immediately, however, *Panther*'s captain would once again prove to be less than cooperative. He refused to provide any assistance from the ship's crew in disembarking the marines and unloading their equipment into the boats, forcing Huntington to leave two companies aboard, one of which was his heavy weapons unit. Again, there occurred difficulty in accessing the battalion's ammunition. Angered, Colonel Huntington appealed to Commander McCalla. The latter's written order on June 12, though couched in Victorian-era civility, reveals a sharp reprimand to Reiter:

> Sir: Break out immediately and land with the crew of the Panther, 50,000 rounds of 6-mm. ammunition. In the future, do not require Colonel Huntington to break out or land his stores with members of his command. Use your own officers and men for this purpose, and supply the Commanding Officer of Marines promptly with anything he may desire.

This directive enabled Huntington to land his remaining two companies aboard *Panther*, and likely some of his remaining field guns.

Perhaps from a combination of gratitude and good politics, the marines' encampment on a hill overlooking Fishermans Point and Guantánamo Bay, surrounded by thick tropical scrub and cactus, would soon thereafter be named Camp McCalla. On the battalion's second night there, acting executive officer Maj. Henry Cochrane, on a prowl around the perimeter, found the security measures in more than one sector far from his liking. He quickly established outposts where needed. The problem lay with the mean age of the company commanders. The combination of the heat and the exertion required was simply wearing out these superannuated officers.

Cmdr. Bowman H. McCalla (1844–1910) was the captain of the unprotected cruiser *Marblehead* and the senior officer present afloat in June 1898, when the First Marine Battalion landed at Guantánamo Bay. The marines appreciated his efforts on their behalf and so named their hilltop position Camp McCalla. He had participated in the Panama landings in April 1885 and would assume a leading role in the failed Seymour expedition to relieve Peking in June 1900, suffering three wounds in the process. (*USMC History Division*)

Two Colt machine guns were landed on the first day and at least one 3-inch field gun went ashore as well. The Spanish attacks on the marine encampment began on the evening of June 11, and on the following day, USS *Texas* (two single 12-inch gun turrets) arrived and landed three additional field guns and two machine guns for Camp McCalla's defenders. Henry J. Whigham, on assignment for the *Chicago Tribune*, in commenting on the assistance given the weary marines by the correspondents present, wrote, "We also helped to get two 3-inch field guns to the top of the hill." How essential a role F Company's 3-inch field guns played in repulsing the attacks by the Spanish army's tough Escuadra de Guantánamo and its Cuban irregulars is questionable, because the shrapnel rounds rarely exploded at the relatively short ranges demanded.

Stephen Crane—whose 1895 novel *The Red Badge of Courage* addressed individual courage and cowardice in the context of the American Civil War—on assignment for Joseph Pulitzer's *New York World*, described the fighting for possession of the hilltop eight months later in *McClure's Magazine*:

> It was my good fortune—at that time I considered it my bad fortune, indeed—to be with [the marines] on two of the nights when a wild storm of fighting was pealing about the hill, and of all the actions of the war, none were so hard on the nerves, none strained the courage so near the panic point, as those swift nights in Camp McCalla. With a thousand rifles rattling; with the field guns booming in our ears; with the diabolic Colt automatics clacking; with the roar of the *Marblehead* coming from the bay, and last, with Mauser bullets sneering always in the air a few inches over one's head, and with this enduring from dusk to dawn, it is so extremely doubtful if any one who was there will be able to forget it easily.

Those clacking Colt machine guns, never the most reliable of automatic weapons, indeed proved prone to malfunction, very likely giving rise to the "diabolic" in Crane's description of them. Their next-morning repair after a night's fighting became an essential routine for the marines on the hilltop. Some of the senior officers complained that the marines' fire was too often panicky and undisciplined. The casualties inflicted by the Spanish nocturnal raids were relatively few, but still significant, including the acting sergeant major and the attached naval surgeon both shot dead.

The fight for Guantánamo Bay—the first land action of the Spanish-American War—ended when Colonel Huntington, having requested that his battalion be permitted to withdraw from the precariously exposed hilltop camp and driven nearly to despair when Commander McCalla refused him, took the offensive on June 14. He sent a mixed force of two marine companies commanded by Capt. George Elliott and fifty insurgents to capture and destroy Cuzco Well, in order to deprive the Spaniards of the only water supply in miles. Lt. Col. Enrique Tomas accompanied the force as the Cuban liaison officer. The marines' volley fire up to 1,200 yards, supported by three machine guns, forced the enemy's withdrawal.

Sgt. John Quick and Pvt. John Fitzgerald won both fame and Medals of Honor by standing on a hillcrest in clear sight of enemy riflemen and calmly wigwagging

American marines and Cuban insurgents gather at Camp McCalla on June 12, 1898, before their assault on Cuzco Well, the main water source for the Spanish forces in the region. Several marines can be seen in tropical khaki uniform, but the one seated behind the Colt-Browning Model 1895 machine gun still wears his heavy blue shirt. The 1897 Pattern blue forage cap, however, was worn for the duration of the fighting at Guantánamo Bay. Both the marines and insurgents are armed with the Winchester-Lee Model 1895 6mm rifle. *History Today*, June 2014.

gunnery directions to USS *Dolphin* (two 4-inch guns). In the absence of the marines' field guns because of the nature of the terrain, the gunboat (originally built as a dispatch boat) lay offshore and fired in support of the assault. In one instance, Quick's fearless signaling averted friendly fire casualties to a marine flanking attack led by 2nd Lt. Louis Magill. The battle ended with the capture and destruction of the well. In October 1903, George Elliott earned promotion to brigadier general, and became the tenth commandant of the corps. Louis Magill died unexpectedly in February 1921, before the Brevet Medal awarded him in June of that year for gallantry at Guantánamo Bay twenty-three years before, could be presented.

Accounts indicate that the marines emplaced four each Fletcher 3-inch field guns and Colt-Browning 6mm machine guns on the beleaguered hilltop. What remains uncertain is the number of each type that came ashore from transport *Panther* and battleship *Texas*. It appears that one or two field guns, probably organic to F Company, were in place atop the hill within a day of the marines occupying the position, and two more were added from *Texas* on the 12th. Those were the pair of guns hauled up by marines and journalists together in harness, as chronicled by newsman Whigham. Other field guns and machine guns were emplaced to defend the base camp below on Fishermans Point, which Huntington improved after he had begun to question if the higher elevation Camp McCalla remained tenable.

After the battle, Huntington criticized the design and quality of the tropical-weight campaign uniforms that eventually reached the marines. For one thing, after a single washing of the items, their color so bleached out it became nearly white, a liability in night fighting. It is not clear in his remarks when precisely these uniforms arrived,

The gunboat *Dolphin*, seen here in the peacetime livery of the early 1890s, supported the marine assault on Cuzco Well with her pair of 4-inch/40cal guns. Built as a dispatch vessel during the pre-wireless age, she had been the "D" of the ABCD warships in the 1880s New Navy. The service later redesignated her as a gunboat. Detroit Photographic Co. (*Library of Congress*)

With the end of the fighting, Capt. Francis Harrington (1843–1906), commanding F Company, poses with the marine defenders of Camp McCalla. The riflemen are equipped with Winchester-Lee Model 1895 6mm bolt-action rifles, and the artilleryman aims a Fletcher Mark I 3-inch field gun. The army-issue 1889 Pattern campaign hats worn here did not become available until after the Spanish forces at Guantánamo Bay had been driven off. (*Henry C. Cochrane Collection [COLL/1], Marine Corps Archives & Special Collections*)

although it is known that the marines had to fight the entirety of the battle in their standard blue garrison caps, likely contributing to the high rate of heat exhaustion during the Cuzco Well foray. The campaign hats—left aboard naval transport *Resolute* (four 6-pounders)—would get there in time for après-combat photos of stern, manly looking marines. Huntington also urged that canister rounds be provided for the field guns, attributing the failure of the shrapnel to explode as being a function of too short a range. Two years later, firing at rather longer ranges showed the problem lay with the design of the fuse.

Actions at Guánica and Cape San Juan, Puerto Rico, and Stirrings in China

In late July 1898, Gen. Nelson Miles, US Army, requested that the navy undertake a reconnaissance of Guánica, on the southwest coast of Puerto Rico. Miles assured the navy that the port lacked harbor defenses and, moreover, that the Spaniards berthed a substantial number of sugar lighters there. The army needed such flat-bottomed craft to ferry its troops and field guns ashore from the transports arriving at the port. If it also captured Ponce 20 miles to the east, a sufficiently large force could land and progress along the main highway to San Juan on the northern coast, cutting the island in two and likely ending the opposition. Accordingly, an American flotilla led by battleship *Massachusetts* (two twin 13-inch gun turrets) arrived off Guánica, and Capt. Francis Higginson warily gave permission to Lt. Cmdr. Richard Wainwright to take his ship, the gunboat *Gloucester* (J. Pierpont Morgan's former yacht *Corsair II*, now armed with four each 6-pounders and 3-pounders), into the harbor. Wainwright not only did so, but he ordered away a thirty-man landing party commanded by his executive officer, Lt. Harry Huse, assisted by Lieutenant Wood, to capture the town.

The landing party debarked from the *Gloucester*'s cutter, replaced the Spanish colors at harborside with the US flag, and soon came under enemy fire from the brush a few hundred yards away. Their cutter provided fire support from its Colt-Browning machine gun, until it jammed and could not be restored to working order. *Gloucester* sent ashore a replacement gun, which the seamen placed behind a wall they had thrown up across the road leading into the north end of town. They also strung two rows of barbed wire across the road, fifty and 100 yards in front of the makeshift wall. When a hostile mounted unit appeared, *Gloucester* opened fire with her 6-pounder and 3-pounder batteries and drove them off. Soon thereafter, the army arrived in strength, and the naval party returned to its ships. In his action report, Lieutenant Huse described the Winchester-Lee navy rifle as abominable, jamming frequently—almost certainly the first report of the widespread unhappiness with this rifle in combat.

A few days later, upon the demand of the warships standing offshore, Ponce and La Playa surrendered without a shot being fired. The action then shifted to the northeast coast of Puerto Rico.

With the intention of making a rendezvous with the transports carrying an army assault force, a small naval flotilla arrived off Cape San Juan, Puerto Rico, on August

1, 1898, and anchored to seaward of the offshore islands. Except for two transports, the anticipated army force failed to show (it landed in Guánica, General Miles having neglected to inform all parties). Capt. Frederick Rodgers, commanding the monitor *Puritan* (two twin 12-inch gun turrets), ordered two boatloads of sailors ashore to reconnoiter, the landing party commanded by Lt. Herman Dresel. They observed Spanish troops on the outskirts of the city of Fajardo, and likely wishing to accomplish something warlike before reporting back to Rodgers, Dresel's men seized a schooner as a prize.

Another party, commanded by Lt. Cmdr. James Selfridge, landed the next morning and occupied the Faro de Las Cabezas de San Juan (the Cape San Juan lighthouse), raised the US national colors, but instructed the keeper of the light to continue with his routine procedure. Three days later, *Puritan*, the collier *Hannibal* (one 4-inch gun), and the two army transports were ordered elsewhere, leaving Capt. Charles Barclay, commanding the monitor *Amphitrite* (two twin 10-inch gun turrets), as the senior officer present afloat. That afternoon, Barclay, Ens. Albert Campbell, some leading citizens of Fajarda, and a fourteen-man landing party boarded the armed tug *Leyden* (two 1-pounders), under the command of Ens. Walter Crosley. Upon docking, they entered the city, and perhaps mostly for its annoyance value, the Americans raised their flag over the customs house and city hall. As a final act before withdrawing his landing party, Barclay organized a citizens' militia to patrol the city.

On the evening of August 6, Barclay ordered a fourteen-man landing party from *Amphitrite* under Ens. Kenneth Bennett to return to the lighthouse and relight its lamp; a Colt-Browning machine gun equipped this party. Very soon thereafter, a second boat holding another fourteen men commanded by Lt. Charles Atwater followed, with Atwater assuming overall command. A number of prominent Fajardans, who feared reprisals from the Spanish because of collaboration with the Yanquis, took shelter with the American sailors at the lighthouse.

Although having withdrawn upon the appearance of the American flotilla, the Spanish army reoccupied Fajardo the following afternoon—this time a force of about 200 men commanded by Col. Pedro del Pino. That same day, protected cruiser *Cincinnati* (one 6-inch gun, ten 5-inch guns) arrived, accompanied by *Hannibal*. Following the relief of half of the shore party, including Ensign Bennett, the bluejackets fortified the lighthouse against the expected attempt by the Spaniards to retake it. They placed their machine gun on the roof in order to achieve a better field of fire. The night of the 8th turned out to be brightly moonlit and the Spanish opened the fight with volley fire from the surrounding forest. Atwater ordered the lighthouse lamp doused as a signal to the warships that his force was under attack, and *Cincinnati* illuminated the area with her searchlights. When a friendly 6-pounder shell penetrated the defenses without exploding, however, Lieutenant Atwater quickly ordered the lamp relighted. Shortly thereafter, firing ceased by both sides. The Americans had expended 1,100 rounds of 6mm ammunition, from both their Winchester-Lee rifles and the Colt-Browning machine gun.

In the early morning of the 9th, Barclay ordered the landing force withdrawn. To cover the withdrawal, thirty bluejackets from *Amphitrite* and thirty marines off

Cincinnati, the latter commanded by 1st Lt. John Lejeune, went ashore. Thereafter the flotilla sailed away, with the refugees most at risk from retribution taken to safety aboard *Leyden*. Both Rodgers and Barclay had rotated the junior officers and men on shore in order to maximize the experience of participating in a landing force in action. In this instance, the experience came with a cost: the revolver carried by Naval Cadet William Boardman slipped out of its holster during the night of the 6th and discharged, with the bullet hitting him in the thigh. He died two days later from blood loss, likely because the projectile had nicked his femoral artery.

Across the world, marines went ashore in China. They did so to protect American lives and property from the widespread unrest directly resulting from the coup de main carried out by Dowager Empress Tzu Hsi against sitting Emperor Guangxu, her nephew. In early November 1898, marine detachments taken off the protected cruisers *Baltimore* (four 8-inch guns), *Boston* (two 8-inch guns), and *Raleigh* (one 6-inch gun, ten 5-inch guns), and commanded by 1st Lt. Robert Dutton and navy Lt. John Gibson, went up the Peiho River to protect the consulate at Tientsin and the legation at Peking. Each detachment took along a Gatling machine gun and plentiful ammunition. The navy pulled out the detachments in mid-March 1899. All too soon, however, American marines and bluejackets would be back.

Unrest in China in the fall of 1898 led to three US cruisers landing two marine detachments, each equipped with a Gatling machine gun, to guard the American legation at Peking and the consulate at Tientsin. The marines are seen formed up in front of the Tientsin consulate, armed with Winchester-Lee 6mm rifles and the encased short Gatling gun. Also shown are the American consul James Ragsdale (1848–1932), his family and staff, and a crowd of curious Chinese. (*Robert Henry Chandless Photograph Collection, University of Washington*)

Joint Intervention in Samoa by the United States and Royal Navies

By the year 1899, the question of who possessed or at least controlled the Samoan Islands had been a matter of contention for a decade among the United States, Great Britain, and Germany. Their considerable value lay in establishing a naval coaling station, and later a wireless station, as a center of influence in the central Pacific, in that period before the construction of the Panama Canal. The United States and Germany came close to war over the islands in 1889, until a March typhoon struck and destroyed in their entirety the German and American battle flotillas sheltered in Apia harbor on the north shore of Samoa's Upolu Island. As a result of either the great wind, the heavy devastation, or the sizable loss of life—or a combination of the three—cooler heads prevailed. Despite the backing off from international conflict, civil war between tribal factions continued on the islands.

Upon the death in 1898 of Malietoa Laupepa, who had been restored as king under a three-power protectorate, his long-time rival, tribal chief Mata'afa Iosefo, claimed the throne, now backed by the Germans. Opposing Mata'afa was Laupepe's son, Prince Tanumafili, supported by Britain and the United States. In early March 1899, the 4,300-ton protected cruiser *Philadelphia* (twelve 6-inch guns), having aboard Rear Adm. Albert Kautz, commander of the US Pacific Squadron, arrived in Apia harbor. On the 15th—the decade anniversary of the destructive cyclone—*Philadelphia* and a flotilla of British warships bombarded hostile strongpoints near the water, followed by the seizure of the town by landing forces from both navies. Efforts by the rebel Mataafans to retake the port town were handily repulsed by the Anglo-American force, which had landed artillery and machine guns. The rebels attacked the American headquarters at the Tivoli Hotel, defended by both American and British naval elements. During the fierce fighting there, the Mataafans temporarily captured a British 7-pounder muzzleloading landing gun, but abandoned it when they were forced to withdraw.

The senior officers presently decided to march on the rebel stronghold based at the coconut plantation at Vailele owned by Deutsche Handels und Plantagen Gesellschaft (DHPG), not far to the east and south of Apia. The operation commenced on April 1. Offshore lay the screw corvette HMS *Royalist* (eight 6-inch/100-pounders), which bombarded the plantation forts before the land battle began. The naval contingent, comprised of 115 to 118 sailors and marines nearly evenly divided as to nationality, and 136 Samoans loyal to Tanumafili, made it to the plantation at Vailele without much resistance. South of that place, the landing force turned back west to return to Apia. Where the road ran between parallel barbed wire fences, a rebel force comprised of several hundred warriors ambushed the landing party and badly mauled it. Very soon into the battle—known as either Vailele or Tagalii—a bullet killed Lt. Angel Hope Freeman, RN, who commanded the expedition.

The Americans had with them a Colt-Browning machine gun mounted on a field carriage, which Lt. Philip Lansdale, the senior American officer, held in high regard. Not for the first time, however, this automatic gun showed its cranky side. It refused to operate at the first attempt to fire it. After being put back into operation, the gun jammed during a critical moment in the battle. With the Anglo-American force now under a terrible fire and confronting one of the barbed wire barriers, Lansdale had no choice but to disable the machine gun and leave it behind.

American Imperial Adventures   159

Lying off Apia harbor, Samoa, in mid-March 1899, the protected cruiser *Philadelphia* pounds hostile strongpoints with her port 6-inch/30cal gun battery. She has a sail set fore and aft to assist in making minor changes in position as the situation ashore requires. Her landing party of sailors and marines will shortly undertake a joint operation with a similar British landing party and a force of friendly Samoans, with disastrous results. (*National Archives and Records Administration*)

Joined by the wife of Luther Wood Osborn, the American consul general, the sailors in the landing party from the protected cruiser *Philadelphia* pose in front of the United States consulate in Apia (the national emblem can be barely discerned between the sailor at the top right and the adjacent pillar). If the photo was taken before the battle at Vailele, there is a good chance the Colt-Browning machine gun shown at the top of the steps is the one that failed during the action on April 1, 1899, and had to be abandoned. The men are armed with the Winchester-Lee 6mm infantry rifle. (*Naval History and Heritage Command*)

Not long thereafter, leading his men through a ravine swept by fire, he was shot in the leg. With a major bone shattered, Lansdale could no longer walk, and he ordered his men to fall back without him. His second-in-command, Ens. John Monaghan, and a few sailors, however, stayed with him. While attempting to carry Lansdale to safety, Seaman Norman Edsall was shot and killed. Despite Lansdale's pleading with his friend "Monny" to leave him behind, Monaghan refused to do so. In consequence, both officers and another American seaman died there. Following the retreat of the combined naval landing force, the rebels beheaded Freeman, Lansdale, and Monaghan, and cut the ears off the British and American ratings. Both of the American officers lay wounded when taken.

Elsewhere on the field, 1st Lt. Constantine Perkins, USMC, collected a mixed party of American and British marines and carried out a methodical fighting retirement to the shore. Passed Asst. Surgeon George Lung, in the absence of line officers, acted to rally and steady the men, and personally carried out a reconnaissance to find a route over which the Anglo-American force could safely withdraw. Neither American officer knew of the deaths of Lansdale and Monaghan until the rear guard reached the beach. *Royalist* lay opportunely offshore and sent reinforcements, but the rebels no longer molested the field force.

The Tripartite Convention, signed on December 2, 1899, and ratified on February 16, 1900, decided the political fate of the Samoas, with the western islands going to Germany and the eastern to the United States. Britain withdrew her claim after trading for some German colonial possessions in the western Pacific and West Africa.

Of the four Medals of Honor awarded for the action on Samoa, one went to Pvt. Henry L. Hulbert, among the last men to pull back and singled out by Lieutenant Perkins for his steadfast bravery. Born in England, he would rise through the ranks to sergeant major, and thereafter be commissioned. He was killed in action during World War I as a fifty-one-year-old first lieutenant, and posthumously awarded the Navy Cross. Four of the men involved in the fight at Vailele, Samoa, would be honored as the eponyms for destroyers USS *Monaghan* (DD 32; DD 354), USS *Lansdale* (DD 101; DD 426), USS *Edsall* (DD 219; DE 129), and USS *Hulbert* (DD 342). As if in some uncanny repetition of the battle more than forty years before, of those four named destroyers that entered World War II at its inception for the United States, first *Edsall*, second *Monaghan*, and second *Lansdale* would all be lost. Only *Hulbert* would survive.

The Seymour Relief Expedition and the Action at Tientsin

The events leading to the relief of the legations at Peking are reasonably well understood, even with the complicated and ever-changing political aims—at times intermeshing, at other times conflicting—of Dowager Empress Tzu Hsi, the Society of Righteous Harmonious Fists (in western parlance, the Boxers), and the Chinese imperial army. With the armies of the various warlords, both close to and far from the imperial city, added into the mix, the situation in China became terribly confusing and challenging to the Western principal figures on the scene. With their legations likely to be besieged and their nationals at great risk, the international powers decided upon a collective response to the emergency.

One of the first warships to arrive off Taku Bar in late May 1900 was the protected cruiser *Newark* (twelve 6-inch guns), commanded by none other than Capt. Bowman H. McCalla of Isthmus of Panama and Guantánamo Bay fame. With him, he brought two contingents of marines, *Newark*'s own commanded by Capt. Newt Hall and the other previously transferred from the battleship *Oregon* (two twin 13-inch gun turrets) under the command of Capt. John Myers, with Myers senior to Hall.

On the 29th, forty-eight marines under the two marine captains went ashore, along with a Colt machine gun and a 3-inch fieldpiece served by a naval gun crew. Three hours later, McCalla himself arrived with a naval contingent consisting of sixty-four officers and men, equipped with a second Colt automatic gun, and joined the earlier landing force at Tongku, the railhead on the Peiho River nearly 40 miles downstream from Tientsin (Tianjin).

Having been refused transport by the Chinese railway authorities, McCalla commandeered a tugboat and some junks and sampans, tied them together, and moved his force upriver to Tientsin, arriving late the same day. The American landing force was soon thereafter followed by naval contingents from several European nations

Standing in front of the Colt-Browning 6mm machine gun on a field carriage that they have just brought up by train, a group of officers pose for the camera at the US legation in Peking. From left to right: Naval Cadet Charles E. Courtney; Capt. Newt H. Hall, USMC; Capt. Bowman H. McCalla, USN; Paymaster Henry E. Jewett, USN; Assistant Surgeon Thomas McC. Lippett, USN; and Capt. John T. ("Handsome Jack") Myers, USMC. McCalla, Jewett, and Courtney would soon return to Tientsin, while the other three would remain to participate in the defense of the international legations against the Boxers. (*USMC History Division*)

and Japan, some bringing along their own landing guns. The total force sent on May 31 by special train to Peking numbered about 450 officers and men. The American contribution slightly exceeded fifty men, mostly marines with a handful of sailors, commanded by Captains Myers and Hall, which included one of the Colt-Browning machine guns. Their subsequent role in the determined defense of the Peking legations has no part in the remainder of this account.

Captain McCalla, accompanied by Paymaster Henry Jewett and Naval Cadet Charles Courtney (USNA class of 1899; commissioned as ensign 1901), who had gone up to Peking with the marines, returned by train to Tientsin on the morning of June 2. The last down train pulled into Tientsin three days later. A council of war on the 9th, in which McCalla—despite having but 112 men, mostly sailors, with a few marine sharpshooters—declared his intention to push on to Peking regardless of what the other nations did, helped to decide the matter. On the following day, the 2,500-man international relief expedition, commanded by Vice Adm. Edward Hobart Seymour, Royal Navy, the senior officer present, set out for Peking in five trains comprising more than 100 coaches and wagons. To protect the expedition's base, a garrison remained in Tientsin.

The relief force reached a point only 25 miles from the imperial city before fierce Boxer resistance halted it. On June 16, low on food and water, substantially reduced in strength by more than 250 casualties, and unable to proceed farther, the expedition began a retrograde movement to Tientsin, but found the track cut behind, as it had been ahead.

A troublesome reality confronted the American sailors: Although their Winchester-Lee bolt-action rifles proved accurate enough, one 6mm round frequently failed to stop a Boxer warrior, more often than not armed with only a sword or spear. Rather, unless the first bullet hit a vital spot, two or more rounds could be required to drop him.

Also on the 16th, by coincidence, the warships of the various navies anchored in the Yellow Sea issued an ultimatum to the Chinese garrisons to abandon the increasingly bothersome Taku forts defending the Hai River, the estuarine part of the Peiho. Early in the following morning, the guns of the forts opened fire on the ships before the ultimatum had expired. Although heavily outnumbered, the ships stripped their crews to deploy as naval infantry, and thereupon assaulted and captured the forts. Casualties were numerous. Only the US Navy did not take part, having been ordered not to make war on the national government of China. The old sidewheel gunboat *Monocacy* stood close in, however, to take on refugees. Later she served to accept wounded men from the contingents of the coalitions, until they could be taken aboard their own warships.

The upshot of the attack and capture of the Taku forts for Seymour's force was that it almost immediately faced large numbers of imperial troops in addition to the Boxers. On the 19th, the would-be rescuers abandoned the trains on which they had been traveling. Now afoot, initially with the Americans at the front, followed by the British, the column faced Chinese harassing fire of ever-increasing intensity. The US Navy 3-inch field gun, firing shrapnel from 1,000 yards down to 700 yards, and a flank attack led by Ens. Daniel Wurtsbaugh, did much in repelling a major enemy assault on the 20th. Falling behind the day after, the American field gun joined with a pair of British muzzleloading 9-pounders as an ad-hoc Anglo-American field battery. In order to come into action, the gunners, dragging their pieces, had to run 300 yards along an embankment swept by hostile fire; the Americans lost two killed and three wounded. A severe leg wound disabled Naval Cadet Joseph Taussig (USNA class of 1899; commissioned as ensign 1901), commanding 1st Company. Later in the day, the American Colt-Browning and the British Maxim guns joined similarly in a combined machine gun battery to provide an hour's steady firing against the hostile onslaught that threatened.

Tormented by the heat, thirst, and hunger, encumbered by wounded men, many in agony, and depleted of ammunition, the expedition faced a precarious prospect. Its fortunes took an upswing on June 23, however, when it was able to take refuge in Hsiku Arsenal. British marines led by Maj. James Johnstone, and supported by the Germans and Americans, captured the arsenal and its fort after the garrison had fired on Seymour's men attempting to pass it peaceably. In addition to safe haven, at least momentarily, the members of the relief force found ample food, medical supplies, and additional weapons there. Many of the American sailors and marines took the opportunity to divest themselves of their 6mm Winchester-Lee rifles and replace them with heavier-hitting Austrian Mannlicher 8mm infantry rifles stocked in the arsenal. The new arms would be needed at once: Gen. Nieh Shih-ch'eng launched his imperial troops in an immediate and determined attack to wrest the arsenal and its matériel from the coalition force. That attack failed, as did the one the following day, this time made by both imperial troops and Boxers.

After the Chinese had handily turned back Major Johnstone's marines intending to win through to Tientsin, in which the British suffered five killed, the expedition settled in to await relief. Chao Yin-ho, the loyal Chinese servant of Seymour's civilian translator, British legation officer C. Clive Bigham, crept out of the beleaguered stronghold and began his lonely and dangerous journey to Tientsin to carry the word of the plight of Seymour's force. In the days previous, that city had seen some significant happenings.

On June 18, 1900, Maj. Littleton Waller arrived on the hospital ship USS *Solace* with a company detached from the 1st Marine Regiment at Cavite on Manila Bay, reinforced by another thirty men off the gunboat *Nashville* (eight 4-inch guns), who brought along their machine gun and 3-inch rifle. The men of Waller's company were the first marines detached from the Philippines, which had been awarded to the United States as a result of the Spanish-American War and where the Americans now had their hands full with a widespread insurrection.

With engine room sailors from *Monocacy* crewing the locomotive, the marines loaded a train with their weapons, as well as extra rails and ties to repair the railway, and set out for Tientsin. A dozen miles shy of the city, the 140 Americans met up with a force of more than 400 Russian troops, and together on the following day, the 21st, despite the doubts expressed by Waller, they continued the march. The Russians put Cossack cavalry in the lead to carry out reconnaissance. Early in the advance, the old 3-inch breechloading rifle accompanying the marines proved to be irreparably defective, and Waller ordered it rolled into the river (or a tributary canal).

On the outskirts of Tientsin, the combined force was stopped cold by overwhelming odds centered at the imperial East Arsenal. During the course of this hard fight, the enemy shot down all but one of the crew of the single Colt machine gun. Because it had jammed several times, its commander, 1st Lt. William Powell, ordered it disabled and abandoned. The hard-pressed coalition force finally withdrew to the former Russian encampment, carrying its wounded but leaving its dead on the field. During that day and the next, a coalition force in excess of 2,000 troops gathered, with Waller's marines attached to the British naval brigade commanded by Cmdr. Christopher Cradock (who as a rear admiral commanding the British cruiser squadron engaged, would be killed in action at Coronel in November 1914). On June 23, this force drove back the Chinese opposition and marched into Tientsin, relieving the foreign concessions adjacent to the treaty port. By that time, its buildings were bullet-ridden and much damaged by shell fire.

In the early morning of the 25th, a mostly Russian and British force of 1,800 men commanded by Colonel Sherinsky started for Hsiku Arsenal to save Seymour and his men (despite more than one close call, Chao Yin-ho had gotten through). They encountered minimal opposition and arrived the same day. The rescuers and the rescued, of whom around 300 were wounded, departed for Tientsin the next morning, blowing up the arsenal on their way out. So many men were wounded or needed to assist the wounded—typically, four men bore a laden stretcher, with two more men required to spell the bearers in the onerous heat—that much equipment had to be left behind. Having disabled the piece by the removal of its breechblock, the Americans abandoned 3-inch field gun No. 31, which had landed from *Newark* a month earlier and thereafter provided effective service.

On June 21, 1898, superior Chinese forces before Tientsin repulsed the advance of 400 Russian troops, including Cossack cavalry, and 140 American marines. The coalition contingent then withdrew to its camp on the Peiho River, awaiting reinforcement. Behind the central figure in the sun helmet can be seen numerous naval infantry. The man standing on the right appears to be wearing the Royal Navy summer-issue straw hat, and the soldier seated on the far right holds a Russian Mosin-Nagant Model 1891 infantry rifle. The following day, this force would fight its way into Tientsin. (*Naval History and Heritage Command*)

The seriously incapacitated men from the Seymour expedition included thrice-wounded Capt. Bowman McCalla, USN, nearly fatally wounded Capt. John Jellicoe, RN, and American Naval Cadet Joseph Taussig, who had a serious leg wound. McCalla became a rear admiral in 1903 and died one decade after the events in China had taken place; Jellicoe survived to command at Jutland in 1916; Taussig, after a lengthy convalescence in the American naval hospital in Yokohama, rose to the rank of vice admiral in World War II. The Americans alone had suffered thirty-one men killed and wounded in this ill-fated attempt to relieve the Peking legations.

But now another problem needed to be addressed. Estimates vary, but at least 20,000 Boxers and 15,000 imperial troops, the latter equipped with Krupp artillery of excellent quality and impressive quantity, defended the fortified Chinese city within Tientsin. Those guns even then bombarded the "foreign devils" inside the more expansive city. In the European concessions, men housed in godowns serving as barracks suffered casualties. On the day following the return of the rescuing force from Hsiku Arsenal, much the same force of 1,800 men, largely Russian but supported by other coalition contingents, assaulted and captured the vital East Arsenal, held by nearly four times the number of imperial troops. A week into July, Waller sent 1st Lt. Robert Wynne, USMC, with a party of sharpshooters composed of American marines and Royal Welsh Fusiliers to the tower of the English missionary college to combat enemy snipers. Two days later, sailors of the Japanese landing force and American marines captured the troublesome West Arsenal, close to the city.

Additional reinforcements began to reach the Americans at Tientsin. On July 11, two battalions of the 9th US Infantry arrived, led by the regimental commander, Civil War veteran Col. Emerson Liscum. The following day, the armored cruiser *Brooklyn* (four twin 8-inch gun turrets) anchored off Taku and unloaded the remainder of the 1st Marine Regiment under the command of Col. Robert Meade, which included the regimental headquarters, an additional infantry battalion, and the artillery company armed with three each 3-inch naval field guns and Colt-Browning 6mm machine guns on field carriages. The 1,000 or so Americans—two-thirds soldiers and one-third marines—were again brigaded with the 2,200-man British force at Tientsin, as were the sailors of the Japanese naval infantry. The total international presence now stood at just shy of 6,000 troops. This force would undertake the reduction of Tientsin's native city.

An outer mud wall, 12 to 15 feet in height, encompassed the entire city of Tientsin. Within this wall lay residential and commercial buildings, both Chinese and European. Religious edifices, some of them grand, had been erected here; the recently captured West Arsenal stood on the south side, near a gate in the mud wall. The Tongku to Peking railway entered from the east and just past the Tientsin station, looped around and exited to the northeast. The Peiho River entered from the northwest, and after taking on a northern tributary, flowed out on the east, where the treaty port and the foreign concessions stood; both moats and canals, fed by the river, saw water traffic. South and west of the river, but otherwise within the center of the outer city, sat the formidable Chinese city. Its massive walls measured 33 feet thick, consisting of masonry and solid earth, with the earth ascending 20 feet and the masonry face another 6 feet in height. The upper stone wall, pierced with loopholes and embrasures, presently bristled with modern Krupp artillery.

To answer the Chinese guns, the Western powers brought a variety of field artillery. In addition to the American marines' 3-inch field guns, HMS *Terrible* (two 9.2-inch guns) provided five 12-pounder QF and two 9-pounder ML naval landing guns, as well as a number of larger quick-firing shipboard guns mounted on improvised field carriages and firing shells filled with lyddite (fused picric acid).

After a council of war decided the issue on July 12, the attack began early on the following morning. The American marines were on the extreme end of the left flank, accompanied by civilian engineer and future US president Herbert Hoover, who was well acquainted with the city. The soldiers of the 9th Infantry directly supported the British naval brigade. Upon crossing the outer mud wall, the attackers advanced in rushes, taking what cover was offered by the drainage ditches and burial plots that broke up the ground. With the sunrise, the temperature soared to a dreadful level, as it had been all week long, and the attacking force was soon assailed by heavy sniper fire. Pursuant to orders from the brigade commander, British Brig. Arthur Dorward, the marines moved forward to attack a troublesome enemy battery and to extend the left flank in order to protect against enemy soldiers attacking that flank. Closer to the Chinese city, the attackers found that the Chinese had sometime earlier diverted water from the canals to flood the area. Now they had to wade slipping and sliding through the foul water, fired at incessantly by the Boxer snipers and the artillery pieces mounted on the city wall.

Company commander 1st Lt. Smedley Butler stopped a bullet here, and after a swig of brandy provided by a British officer, he had to be removed to the aid station. His steadfast bravery won him a brevet captaincy, and in 1921 the award of the Brevet Medal. At that time it was the second highest decoration bestowed on marine officers for extreme gallantry, surpassed only by the Medal of Honor. The corps conferred only twenty-three Brevet Medals during the two decades the award remained in being.

Capt. Ben Fuller's artillery company had been active, providing 3-inch covering fire against the stone wall 2,200 yards distant, and also firing into more distant Chinese positions within the native city, until forced back by counterbattery fire from those all-too-accurate German artillery pieces. Having exhausted their 130 rounds of shrapnel—which proved to be mostly duds—the marine gun crews picked up their rifles and went into action as infantry. The crews of the machine gun section joined D Company, commanded by 1st Lt. Robert Dunlap, and moved to the far left flank to resist enemy pressure there. Capt. Charles Long nominally commanded D Company, but upon being appointed commanding officer of the provisional second battalion, his executive officer, Lieutenant Dunlap, assumed command of the company. One of the Colt machine guns followed the company, but soon after resuming fire, its operating bolt broke. Somewhat later, Brigadier Dorward sent the field gun crews forward to support the hard-pressed 9th Infantry.

The marine infantry had gone into battle with plentiful ammunition, each man with 100 rounds in his ammunition belt and eighty additional rounds in his haversack. As the battle wore on, Colonel Meade became concerned that even that supply would not be enough, and a desperate last-gasp fight might ensue, with fixed bayonets the last resort. The members of D Company, as previously reinforced, climbed onto roofs and other vantage points, and provided a crossfire helpful to those closer to the inner-city walls. Although assailed by heavy fire and continuing to take casualties, the marines held.

At day's end, the Anglo-American brigade pulled back through the murderous fire to lick its wounds. Both the American marines and soldiers had suffered serious casualties, as had the Japanese and British contingents. The 9th US Infantry had been especially badly punished; its casualties included its stouthearted commanding officer, Col. Emerson Liscum, who died exhorting his men to keep up their fire.

That night the international force hunkered down back behind the mud wall, wondering what the morrow would bring. The Imperial Japanese Army, however, did not spend its night in such pondering. In the wee hours, its soldiers blew in the outer south gate, and with bayonets fixed on their Murata rifles, poured through the gateway, scaled the inner gate, and threw back the defenders. The remainder of the international force followed—the Russians entered independently through the east gate—and soon after daybreak, the native city lay in coalition hands. In an unfortunate collapse of discipline, almost immediately the victors committed atrocities and other reprisals, and engaged in wholesale looting. This was the last action seen by the American marines' artillery during the relief expedition.

The US Navy and Marine Corps next tested the mettle of the Japanese during World War II, when the two nations, no longer allies, met in combat. The Americans would find the members of Japan's armed forces just as tough as they had been in China some forty years before.

Taking Stock: Action Reports, Postmortems, and Analysis

Following the campaign to relieve the Peking legations, more than a few field- and company-grade officers of the army and marines vehemently criticized the issuance of the dark blue fatigue shirt for combat, pointing out they were hot and made too nice a target. They argued instead for all-khaki campaign uniforms, such as had been worn by the military and naval troops of most of the other nations involved in the expeditions. The marines and navy were equally vociferous in their demand to hasten the issuance of a shoulder arm firing a heavier bullet than the 6mm Winchester-Lee, in order to provide better man-stopping capability. They urged that the adoption of the Krag-Jørgensen .30-40 rifle by the naval establishment, agreed upon two years before, be implemented promptly, so that all three services carried the same basic infantry weapon. Both recommendations would come to fruition within the next few years, with the marines in the Far East stations getting theirs by year's end.

In his report dated June 16, 1900, three days after the assault, Capt. Ben Fuller offered no complaint about the 3-inch field gun itself, but described its ammunition as "poor," resulting not only in relatively few shrapnel rounds exploding but making proper fire control next to impossible. After all, the British had not had trouble with their shells, one of which had even blown up a Chinese magazine. On the same day, one week before a medical board invalided him to the naval hospital on Mare Island because of acute rheumatism, Col. Robert Meade, commanding the 1st Marine Regiment, filed a long report on the action at Tientsin. In it, he was critical of the new naval landing gun in any putative role as a general field gun, because of its trail wheel, arguing that gunners could not depend upon the ground underneath being hard enough to make that wheel advantageous. He stated that in the Tientsin action, the soft earth had fatigued the gun crews enormously in moving the guns across obstacles such as dikes and ditches, thus impairing the men when finally they did get into position to fire. In order to improve the portability of the field gun, Meade opined, it needed to be hitched to a limber and divested of its trail wheel.

The performance of the Colt-Browning machine gun, at least in the 1890s, can only be described as faulty and unreliable. It hints at the predilection of the navy's Bureau of Ordnance to choose, all too often, inferior weapons to be its standard pieces, when tried and true alternatives were readily available. One may honestly protest the inexplicable decision for the Colt-Browning "potato digger" in the face of the reliable Maxim gun, even factoring in the difference in weight. Or of opting for the Fletcher landing gun, when either Hotchkiss or Driggs offered a better and proven product. The Fletcher gun, in its limited field service so far, had not yet offered a hint of the monster that would rear its ugly head as soon as the navy sailed into the twentieth century. Its shrapnel did not work very well at Guantánamo Bay or Tientsin. No fault of the gun, to be sure, but again, that dereliction is a reflection on the bureau's lack of attention to matters that lay firmly within its area of responsibility. Thorough testing of matériel, particularly that manufactured by contractors, remained clearly essential.

There resulted at least one positive development. As had been done in the Spanish-American War, during the war emergency of 1900, the Marine Corps appointed from civilian life and commissioned directly a number of second lieutenants. This time, such men were enrolled promptly into the School of Application, established at the Washington Navy Yard on November 5, where they undertook both classroom studies and field exercises. The corps expected these new junior officers to acquire a thorough knowledge of light artillery, with the guns to be mastered being the 3-inch field gun and both the Hotchkiss and Driggs-Schroeder 6-pounders on field mounts.

One last point needs addressing. Much was made over the abandonment of the Fletcher field gun on June 25, 1900, the day before the Seymour and rescuing expeditions withdrew from Hsiku Arsenal. Indeed, the loss of gun No. 31, manufactured at the Naval Gun Factory in 1897 and thereafter assigned to *Newark*, is chronicled in more than one place and details are readily found in the US National Archives. No other Fletcher gun appears therein as having been lost in combat. Some of the old 3-inch BL howitzers still saw sea service in the final years of the nineteenth century, and the gun taken off *Nashville* and "rolled into the river" on June 21, during the failed Russo-American attempt to fight its way into Tientsin, is described in reports as a 3-inch rifle. One must therefore conclude it was one of those older steel pieces of 1870s design. Maj. Littleton Waller recounts its loss in his report dated June 22, 1900: "The 3-inch rifle was defective, so I was obliged to disable it and hide it in a canal." Perhaps if the canal was shallow enough, this landing gun eventually may have been recovered by American or other coalition personnel, and would readily explain why so few details about its loss exist in then-contemporary naval ordnance records.

The Fletcher Mark I 3-inch field gun passed into the twentieth century with its serious problems largely undiscovered. A highly polished example stands on the deck of armored cruiser *New York* in 1899. In front of the gun, the ship's goat mascot Pitch and Adm. William Sampson's nine-year-old son Harold have hit things off nicely. Both he and his older brother Ralph followed their father into the US Naval Academy, but only Ralph graduated. Detroit Photographic Co. (*Library of Congress*)

References Consulted

Abdow, E., "The Boxer Rebellion: Bluejackets and Marines in China 1900–1901," Naval History and Heritage Command, 2022 (online).

Clifford, J. H., *The History of the First Battalion of U.S. Marines* (privately published, 1930).

Coletta, P. E., *Bowman Hendry McCalla: A Fighting Sailor* (Washington, DC: Univ. Press of America, 1979), pp. 105–20.

Crane, S., "Marines Signaling Under Fire at Guantanamo," *McClure's Magazine*, February 1899, pp. 332–36.

Crowe, G., *The Commission of H.M.S. Terrible, 1898–1902* (London: George Newes, 1903), pp. 243, 250–54.

Davidson, W. C., "Operations in North China," US Naval Institute (hereafter USNI), *Proceedings*, 1900, 26(4): 637–46.

Dictionary of American Naval Fighting Ships, Naval History and Heritage Command (online). Consulted for each US naval warship and auxiliary mentioned in this chapter.

Ellsworth, H. A., *One Hundred Eighty Landings of United States Marines 1800–1934*, History and Museums Division, HQ, US Marine Corps (Washington, DC: Govt. Print. Off., 1974 [originally printed 1934]), pp. 32–39, 146–49.

Heinl, R. D., "Hell in China," *Marine Corps Gazette*, November 1959, 43(11): 55–68.

Leonard, H., "The Visit of the Allies to China in 1900," Mil. Hist. Soc. of Massachusetts, *Papers*, 1918, 14: 296–318. This paper was originally read before the society on December 4, 1900.

McAulay, J. D., *Rifles of the United States Navy & Marine Corps 1866–1917* (Woonsocket, RI: Mowbray Publishing, 2017), 93–95, 101–103, 146–47, 149–54, 158–59, 167, 174–78.

Millett, A. R., *Semper Fidelis: The History of the United States Marine Corps*, Chapter 5, "The Marine Corps and the New Navy," pp. 115–44 (New York: The Free Press, rev. edit. 1991).

Multiple authors, *Marines in the Spanish-American War 1895–1899: Anthology and Annotated Bibliography* (Washington, DC: US Marine Corps, 1998).

Nalty, B. C., *The United States Marines in the War with Spain,* rev. edit., Historical Branch, HQ, US Marine Corps (Washington, DC: US Marine Corps, 1967), pp. 8–11, 14–17.

Preston, D., *The Boxer Rebellion: The Dramatic Story of China's War on Foreigners That Shook the World in the Summer of 1900* (New York: Walker Publishing Co., 2000).

Ryden, G. H., *The Foreign Policy of the United States in Relation to Samoa*, originating from his 1928 doctoral dissertation, Yale University (New Haven: Yale University Press, 1933).

Schlieper, P., "The Seymour Expedition," typescript compiled and translated from this German officer's personal journal, June 1900. In Phillips Library, Peabody Essex Museum, Salem, MA.

Taussig, J. K., "Experiences during the Boxer Rebellion," USNI, *Proceedings*, 1927, 53(4): 403–20.

US National Archives, Record Group 74, Records of the Bureau of Ordnance. See especially Entry 112, Ordnance Registers, 1842–1900, Washington, DC.

US Navy Dept., *Annual Report of the Secretary of the Navy, 1899,* pp. 3–6.

———, *Annual Report, 1898,* Bureau of Navigation, p. 324; Appendix, pp. 491, 635–57.

———, *Annual Report, 1899,* Bureau of Ordnance, p. 26.

———, *Annual Report,* Marine Corps, *1898,* pp. 822–26, 831, 836–48; *1899,* pp. 935–45; *1900,* 1116–29, 1148–66; *1901,* 1273–75.

Venzon, A. C., (ed.), *General Smedley Darlington Butler: Letters of a Leatherneck, 1898–1931* (Westport, CT: Greenwood Publishing Co., 1992), p. 8.

Wieand, H. T., "The History of the Development of the United States Naval Reserve, 1889–1941," doctoral dissertation, University of Pittsburgh, 1952, pp. 58–63.

Wingfield, T. C., and Meyen, J. E., (eds.), "Lillich on the Forcible Protection of Nationals Abroad," Appendix I. "A Chronological List of Cases Involving the Landing of United States Forces to Protect the Lives and Property of Nationals Abroad Prior to World War II," *International Law Studies* 77: 138–42.

Wurtsbaugh, D. W., "The Seymour Relief Expedition," USNI, *Proceedings*, 1902, 28(2): 207–19.

8

Landing Guns for the Twentieth Century

The US Navy entered the first year of the twentieth century, 1901, with mostly battleships on its mind. The world would soon be awash with them, and the United States wanted its share. When in the odd moment, the American navy did cast its attention on landing guns, it reacted with annoyance and exasperation. The service had purchased 100 Mark I 3-inch field guns, which suffered from a fundamental design problem that interfered with their effective use. Although it subsequently ordered sixty-two somewhat more modern Mark I mod 1 field guns, also with a Fletcher breech mechanism, they revealed the usual teething problems at the Naval Proving Ground. For several years, the US Navy would find itself nearly devoid of field artillery to support its landing forces.

In this first year, however, not everything went awry. The Naval Gun Factory did successfully design a new water bucket for its field guns, with a leather bottom replacing the old wooden one, which had been prone to cracking and breakage. The new model bucket also came with a watertight lid, preventing the water from sloshing out as the gun moved across rough country. The gun factory pegged its price at $7.50, and the stout new bucket duly saw adoption.

Trying to Fix the Fletcher Mark I 3-inch Field Gun

The first hint that severe problems awaited the Mark I field guns presented itself in a June 1901 report from the proving ground. During the test firing of gun No. 76 mounted on field carriage No. 76, the cast-steel recoil cylinder failed at a blowhole well rearward of the trunnions, permitting the recoil fluid to run out. The proving ground staff had already discovered that this gun did not return to battery after each firing, which they attributed to a weak counter-recoil spring within the cylinder. Examination, however, showed a permanent swelling in the internal diameter of the cylinder, uneven throughout the cylinder's length, which had caused twisting and distortion along the piston. After the cylinder on a second gun cracked along its length, again dumping its recoil fluid, it became clear that the problem was not a random one simply of steel quality. Therefore, the Bureau of Ordnance strongly suggested the immediate replacement of the faulty steel cylinders by ones made of manganese bronze.

By November 1, seventeen bronze cylinders had been manufactured and were ready to ship. The gun factory carefully pointed out that the new cylinders lacked registration numbers, but that a number could be stamped on each one to match that of the specific gun, whether aboard a ship or at a naval station. In its instructions to install the new cylinders, the gun factory included a warning to exercise caution when installing the counter-recoil springs. At mid-month, the first of the bronze cylinders began shipping to individual ships and stations, with the directive to turn in the old cast-steel cylinders to store at the major navy yards. During the course of re-equipping the battleship *Illinois* (two twin 13-inch gun turrets) with the new cylinders, the gun factory noted that the funnel for filling the original recoil cylinder did not fit the new one. As the yard had only ten new filling funnels in store, the superintendent recommended the provision of forty more.

The bronze cylinders went to the various navy yards on both coasts over the next six months, as well as to the Cavite Naval Station. Personnel fitted them to the intended guns in lieu of the original steel cylinders; some of the latter remained in store at the particular navy yard, and others got shipped back to the Washington Navy Yard. At the same time, new counter-recoil springs were installed. The steel twist comprising the coils was round in cross-section, however, rather than square as in the originals, and they were more flexible.

Near the end of February 1902, fifty bronze cylinders, fully half of those required for the Mark I field gun, had been supplied. The gun factory urged the manufacture of fifty more to complete the substitution for the 100 guns in service. As the facility tooled up to finish the job, a jarring report arrived from Puerto Rico. In late April, the battleship *Massachusetts* (two twin 13-inch gun turrets) landed its two field guns re-equipped with bronze recoil cylinders on the island of Culebra in order to field-test them firing shrapnel. With the first round for each gun, it became apparent that a problem existed: neither piece returned to battery, and had to be drawn back out manually. The bronze cylinder of one gun split after six rounds, that of the other gun after twenty-four rounds. Examination on the spot revealed that one section of the counter-recoil spring had overridden the other, with the comment made that such a fault had not been possible with the older springs, having the square-cross-section steel coil and greater rigidity.

On May 1, the Bureau of Ordnance at Washington directed the proving ground to repeat the tests made by *Massachusetts*, and on the following day asked the navy yard in the city to suspend all work on the bronze recoil cylinders. During the next few days, cables and telegrams flew out, instructing the various stations and ships to cease the installation of the bronze cylinders, or if installed, to remove them from the guns, and return them to the Washington Navy Yard. The bureau suggested that the cast-steel cylinders previously removed could be refitted, but left the decision to the discretion of the commanding officer on the scene.

Landing Guns for the Twentieth Century

Wonderfully complementing each other, the battleship *Iowa* steams down the East River and under the Brooklyn Bridge after her 1903 refit at the New York Navy Yard. Armed with four 12-inch/35cal main guns and twice as many 8-inch/35cal secondary ones, she also carried a multitude of shipboard rapid-fire pieces and two Mark I 3-inch field guns. Detroit Photographic Co. (*Library of Congress*)

Adjacent to one of *Iowa*'s main turrets mounting twin 12-inch guns and under an awning rigged to provide some shade, this grizzled seaman wearing whites grimy from the burning of coal poses behind a Mark I 3-inch field gun. The larger diameter recoil sleeve girdling the gun barrel can be seen to be the original configuration, whether of steel or bronze, dating the photograph close to the turn of the century. Detroit Photographic Co. (*Library of Congress*)

Similar trouble was encountered in the proving ground tests of gun and carriage No. 76 fitted with a new recoil sleeve. Of the fifty-three rounds fired, on the first round, the gun failed to return to battery and then performed well for the next thirteen rounds. When the gun failed to return to battery two more times, the gunners opened the filling port and allowed a quantity of the glycerin-and-water recoil fluid to spill out. From the nineteenth to thirty-third rounds, the gun returned to battery, but once again, it failed to do so during the final twenty firings. The reduction in the expanded recoil fluid averted permanent damage to the gun, and the officer in charge saw no reason not to continue the tests if so desired.

The results showed that at least part of the problem lay with the excessive pressure generated by the recoil fluid from its heating by the gun barrel upon repeated firing. The gunners realized that the recoil cylinder surrounding the barrel and lying in close contact with it only hastened and worsened the difficulty. Virtually every naval officer had knowledge of the dictum of physics extrapolated from that for gases: For almost all fluids contained in a closed vessel of fixed volume, an elevation in temperature results in an increase in pressure of the fluid so restrained. In the Mark I recoil apparatus, if no release of the pressure took place, the recoil cylinder swelled, bound up the barrel, and shortened the recoil. The overriding of the recoil springs only compounded the problem. The question then became what to do about it.

The answer did not lie simply in reducing the recoil fluid to less than full volume in the cylinder at the outset of firing, because all rounds fired required a full volume of fluid in the closed system. The answer also hardly lay in spilling off fluid during an expedition after firing had commenced, possibly at a critical point in the action, when the artillery fire would be most needed and the gun crew be at the greatest risk in ceasing fire in order to draw off the recoil fluid.

Significantly, in late May, Capt. Henry Manney, the commanding officer of *Massachusetts*, suggested three possible solutions to the difficulty arising from the increased temperature and pressure. He first contemplated either a safety valve, opening automatically, or a pressure gauge and cock, either of which would operate to draw off the expanded fluid once a given pressure was obtained. His third alternative differed: an air chamber that could hold a quantity of fluid equaling the expansion in the cylinder due to the increased temperature and pressure. If the bureau opted for the last possibility, wrote Manney, the chamber would be fitted with a cock at its junction with the recoil cylinder, to open at a set temperature or pressure and allow the expanded fluid to empty into the relief chamber, where it would be preserved until the gun cooled down.

The hypothesis that increased temperature and pressure caused all of the other problems experienced by the Mark I field gun met resistance. Critics pointed out that those factors could not explain the failure of the gun in more than one instance to return to battery on the first round or first several rounds fired. They postulated that the overriding of the coils of the recoil springs or the sticking of the packing in the stuffing boxes could be coming into play. Those fears were underscored with the discovery that the bronze recoil cylinder on gun No. 76 at the proving ground had in fact suffered enlargement, in spite of the initial perception that it had not. Some deformation of the counter-recoil springs had also taken place. Therefore, the Bureau of Ordnance determined to carry out suitable experiments to seek a solution.

In August 1902, a sealed oil cup having a half-pint capacity and fitted with a nipple was screwed onto the front of the recoil cylinder, and steam pressure used to heat the gun to about 300 degrees Fahrenheit (149 degrees Celsius). The resulting pressure forced about a quarter pint of fluid into the cup. Having satisfied itself that the relief cup worked in principle, the proving ground fired the gun about fifteen times, at the end of which, it found that the fluid had expanded and forced its way into the cup. To counter fears that the obvious location of the relief cup made it vulnerable to hostile rifle fire, in lieu of the cup, the proving ground removed about four tenths of a pint of fluid at the outset and fired the gun six times. The testing staff discovered a through-crack extending 6 inches longitudinally along the recoil cylinder and deformation of the cylinder. Tests were made immediately to determine if the character of the metal had led to the cracking. The officer in charge, however, believed the problem arising so repeatedly had to be inherent in the design of the recoil system. Virtually no one senior to him wished to believe that supposition.

The tests on strength and elastic limit showed the metal quality in the vicinity of the crack to be porous and weak, but to be up to specification farther from the crack. Independently, the gun factory subjected a test cylinder to a pressure of 1,800 pounds per square inch (psi), without enlargement, but upping the pressure to 2,100 psi did produce a permanent distortion. In what may fairly be described as a hope-springs-eternal moment, the navy decided the design of the recoil cylinder to be fundamentally sound, and it proceeded to manufacture more than 100 new cylinders with the expansion chamber included as part of the casting, with its volume specified as one half pint. That decision would consume enormous amounts of time, effort, and money for the remainder of the decade.

Two months after the Bureau of Ordnance had circulated a general directive to cease issuing the bronze recoil cylinders to the fleet and turn them in to store (with the decision to reinstall the original cast-steel cylinders left up to the individual yards and stations), the government let a contract with a specialty steel firm, I. G. Johnson & Company, located in the Spuyten Duyvil section of the Bronx, for five cast-steel recoil cylinders having the relief chamber integral with the cylinder. To facilitate their manufacture, the Washington Navy Yard shipped a pattern for the new model cylinder to the Johnson works. The company acknowledged the receipt of this pattern on October 22, 1902, and promised that it would start work on the five new steel cylinders at once. The bureau spent the balance of the last two months of the year responding to the clamor from warships and yards that their field guns remained out of commission, by reminding them to ship by the first available transport all of the interim bronze cylinders to the Washington Navy Yard and reassuring them that the new steel replacements should be expected in six months.

Early in 1903, it became clear that much of the fleet lacked operative field guns for shore expeditions. Soon, a year had gone by, and then another. Although the bureau equipped some warships with the new Mark I mod 1 field gun, the navy had ordered only sixty of them initially, and a delay also ensued with their manufacture and distribution. Increasingly insistent reminders from the fleet reached the bureau, declaring that a lengthy delay had transpired since their landing guns had been operative, and some of them were now stored in precarious places, susceptible to damage. Exasperated captains were asking what to do about it.

Meanwhile, the bureau turned its attention to the counter-recoil springs, which were quite evidently faulty, given that the Mark I field gun often refused to return to battery on the first rounds, before the gun had heated up sufficiently to account for this malfunction. In April, tests proved that the springs were not strong enough, on both the Mark I and Mark I mod 1 field guns. Cleverly, the proving ground set up a Mark I gun lacking a spring but with its recoil cylinder full and its stuffing boxes packed, and then, manipulating the gland, subjected the gun to ten firings. From that experiment, the officer in charge concluded that a counter-recoil spring needed 500 pounds initial pressure to operate properly. Once again, however, the problem of the coils overriding each other had been manifest, and the suggestion was made to return to coil steel of square cross-section. The Bureau of Ordnance thereupon instructed the Railway Steel Spring Company in New York City to discontinue the manufacture and shipment of counter-recoil springs for the 3-inch field carriage until further advised. The delay in getting these guns back to the fleet thus became even more prolonged.

Worse news came one year later, when upon firing only the third round during tests at the Naval Proving Ground, one of the five newly manufactured steel recoil cylinders cracked closely along a seam on its underside. Questions arose about the need for yet another change in the design of the cylinder, but once again, initial examination showed flaws in the metal in this particular cylinder. The gun factory refitted the test gun, No. 100 (by coincidence the final gun in the sequence provided in the late 1890s), with a new Johnson-made recoil cylinder and this time successfully fired the piece forty-nine times without mishap. It was noted that the recoil fluid had warmed sufficiently during the firings to ooze past the packing of the glands in the stuffing boxes.

Then came word in late April 1904 that should have gotten the bureau's closer attention. Follow-up metallurgical tests on the cracked cylinder showed the tensile, elastic, and elongation data met specification, and thus the quality of the metal had not been compromised. However, Capt. Edwin Pendleton, superintendent of the Naval Gun Factory, stated his belief that no additional design changes would be necessary. Without additional firing using the recoil cylinder currently in place at the proving ground, or installing for test a third cylinder from the five manufactured by I. G. Johnson—after all, one had passed and one had not—the bureau authorized the production of another ninety-six steel recoil cylinders incorporating the relief chamber. It suggested these new cylinders could be cast at the gun factory. This grasping at straws, it appears, existed at more than one level.

USS *Panther* (six 5-inch guns), which had taken the First Marine Battalion to Guantánamo Bay in 1898, became the epitome of the problem that faced so many American warships in 1904. Currently out of commission at League Island Navy Yard near Philadelphia, she had waited for her Mark I field gun to be completed for at least two years. Heedful of the demand for landing guns in the fleet and thus needing to lessen the difficulty in machining the cylinders, the gun factory suggested that the cylinder's recoil grooves be increased from two to four. This was duly approved by the bureau.

In early September, in response to a query from the bureau of when the recoil cylinders and counter-recoil springs for the Mark I field gun would be available for issue, the gun factory admitted that the drawings for the new cylinders had not been completed and would need another week to complete. It estimated that the components desired would be ready for shipment within ten days after the receipt of the needed drawings. One month later, the drawings had still not been finished, with the complaint that the curve for the cylinder rifling remained too great for the rifling machine at the gun factory. The bureau wished to know specifically if the cylinder and spring for *Panther* would be ready to ship in the promised seventeen days, and generally when the entire lot would be available. A few weeks later the gun factory still did not have the drawings at hand. The estimated availability of two cylinders for *Panther* remained the same after the drawings did become available, and the completion time for the rest of the ninety-six cylinders was estimated at five months, both projections being contingent upon the cylinders passing the hydraulic water test. The bureau expressed the fear, however, that a great many cylinders would fail that test and be rejected. Unfortunately, that prediction came to pass.

The delivery of the counter-recoil springs had also been delayed, with only one pair received to date. The order for 125 springs had been made in May 1904 to the Railway Steel Spring Company, with shipment to the gun factory stipulated within two months, but by the fall of that year the shipment was well past due. The Bureau of Ordnance urgently requested the Bureau of Supplies and Accounts to expedite the delivery of the springs, reminding all parties that the 3-inch field guns for which the components were intended had been useless for several years. In late November, the gun factory sought transportation for the two cylinders earmarked for *Panther*, but five months later, the Bureau of Ordnance still wondered about the status of the recoil cylinders and counter-recoil springs.

The gun factory conceded in May 1905 that it was experiencing trouble in the production of the recoil cylinders, explaining that the casting remained a particularly difficult one, with a great many rejected after machining because they had failed the hydraulic water test. Accordingly, the facility asked permission to make two from manganese bronze. If the request was refused, then the gun factory wished the remainder of the cast steel cylinders contracted to a commercial firm, because the capacity at the navy factory had become overtaxed. Meanwhile, one piece of good news lay in the receipt of all of the counter-recoil springs. The Bureau of Ordnance authorized the fabrication of the two, later increased to three, cylinders made of manganese bronze. They were contracted to Parsons Manganese, Paul Reeves & Son Manganese, and Ajax Manganese, all with plants in Philadelphia. Two months later, the first one completed, from Reeves, withstood a pressure of 2,000 pounds; the other two tested successfully to 1,800 pounds. The gun factory therefore sought authorization to build the remainder of the recoil cylinders of manganese bronze, predicting they would be very much cheaper than steel. For one thing, rifling the bronze cylinder required two days, versus six days for steel.

When testing the bronze cylinders at the end of July, the proving ground found them to be sound, and within two weeks the bureau approved the manufacture of such cylinders for the battleship *Alabama* (two twin 13-inch gun turrets), whose captain had been particularly insistent about obtaining replacements for her long-disabled field guns. The commanding officers of a good many other warships were only slightly less dogged.

As 1905 came to an end, the bureau reminded the commandant of the Washington Navy Yard of its intention to place in service as soon as practicable a number of "old" 3-inch field guns for which new recoil cylinders were currently under manufacture at the gun factory. The bureau needed to know precisely how many recoil cylinders had been completed and were ready to ship. It also sought the opinion of the commandant whether it would be more sensible to ship these replacement cylinders to the sundry navy yards and ships for installation, or to have the guns themselves shipped to the Washington Navy Yard to do so. The bureau also queried him about the mark to be assigned to the new cylinder and whether it would change the mark of the gun or field carriage. The latter question is curious, as that designation would seem to have been a function of the Bureau of Ordnance. Whatever the case, the modified cylinder became the Mark I mod 2.

A Mark I 3-inch field gun displays the replacement Mark I mod 2 recoil cylinder, topped by its distinctive expansion chamber to draw off the recoil fluid hot from firing. All attempts to achieve a reliable and safe recoil of the piece failed, and the navy finally threw in the towel at the end of the first decade of the twentieth century. The awkward elevating hand wheel is clearly seen beneath the breech of the gun, on display in West Carrollton, OH. (*Nelson H. Lawry*)

The distribution of the new cylinders began in January of the following year, to replace the previous bronze ones lacking the expansion chamber, most of which had been returned to the Washington Navy Yard months to years before. The early recipients included the battleship *Massachusetts*, two guns, and Honolulu Naval Station, three guns. Six months later, the yard in Washington confirmed that 101 cylinders, Mark I mod 2, had been ordered, of which forty had so far been built (one had of course failed at the proving ground). Of those last, thirty-three were available at the yard, nineteen steel and fourteen bronze, for which no shipping orders existed. In its mid-June letter, the gun factory declared that, inexplicably, shipping orders for the new carriages were seldom received. Its superintendent expressed the hope that it was not the bureau's intention to continue the manufacture of the recoil cylinders and have them accumulate in store. This thinly veiled swipe at the Bureau of Ordnance may have pointed out the lack of enthusiasm by that office in distributing what it perceived very likely to be a flawed device.

Later in the month, thirty-seven cylinders went out, twenty-two to the Cavite Naval Station, five each to the New York and Norfolk Navy Yards, four to the Portsmouth Navy Yard, and one to the Puget Sound Navy Yard. Two of the cylinders sent to the New York yard were earmarked for the long-suffering *Alabama*. At the beginning of 1907, the Naval Gun Factory had completed twenty-three more bronze recoil cylinders, and by May, the total had become twenty-six, of which only two were made of steel. In response on both occasions for shipping orders, in mid-May the bureau instructed twelve sets be sent to the US Naval Academy (plus another one week later), six sets to the Naval Training Station, Newport, Rhode Island, and five sets each to the Boston and Mare Island Navy Yards. Two months later, the bureau instructed the naval academy to survey and sell the previous type cylinders, but by September, it had changed its mind and instructed that they be sent to the Norfolk Navy Yard. Those allocations by and large ended the bureau's efforts to distribute the Mark I mod 2 recoil systems, and for all intents its dwindling interest in them.

In March 1908, the storeship *Monongahela* burned to destruction at Guantánamo Bay, Cuba. Built during the Civil War as a steam sloop, she had been converted to a sailing ship in the early 1890s, and served as a training vessel at Annapolis and Newport until relegated to her reduced status in 1904. Mark I landing gun No. 88 had remained aboard for no evident reason, perhaps forgotten, and the fire destroyed it.

Exhaustive firing tests in late 1911 showed the gun and its recoil system little improved, and despite all the money and effort expended to make them right, very much wanting. During a comparative series, whereas the much-modified Mark I mod 1 fired repeatedly with no appreciable heating of its recoil cylinder, the Mark I not only experienced substantial heating of its cylinder, but too frequently failed to return to battery on its carriage. These problems were exacerbated by flying dust and dirt, which could be expected when firing was undertaken on a sandy beach or dirt surface. Consequently, in early January 1912, the Bureau of Ordnance decided to remove from fleet service the Mark I gun and Mark I field carriage (sometimes referred to as the Mark I mod 2 carriage, although strictly speaking that designation had been intended for the recoil sleeve). A few exceptions continued to sailor on for a few years more in monitors, as well as older gunboats serving in the Far East.

From examples of Mark I field guns surviving into the twenty-first century, both displayed in public places and held less prominently in the hands of private collectors, it remains clear that the navy did not replace all of the original recoil cylinders by the Mark I mod 2 type having the expansion chamber. The basis for such uneven treatment has not been revealed.

In the intervening years of the early twentieth century, the bureau frantically sought to provide surrogates for what had been increasingly perceived during the lengthy and frustrating process as an unsatisfactory landing gun, indeed even a dangerous one to fire.

The Naval Gun Factory Mark I mod 1 3-inch Field Gun

In 1901, the Bureau of Ordnance developed, built, and tested a 3-inch field gun and carriage of much improved design, as well as a sorely needed ammunition limber. Initially, the bureau designated the gun as the Mark II, but because a Mark II 3-inch shipboard gun also existed and thus made for confusion, the bureau later respecified the gun as the Mark I mod 1. The field carriage's designation remained the Mark II, with its later minor modification the Mark II mod 1. As with the earlier Mark I gun, the new one was 21 calibers long and single-forged, but with a classic muzzle swell in lieu of the older muzzle ring. It had a modified Fletcher breech mechanism, styled as was the gun, the Mark I mod 1. As with the Mark I gun, its successor generated a low muzzle velocity of 1,150 feet per second and demonstrated a gun elevation of 10 degrees.

Forged on the underside of the breech, a heavy lug allowed the attachment of the piston rod of the recoil apparatus. The gun recoiled through a bronze sleeve, cast as one with the recoil cylinder beneath the barrel. A key on each side of the barrel fitting a matching groove in the slide prevented the gun from turning by the action of the rotating projectile. The hydraulic recoil cylinder included counter-recoil springs divided into three parts and separated by disks, and with a forward-located checking device. The slide, rather than trunnions, secured the gun to the carriage. Much like the predecessor Mark I field gun, the resulting short recoil (about 15 inches) did not permit the absorption of the entire recoil by the field carriage. Although the wheel brakes and the trail spade diminished the force of the kickback, the over-ground portion of the recoil consigned the gun to obsolescence upon arrival.

Improvements in the carriage design, however, had been made. The elevating shaft and its operating wheel were located on the left side of the gun, rather than thrusting awkwardly from under the breech as in the Mark I gun, which had prevented the elevation of the gun and the service of the breech at the same time. That shaft operated through a bevel gear, with a collar surrounding the axle for the wheels swinging the gun in elevation and depression. The shaft for the training wheel, also on the left side, turned in bearings and worm gears in an arc on the cylinder and slide casting, accomplishing a limited gun traverse of 3 degrees each, left and right. To achieve greater traverse, the gun had to be moved by the use of the trail stave.

From 1905, the modified Fletcher breech mechanism contained a Tasker firing lock, which differed from the Badger lock in the Mark I gun in that the firing pin actually

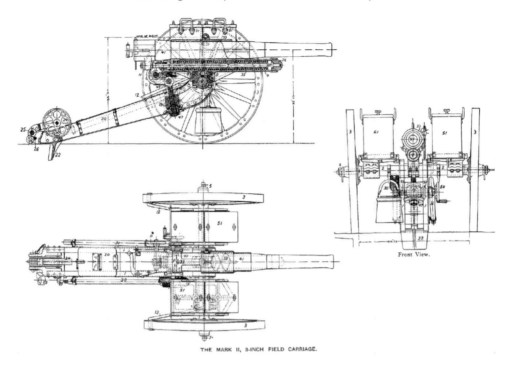

THE MARK II, 3-INCH FIELD CARRIAGE.

Drawings of the Mark I mod 1 3-inch field gun reveal the substantial improvements from the original Fletcher Mark I landing piece. The redesigned recoil cylinder occupies a more traditional position under the gun barrel and the elevating hand wheel is sensibly located on the left side of the carriage. The gun attaches to the carriage through a slide, not trunnions, and its forepart displays a graceful muzzle swell. The positive camber of the wheels is noticeable. Fullam and Hart, *Text-Book of Ordnance and Gunnery*, 1903. (*Eugene L. Slover, online*)

struck the primer on the cartridge case. A sear restrained the firing pin in the cocked position, with the mainspring slack. When the cocking lever was pulled rearward, the sleeve surrounding the firing pin was forced forward, so compressing the spring and tripping the sear, allowing the firing pin to fly forward against the primer. Upon the release of the lanyard, the mainspring brought the entire mechanism to its previous position at the rear, the cocking lever upright and the sear re-engaged with the firing pin. As this firing lock also permitted the premature discharge of the round before the breech completely closed, it too required the retrofitting of the Graeme safety device. The late 1911 firing tests mentioned previously showed that both the Badger and Tasker firing locks performed well, and the Graeme device proved equally effective in these locks. The extractor in both instances only loosened the spent cartridge case, however, requiring that the case be withdrawn further by a crew member's fingers clamped onto the rim. Clearly a proper extractor would have fully ejected the case.

Testing of the pilot gun and carriage revealed the usual teething problems, particularly the propensity for the wheel brakes to fly off during firing, with the consequent increase in over-ground recoil. Making things worse, neither the trail in its original configuration nor the trail spade after it had been added, dug in well when firing the piece on hard earth. Moreover, the gun suffered from excessive breech drop that proved very difficult to correct, and when the gun fired from extreme train at 3 degrees, it jumped several inches into the air. The gun factory was still at work on these problems as 1901 came to an end, and indeed they continued thereafter.

In an apparent response to the criticism that the Mark I field gun had not been provided a limber and was thus limited to four ammunition boxes containing a total of thirty-two rounds carried on the field carriage, the Bureau of Ordnance designed a limber carrying nine identical boxes. When that proved too heavy and cumbersome, the bureau reduced the number of boxes to six, thus loading forty-eight rounds additional to those conveyed on the carriage. At that time, the Fletcher field guns were provided only shrapnel projectiles, but during the field testing of the limber, the question arose if shell was also needed. The bureau confirmed its desire to issue shell to the 3-inch field guns and acknowledged the need for either projectile to fit within the ammunition box such that the lid could be fastened securely, which at that time it did not.

The bureau requested an order to the Naval Gun Factory in September 1901 to make sixty 3-inch field guns of the new type, pursuant to which the navy let a contract on November 11 to the Fore River Ship and Engine Co., Quincy, Massachusetts, for that number of nickel steel forgings for the tubes and breech plugs. Things went well enough initially. Fore River delivered twenty-four sets of forgings in 1902 and another dozen the following year. Thereafter, work at the gun factory on the new fieldpieces ground to a halt, with the annual report of the chief of the Bureau of Ordnance for

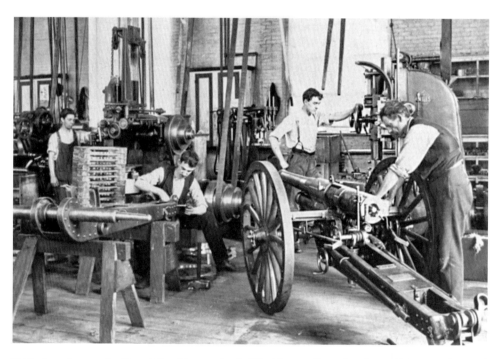

Skilled workers at the Naval Gun Factory in the Washington Navy Yard assemble a Mark I mod 1 3-inch field gun, while another man completes a second carriage trail to the left. During the early twentieth-century period shown, much machinery, belt-driven from overhead line shafts, remained in use for gun production. (*Naval History and Heritage Command*)

1904 reporting that no additional forgings for the new field gun had been received that year, and two dozen forging sets were at that time outstanding.

As so many slides had been rejected because the bronze cylinder had failed the hydraulic pressure test—a problem hardly unique to the newest cylinders—in March 1903, the bureau approved a request by the gun factory to fabricate a bronze slide with a wrought-steel cylinder. On September 9, the gun factory reported that the newly built slide had performed satisfactorily at the proving ground and requested permission to build thirty more of this design. The following day, the chief of bureau turned down the request because of the difficulty at sea in preventing the corrosion of steel parts. He insisted that bronze cylinders be used because such members were not exposed to particularly great pressure. The chief believed that ordinary bronze would serve, but left it to the gun factory to choose manganese bronze if the facility considered it more desirable.

The proving ground tests continued to show problems with the brakes flying off, the sight mass becoming loose or bending, and the lever locking catch breaking during firing. Perhaps of greatest concern, firing at extreme train required the gun and carriage to be put back into position after each round. When firing at any depression, great effort was needed to elevate the gun again. A board appointed in early November 1903 to examine the Mark II field carriage and Mark I limber reported out in July of the following year, recommending several changes to both wheeled vehicles. Among those changes, the board suggested that the trail wheel be enlarged from 10 to 18 inches in diameter, with which the bureau concurred.

Meanwhile, the Fore River Ship and Engine Co. was in trouble. Although it had punctually delivered the first two dozen of the specified sixty forgings, unaccountably it experienced difficulty with the remaining three dozen. In the early summer of 1903, the company appealed to the bureau to substitute carbon or ordinary gun steel for nickel steel, but without success. In response to a query from the Washington Navy Yard in late August, the company reassured Lt. Cmdr. Edward Scribner, the naval inspector at the plant, that it continued to make every effort to complete the contract and deliver the balance owed within five months. Scribner passed those assurances on to the chief of bureau.

The chief demanded an explanation for the delay in early December, reminding Fore River that the navy urgently needed those forgings. Soon thereafter, as already written, the company delivered another dozen of them, making a total of thirty-six forging sets furnished on the contract. The company explained that it was having extreme difficulty in obtaining the steel quality called for in the specifications, despite having made twelve heats and more than fifty forgings. Fore River Ship and Engine Company had thereby suffered substantial financial loss and it requested that the contract be cancelled.

While the navy pondered the matter of penalties to Fore River in January 1904, wishing to ensure that the government's interests did not suffer, the company advised the navy that it had twenty-five breech plugs at hand that had passed inspection. The company inquired if the navy would accept them before the contract was closed, no doubt to lessen the financial damage to itself. The Bureau of Ordnance promptly agreed to accept twenty-four plugs (it may have subsequently changed its mind and taken the additional one as a spare). In the interim, the bureau contacted Midvale Steel regarding the possibility of manufacturing twenty-four nickel steel gun forgings. Midvale quoted a price of 50¢ per pound and estimated it would deliver the first set in thirty days and complete the contract in seventy days. In the end, the navy refused to authorize the cancellation of the Fore River contract, but declared it would be pleased to receive further suggestions on the matter. At the same time, the bureau informed Midvale Steel it had merely floated the notion for informational purposes.

Five months later, Lieutenant Commander Scribner confirmed that Fore River had failed to respond to the request for suggestions to ease the requirement to complete its contract. In September 1904, Fore River Shipbuilding Company succeeded Fore River Ship & Engine Company and assumed the 3-inch gun forging contract, to which the sureties had assented. The two dozen gun forgings outstanding very likely bore the final twenty-four registration numbers, Nos. 399 through 422. Including the pilot gun, the entire range of the Mark I mod 1 3-inch gun comprised Nos. 362 through 422, with the Mark II field carriages Nos. 101 through 161. Before decade's end, the service added two more units to complete naval militia needs, guns Nos. 906 and 907 and carriages Nos. 187 and 188.

That the new company had solved the steel quality problems and had begun to deliver the last two dozen forgings was demonstrated in late July 1905. At that time,

Less than a decade into the twentieth century, this Mark I mod 1 3-inch field gun appears next to one of the secondary twin 8-inch gun turrets on the battleship *Kansas*. The habitual water bucket is slung under the carriage. A sister battleship of the Connecticut class appears off the starboard bow, displaying pole masts fore and aft, thus prior to the installation of the imperfect lattice (or cage) masts in 1908 or 1909. *The United States Navy Illustrated*, C. S. Hammond Co., 1912. (*Buffalo and Erie County Library, NY*)

the Bureau of Ordnance assigned Mark I mod 1 3-inch field guns Nos. 402 through 407, inclusive, along with their carriages and limbers, to the recently launched semi-armored cruisers *St. Louis*, *Milwaukee*, and *Charleston* (each with fourteen 6-inch guns). Those field guns had originally been intended for the pre-dreadnoughts *Virginia*, *New Jersey*, and *Rhode Island* (all with two twin 12-inch gun turrets, each with a superimposed twin 8-inch gun turret), but then these battleships and their sisters were slated for substitution by landing guns provided by the Bethlehem Steel Company, which the navy had turned to in the interim to fill its needs. The Mark I mod 1 would also see service aboard American gunboats on the China station, such as *Monocacy* and *Palos* (each with two 6-pounders), and *Asheville* and *Sacramento* (each with three 4-inch guns).

The Mark III Heavy Field Carriage to Mount the 3-inch/50-caliber Shipboard Gun

At the same time that the navy realized its need for a new light field gun, it decided upon the provision of more powerful guns on shore. It would achieve this by removing existing 3-inch/50-caliber deck guns, landing them, and then mounting them on heavy field carriages brought ashore at the same time. The Bureau of Ordnance intended a pair of heavy field carriages each to be issued to battleships and armored cruisers mounting such deck guns. In the end, these pieces turned out to be another bad idea.

The Naval Gun Factory designed and began construction of such an experimental field mount for the long 3-inch gun in 1900. The facility completed the original carriage one year later and sent it on to the proving ground, where the gunners mounted a 3-inch/50-caliber gun and fired it many times. Weighing a little more than a ton, the carriage was found to be sufficiently heavy to retard recoil and amply strong to allow rapid firing. Without an on-carriage training device, however, shifting the heavy gun in train was extremely difficult because of the great weight of the trail. With the trunnions of the gun placed over the carriage axle and the trail thus unbalanced, shifting the gun needed eight men. Extension of the trail piece forward of the axle achieved near balance of the trail, and then the task could be done by two men. The total weight of just the gun and carriage at nearly 2.5 tons had not changed, however, and the fieldpiece remained terribly difficult to manhandle on shore. The task became all the more onerous because the party also needed to haul the ammunition-laden limber, another ton and a half, which began testing in 1902. By 1904, the navy had under construction twenty Mark III 3-inch carriages that were 90 percent complete and another ten essentially just begun.

In January 1905, the bureau declared its intention to replace the slide and sight on the carriage, both current components having been found to be unsatisfactory. The 8-inch recoil of the 3-inch/50-caliber gun on its Mark III carriage caused significant stress, and the bureau recommended that the recoil be doubled to 16 inches. The new components for the replacement slide included steel extension cylinders with new rifling grooves, recoil piston rods, and possibly a bronze bracket to stiffen and support the extension cylinders.

In 1901, a sailor assigned to the Naval Proving Ground prepares to load a round into a 3-inch/50cal gun mounted on the Mark III field carriage. The original idea was to dismount shipboard guns of this size and marry them to the carriages ashore, to be used as more powerful landing guns with longer range. The navy stored the thirty carriages in two shore depots, one on the east coast of the continental US and the other on Subic Bay, Philippines. The latter carriages saw brief use in coast defense batteries on the island of Guam, but overly heavy, unwieldy, and obsolete, all were eventually scrapped. *Annual Report of the Navy Department, 1901.*

Capt. Edwin Pendleton, superintendent of the Naval Gun Factory, recommended a competitive test between guns fitted with the old and new slides be made at the proving ground. Five months later, in June, the navy got around to that comparative firing. The recoil of the gun equipped with the short cylinders was described as violent, having a racking effect on the field carriage. The design and working of the long cylinders won praise from the officer in charge of the proving ground. Near month's end, the bureau recommended the incorporation of the long recoil cylinders into the new slides.

During the next two years, the gun factory made additional improvements—the pair of recoil cylinders now being made of nickel steel rather than the previous wrought steel, the addition of the army Model 1902 panoramic sight (later to be replaced by the Mark VIII panoramic sight), the repositioning of the elevating hand wheel so not to interfere with the new sight—and the bureau ordered up at least one more test carriage to incorporate them. In March 1907, the end of the slide was cut off, saving 24 pounds in front of the trunnions and nearly correcting the previous muzzle-heavy imbalance. The intention existed, as always, to send the new pattern carriage to the proving ground for thorough testing, but abruptly in early April, the bureau cancelled those tests as unnecessary. Instead, it directed that the manufacture of the new slides proceed as soon as practicable, accompanied by the requisition of the desired number of sights from the army's Ordnance Department.

It took another two years for the first remodeled carriages to appear, but then the navy ran into a logistical difficulty: there remained no more storage space under cover, so the new carriages began to accumulate in the open. Because of their weight—a hefty 4 tons when combined with the ammunition-laden limber—and the need, once ashore, for animals to haul them, the navy decided not to issue the Mark III carriages to warships, but rather to distribute them to certain navy yards for storage awaiting expeditionary deployment. At the end of April 1909, the first six carriages were sent from the Washington Navy Yard to the Norfolk Navy Yard. Two months later, fifteen more carriages had been completed, and Rear Adm. Eugene Leutze, wearing two hats as superintendent of the gun factory and commandant of the navy yard, had to remind the bureau that they remained in the open, exposed to every kind of weather.

By mid-1909, it had been decided to allocate the Mark III carriages and their Mark II limbers to the Marine Corps advance base force. Because of the very near war scare with Japan two years before, the fifteen carriages referred to previously were sent for that purpose to Olongapo Naval Station, Subic Bay, Philippines. Confusion arose when the carriages arrived there without their 3-inch/50-caliber guns, and it became necessary to confirm that such issuance had never been intended, but rather when needed, the guns would be landed from fleet units.

The gun factory recommended in September that instead of manufacturing a new Mark III carriage as the bureau desired, the original type carriage rather should be fitted with a slide of 16-inch recoil. The original intention had been to land these carriages using the slides from the ships' 3-inch/50-caliber batteries, so the pilot carriage still lacked one. The gun factory assured the bureau that if so fitted, along with an army sight, the carriage would be alike in all respects, and the bureau agreed to the recommendation.

By year's end, the gun factory informed the bureau that ten carriages and limbers, ordered more than five years before, stood ready for shipment. They were sent to Norfolk Navy Yard in early 1910, to join the six already sent for the advance base force there, for a total number of sixteen. The pilot carriage fitted with the modern recoil sleeve very likely saw additional testing at the proving ground. A few odds and ends forgotten in the shuffle had to be acquired and distributed; for example, thirty-one each spring compressors and open-end wrenches to manipulate the recoil cylinder glands for those carriages sent as part of the advance base outfits.

At decade's end, the thirty-one heavy Mark III 3-inch carriages resided without guns at two advance force bases: sixteen at Norfolk Navy Yard and fifteen at Olongapo Naval Station. They bore serial numbers 190 to 220. Within a year, fifteen of those stored at Norfolk would be transferred to the Philadelphia Navy Yard as part of the advance base force projection, with one sent to the Naval Proving Ground for the necessary test work there. In 1914, the fifteen Mark III carriages at Olongapo were removed from the advance base force inventory and shifted to the defenses of Guam.

American & British Manufacturing Company Submits Its 9-pounder Landing Gun

Likely because of the ongoing difficulties in equipping the fleet with landing guns, Rear Adm. Charles O'Neil, past chief of the Bureau of Ordnance, suggested to Charles Gulick, the manager of the American & British Manufacturing Company—successor to the American Ordnance Company of Bridgeport, Connecticut—that his firm develop a light artillery piece to fill that role. The company did so, and after testing the 9-pounder gun at its own proving ground, Gulick informed the bureau in late March 1904 that his firm had shipped the new landing gun, carriage, and boat stand, along with a small amount of ammunition, to the Washington Navy Yard for forwarding to the Naval Proving Ground at Indian Head, Maryland.

The low-power landing gun generated a muzzle velocity of only 1,000 feet per second, less than the Fletcher field guns then encountering difficulties. Because the Tasker breech mechanism and firing lock equipped the gun, the United States Ordnance Company quickly inserted itself into the negotiations. This firm was a shell company in existence solely to hold the Tasker patents and look out for the inventor's interests, especially in the assignment of royalties. Indeed, it requested its own copy of the proving ground's report on the Tasker mechanism and of the comparative report on it and that of the present navy 3-inch field gun. Gulick assured the bureau that if the navy was not happy with that breech mechanism for any reason, his company was prepared to substitute another type. For its part, the bureau promised that the gun, field carriage, and accessories would be given a fair and rigorous testing protocol that would simulate actual service.

As issued in early June 1904, the proving ground's report on the 9-pounder was in general a positive one, pointing out its advantages and some perfunctory problems. Probably because it had other fish to fry, the proving ground made the suggestion to pass the gun on to the nearby naval academy for additional field testing both ashore and afloat, by either the midshipmen or the marines assigned there. Those follow-up tests happened in late October, and again garnered favorable comments on its reliability and ease of operation, particularly given its size and weight vis-à-vis the standard navy field gun in use. The prospect of reducing the size of the gun crew represented a promising direction. Despite the favorable results garnered in both the proving ground and academy tests, the bureau lost whatever interest it had originally. Although not precisely stated, one reason probably lay in the gun's short recoil and over-ground recoil component, which by 1904 was viewed as an obsolete design.

The navy's lack of interest notwithstanding, it seemed that it wanted to keep its hands on the 9-pounder landing gun. Two years after the academy tests, Manager Gulick wrote to remind the chief of bureau that the American & British landing gun still remained in navy possession. The acting chief responded in early February 1907, declaring that the bureau's chief objection lay in the odd (i.e. non-standard) type of gun submitted. Later in the month, the gun, its components, and what remained of the ammunition originally sent were returned to Bridgeport.

The Bethlehem Steel Company Mark IV 3-inch Landing Gun

A January 1904 letter from the Bethlehem Steel Company, following up a preliminary conversation between the company and the Bureau of Ordnance, confirmed Bethlehem's ability to provide twenty-five 3-inch naval landing guns, carriages, and outfits at $5,300 each, FOB at South Bethlehem, Pennsylvania. With some minor modifications—for example, to use ammunition consistent with that already standard in the US Navy—these guns were essentially off the shelf, that is, of existing Bethlehem design. The immediate need was to fill the landing gun gap then existing, with the general intention to put such guns aboard the navy's late pre-dreadnought battleships. Once again, because of the exigency of the moment, the navy had to accept fieldpieces having an outdated design, with some over-ground recoil component, despite the two long recoil/counter-recoil cylinders affording 30.5 inches of on-carriage recoil. The company anticipated the first deliveries in one year, with the completion of the contract in two years. The government duly let the contract with Bethlehem Steel in mid-September of that year.

The Mark IV field carriage permitted 10 degrees traverse and 15 degrees elevation, providing a gun range of 5,500 yards. The Mark IV single-forging gun weighed 576 pounds, with two hydraulic cylinders beneath the 23.5-caliber barrel designed as a whole with the gun, and interconnected by an equalizing tube. Adding the 1,141-pound carriage gave a total weight of piece of 1,717 pounds; the Mark III limber weighed 620 pounds without ammunition. The carriage carried two ammunition boxes and the limber another two, each with a capacity of twelve rounds. Another dozen rounds were stowed in compartments in the limber. With each complete round weighing 16.25

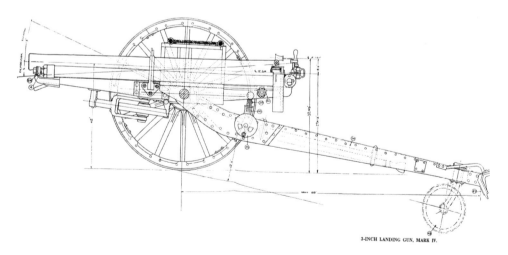

The drawing shows the left side of the Mark IV 3-inch naval landing gun, a catalog design built by the Bethlehem Steel Company. Its pair of long hydraulic recoil/counter-recoil cylinders did not absorb the entirety of the recoil and its breech mechanism offered the least strength of all the modern US Navy landing guns. *Handbook of 3-inch Field and Landing Guns and Their Ammunition*, Ordnance Pamphlet No. 146, Sept. 1915. (*National Archives and Records Administration*)

pounds (later specified as 16.5 pounds), the full sixty rounds amounted to 975 pounds. The entire weight to be hauled amounted to 3,446 pounds, well over a ton and a half.

The breechblock was slightly conical in shape, with interrupted involute (spiral) threaded collars, whose plane lay perpendicular to the long axis of the block. Taken together, the two threaded sectors occupied 240 degrees of the breechblock's circumference, separated by two cylindrical blank sectors of 60 degrees each. The cross-section of each collar resembled an inclined V thread, with rounded top and bottom, which provided no translational motion (withdrawal) while rotating. The operating lever pivoted to the top of the carrier hinge pin, so that a continuous motion of the lever rotated the block by means of the threaded carrier hub. The rotation easily caused the collars to disengage from their seats, swinging the conical block clear of the breech recess. From almost the outset, the breech mechanism occasioned questions about its strength.

In April 1905, the company advised the bureau that the brass cartridge cases supplied by the navy for the first firings fit rather tightly, but it desired a dozen more before making a design change to the powder chamber of its gun. Before complying with that request, Captain Pendleton at the Naval Gun Factory wished to have more details of the company's firing tests. Within a day, the bureau took the decision out of his hands and agreed to ship Bethlehem the dozen rounds requested, and a week later upped the number to twenty-five cartridge cases to be sent. In mid-June of that year, in response to Bethlehem's query, the bureau informed the company that the guns would be bronzed and the carriages painted metallic brown.

Soon it became clear that Bethlehem Steel would not be able to meet its delivery promise. In February 1906, the company still struggled with a design error in the hinge lugs on the breech. The bureau approved its suggested method of correction, but that problem proved not to be the only one or the major one. In August 1907, the bureau still had not settled upon the manner in which to attach the bracket for the army panoramic sight in order to ensure its stability. In the end, the bureau decided to stay with the gun's original peep sight. The bureau also expressed unhappiness with the Bethlehem spiral collar breech mechanism, which it found to be not only difficult and outmoded, but having the least strength of any US Navy landing gun breech mechanism.

The problem of most immediate concern, however, which first manifested itself in the initial proof tests during 1906, lay in the proportions of the gun's powder chamber: it was too short for the brass cartridge containing the standard propellant in naval service. Designed for nitroglycerin-based gunpowder, the gun had performed well at the Bethlehem test range, but once at the Naval Proving Ground and using navy bagged powder, it failed to generate much more than 1,650 feet per second, rather less than a gun of its strength should have been capable of. Even when the gun did fire at that muzzle velocity, however, jump occurred and the sight bar went awry, with accuracy suffering.

The fleet's need for landing guns had become so acute that four-fifths of the Bethlehem Mark IV guns had been placed on board the recipient battleships before the results of the more exhaustive proof tests done two years later had come in, and those results were not good. In official parlance, the chief of bureau deemed the proof of the Bethlehem Steel Mark IV landing gun unsatisfactory. Taking into account the short length of the powder chamber, the charge necessary for the maximum efficiency, sufficient muzzle velocity, and proper functioning of the gun challenged the bureau. The task became all the more difficult because the bureau attributed the deficient recoil and bending of the slide to the improper design of the recoil cylinders and/or pistons. The question arose whether it was practicable to achieve a standard muzzle velocity of even the sought-for 1,650 feet per second. If the bureau desired that as an acceptable minimum, then a charge compatible with the size of the too-short powder chamber had to be determined. Toward that end, the bureau ordered a quantity of smokeless powder from the magazine at Fort Mifflin, near Philadelphia, sent to the proving ground.

So far, the problems described were those restricted to the gun and its recoil components. At the same time, the Bethlehem Mark IV field carriage was put through a series of arduous tests and was found to be defective in its design. The carriage permitted the gun an acceptable train of 10 degrees on either side of its centerline, but

On board the Virginia-class battleship *Georgia*, the main (twin 12-inch/40cal) and secondary (twin 8-inch/45cal) batteries dwarf the Mark IV 3-inch landing gun. The navy considered this landing gun so lackluster that it declined to order any more after the initial twenty-five. The secondary gun turret superimposed on the main gun turret hardly represents one of the navy's better notions, as such arrangement led to more than one difficulty, including that of obtaining accurate fire control. The navy did not repeat the design in its following classes. (*Leslie Jones Collection, Boston Public Library*)

only 15 degrees in elevation. The most serious problem lay in its insufficient girder strength, which led to the jamming of its parts and an overly long over-ground recoil. Little wonder that in early August 1908, the bureau, while conceding that all of the Mark IV landing guns had been assigned to service in the fleet, wrote, "As this gun has not proved satisfactory, this Bureau does not contemplate the purchase of any additional guns." Not a few officers in the fleet believed that the Mark IV landing gun left a lot to be desired and perhaps hoped that it never saw action. If so, those hopes would be dashed in the middle of the following decade.

Entering the Second Decade of the Century

In October 1911, the reality of a limited number of landing guns confronted the US Navy. The Fletcher Mark I 3-inch field guns had proved thoroughly deficient, if not dangerous, and within four months were removed from the fleet. The navy refurbished them and thereafter assigned more than 85 percent of the surviving pieces to naval militia units, training establishments, and training vessels. The sixty-three Naval Gun Factory Mark I mod 1 3-inch field guns were adequate but decidedly obsolescent, although with a new optical sight replacing the original flawed one, they would see decades more service. The twenty-five Bethlehem Steel Mark IV 3-inch landing guns had been purchased as stopgap weapons and had performed as expected, which is to state rather indifferently, but they too would see use into the 1930s. The fifty new Ehrhardt Mark VII 3-inch landing guns being built soon came to be considered as not ideal for fleet use and were duly conveyed to the Marine Corps as light field artillery. For this reason, coupled with the perceived need ashore, on October 19, 1911, the navy reduced the number of 3-inch landing guns aboard its large capital ships—for the most part battleships and armored cruisers—from two to one. In the meanwhile, it sought a superior and more modern landing gun. The unmistakable trend in progressively heavier landing gun outfits, however, would continue.

Action in the Philippines and a False Alarm in Tangier

At the end of the brief Spanish-American War, the United States became the long-term landlord of the former Spanish colony in the Philippine Islands. To the native Filipinos, one master from across the seas was as bad and unacceptable as another, and almost immediately they rose up in what is variously known as the Philippine Insurrection or the Philippine-American War. Even after the government of the First Philippine Republic had capitulated in July 1902, the war continued for another decade, largely against the Moros, the indigenous Muslim people of the Philippines, whose warriors all too frequently lived up to their reputation for fearlessness.

The land war proved to be a grim and terrible one, marked by atrocities, into which guerrilla wars too often descend. "Civilize 'em with a Krag" became the watchword among the American forces, referring to the Model 1898 Krag-Jørgensen caliber .30-40 infantry rifle. Although the round fired proved more powerful than the 6mm one, the rifle's design represented a backward step from the Winchester-Lee, having a side-mounted box magazine with a capacity of five rounds. The action included a magazine cutoff, which essentially made the rifle a single-shot piece under most circumstances. Both the predecessor and successor to the Krag-Jørgensen had charger-loaded magazines: an en-bloc clip for the Winchester-Lee Model 1895 6mm rifle and a stripper clip for the Springfield Model 1903 .30-caliber rifle.

Armed vessels of both the army and navy ably supported the American regular and volunteer infantry in action against the Filipinos. In the fighting around Manila in early 1899, the 125-foot sidewheel gunboat *Laguna de Bay* took part. She had been a passenger steamer plying the Pasig River, but recently purchased for the army by Maj. Gen. Elwell Otis, and armored by soldiers of the Utah Volunteer Light Artillery commanded by Capt. Frank Grant. The eight guns aboard included Mark I 3-inch field gun No. 11, transferred from the navy after brief service on the revenue cutter *McCulloch* (four 6-pounders). Notable in *Laguna de Bay*'s service was the support fire she provided during the March 13, 1900 ground attack on the church and convent of Nuestra Señora de Guadalupe, which the Filipino insurgents had fortified.

Years later, a mystery unfolded, which only in hindsight becomes amusing. The navy had lost track of the gun once aboard *Laguna de Bay*, along with its sister, No. 4. In response to repeated inquiries from the Bureau of Ordnance, in 1913 the War Department reported that those guns had gone first to the Manila Ordnance Depot, then across the Pacific to Benicia Arsenal northeast of San Francisco. Several years later, without seeking the navy's permission, the army had condemned the guns and sold them to Francis Bannerman, a well-known armaments dealer in New York City. At that point, the bureau seems to have pursued the matter no further.

At sea, inadequate charts did what the Spanish navy could not. In early November 1899, the protected cruiser *Charleston* (two 8-inch guns; not the later St. Louis-class semi-armored cruiser mentioned previously) grounded on an uncharted reef off Camiguin Island, north of Luzon. Wrecked beyond salvage and abandoned by her crew, she was accompanied in her destruction by Mark I landing gun No. 9. But elsewhere in the Philippines, particularly during the extended war following the 1902 capitulation, an air of adventure prevailed. A number of small gunboats, the majority acquired from Spain, did useful service in patrolling lesser rivers and shoal waters, transporting troops, giving fire support, landing shore parties, fighting pirates, and disrupting the efforts of gun runners and slave traders. Typical of such warships were the five 243-ton gunboats of the Arayat class—*Arayat, Callao, Pampanga, Paragua,* and *Samar*—built in Manila for Spain in 1887–88. They were not only small, but fast and maneuverable, and armed with 6-pounder and 3-pounder RF guns and machine guns. Ens. William Davidson, who would not long thereafter see service in China during the Boxer Rebellion, became the first American commanding officer of *Paragua*.

When even those small gunboats proved too large for the task, 60-ton steam launches armed with 1-pounder RF boat guns and Colt-Browning machine guns saw inshore use. Midshipman Allen Reed, executive officer of *Paragua*, took command of one of these launches when the navy sent its small warships to cooperate with army operations on Samar in June 1905. Ashore, his judgment and coolness—a favored term for courage under fire—won praise on the fitness report written by his commanding officer, Ens. Charles Kerrick, whose command of *Paragua* garnered tribute in turn from Maj. Gen. Leonard Wood, commanding the army expedition on the island. The day following his commissioning as ensign in February 1906, Reed assumed command of *Paragua*, which operated out of Cavite for the next ten months. As a lieutenant and the executive officer of USS *Denver* (ten 5-inch guns), Reed would command that ship's landing party put ashore at Corinto, Nicaragua, in late August 1912.

The assignment to a gunboat in the Philippines was clearly a junior officer's dream-come-true, offering the opportunity for unprecedented freedom, independent action, and feats of derring-do, far from the watchful eye of his superior officer. Among the young stalwarts cutting their teeth in the Philippines during that decade, who would go on to win fame as flag officers in the 1930s or World War II, may be counted William Halsey, Frederick Horne, William Leahy, John McCain, Sr., Chester Nimitz, and Yates Stirling, Jr.

Despite his great success as commander-in-chief Pacific (CinCPac) in that later war, in a 1950s response to a letter written by a USNA midshipman (later a US senator representing Arizona), Admiral of the Fleet Chester Nimitz looked back fondly to 1907, decades before, when he was a recently commissioned ensign. In the rosy hue of misty memory, his command of the 162-ton gunboat *Panay* (one 6-pounder), with his trusty exec Midshipman John McCain at his side, loomed as the highlight of his naval career. The John McCain in question happened to be the grandfather of the midshipman Nimitz was replying to. Both swashbuckling young officers of that bygone era doubtless appreciated that it took a pirate to catch a pirate.

Newly commissioned Ens. Chester W. Nimitz (1885–1966) looks very smart in his blues in 1907. As a midshipman shortly before his commission, Nimitz took command of the former Spanish gunboat *Panay*, with Midshipman John S. McCain (1884–1945) as his loyal executive officer. Far from the vigilant eye of their superior officer, they enjoyed some heady adventures in the waters of the Philippine Islands. Although Nimitz suffered a setback during his next command, that of destroyer *Decatur*, great things lay ahead for this young officer and his protégé. (*Naval History and Heritage Command*)

Ironically, in May 1904, a landbound pirate did cause consternation in the United States and precipitate the demand to land marines in a foreign place. Following a string of such incidents, Berber brigand Mulai Ahmed er Raisuli (or Raisuni) kidnapped Ion Perdicaris, putatively an American citizen, and his British stepson, Cromwell Varley, from the elder man's villa near Tangier, Morocco. Records subsequently disclosed—though not to the public—that Perdicaris had renounced his American citizenship for Greek forty years before. President Theodore Roosevelt nonetheless ordered the South Atlantic Squadron, commanded by Rear Adm. French Chadwick in the armored cruiser *Brooklyn* (four twin 8-inch gun turrets), to Tangier. Those ships were reinforced within days by three warships from Rear Adm. Theodore Jewell's European Squadron. Although highly placed Americans on both sides of the Atlantic engaged in idle talk to land marines to secure the customs house, wiser heads prevailed and the marines stayed away from that office. A small detail of marines, commanded by Capt. John Myers of Peking fame, did proceed ashore to guard the US consulate, where they remained for nearly a month until the dust settled.

The most significant victory in the incident turned out to be Roosevelt's. Facing strong opposition from members of his own party, the president took no chances at the Republican national convention in late June. Although Secretary of State John Hay penned the telegram with the trenchant admonition, "This Government wants Perdicaris alive or Raisuli dead," the delegates correctly perceived Hay as TR's staunch surrogate. Such bully sentiment clinched the president's nomination for a second term. By coincidence, Raisuli released the captives almost immediately thereafter. Seventy-one years later, the incident provided fodder for the fanciful movie, *The Wind and the Lion*, in which marines and bluejackets do land in strength, and without provocation or warning, slaughter the bashaw's palace guard. Later, in rescuing the wife of Perdicaris, who had in fact *not* been abducted, the marines battle German troops amidst the desert sands. At least the ordnance in use and other technical details are pretty much correct.

References Consulted

Bethlehem Steel Company, *Mobile Artillery Material*, undated, pp. 14–15, and drawing.
Bylines not specified, *New York Times*: "The Paragua shells a town," Sept. 15, 1899; "Gunboat beats off Filipinos," Sept. 20, 1899; "Smuggling on Panay Island," Jan. 3, 1900; "The United Service–Navy," Dec. 14, 1900; "Operations in Samar Island," July 18, 1901; "The United Service–Navy," Dec. 28, 1901; "Reducing the Asiatic Fleet," July 13, 1902; "Our Forces Ready for Chinese Rising," Sept. 30, 1910.
Dictionary of American Naval Fighting Ships, Naval History and Heritage Command (online). Consulted for each US naval warship and auxiliary mentioned in this chapter.
Dumindin, A., *Philippine-American War, 1899–1902*, 2006 (online).
Ellsworth, H. A., *One Hundred Eighty Landings of United States Marines 1800–1934*, History and Museums Division, HQ, US Marine Corps (Washington, DC: Govt. Print. Office, 1974 [originally printed 1934]), p. 8.

Fullam, W. F., and Hart, T. C., *Text-Book of Ordnance and Gunnery*, 2nd ed. (Annapolis: US Naval Institute, 1905), pp. 139–41, 303–09.

McAulay, J. D., *Rifles of the United States Navy & Marine Corps 1866–1917* (Woonsocket, RI: Mowbray Publishing, 2017), pp. 104, 167, 169–74, 181–88, 191–94, 199–200, 202–03, 208.

McCain, J., and Salter, M., *Faith of My Fathers* (New York: Random House, 1999), pp. 133–34.

Smith, J. E., *Small Arms of the World*, 10th rev. edit. (Harrisburg: Stackpole, 1973), pp. 63–65.

Tuchman, B., "Perdicaris Alive or Raisuli Dead," *American Heritage*, August 1959, 10(5): 18–21.

US National Archives, Record Group 74, Records of the Bureau of Ordnance; Washington, DC.

US Naval Academy, "Quarterly Reports on the Fitness of Midshipmen" for Midshipman Allen B. Reed, April 19–June 30, 1905.

US Navy, *Handbook of 3-inch Field and Landing Guns and Their Ammunition*, Ordnance Pamphlet No. 146, September 1915, pp. 11–12.

US Navy Dept., *Annual Report of the Secretary of the Navy, 1898*, p. 54.

————, *Annual Report*, Bureau of Navigation, *1900*, pp. 503–04, 548–50; *1905*, pp. 388, 484–86; *1906*, pp. 399–400, 477; *1907*, pp. 368, 432–36.

————, *Annual Report*, Bureau of Ordnance, *1900*, pp. 570, 603; *1901*, pp. 633, 635, 688, 691–92, 700–02; *1902*, pp. 490–492, 505–06; *1903*, pp. 672, 727, 730; *1904*, pp. 580, 614, 618, 626, 635; *1908*, p. 436; *1910*, p. 347; *1911*, p. 222; *1912*, p. 213; *1913*, p. 179.

Williford, G. M., "The American Defenses of Guam 1909–1932," *Coast Defense J.*, August 2017, 31(3): 1–23.

Wingfield, T. C., and Meyen, J. E., (eds.), "Lillich on the Forcible Protection of Nationals Abroad," Appendix I. "A Chronological List of Cases Involving the Landing of United States Forces to Protect the Lives and Property of Nationals Abroad Prior to World War II," *International Law Studies* 77: 146–47.

9

The Marines Get Their Share

The Greer Board and Its Aftermath

The United States Marine Corps evolved so substantially during the twentieth century that most officers from the previous one would no longer have recognized the corps or appreciated its new roles. Likely, however, those officers would not have included the likes of Samuel Nichols, Presley O'Bannon, Archibald Gillespie, McLane Tilton, Robert Huntington, Henry Cochrane, John Myers, Newt Hall, or any other marine officer involved in amphibious landings, whether of large-scale or small, or in independent operations ashore. The problem lay with what the corps itself saw, almost always through the eyes of its commandant, as its primary role: that of a ship's guard to protect the naval officers aboard from their crew. The crews of the vessels of the New Navy, however, had been recruited, not press-ganged, and were not unreasonably resentful of the implication in having such policemen aboard. The distasteful policy endured that no matter how well trained, thoroughly disciplined, and otherwise courageous, the sailors could not be trusted. Moreover, the members of the marine detachment all too often acted in a manner suggesting superiority to the naval crew, with the result that the sailors came to despise them.

In 1889, the newly installed secretary of the navy, Benjamin Tracy, commissioned the Board of Organization, Tactics, and Drill, headed by Commo. James Greer, and thus all-too-soon known informally as the Greer Board. One of the board's important functions, as defined by Tracy, lay in defining tactics of the US Navy, including landing operations. The board included Lt. (j.g.) William Fullam and Lt. Seaton Schroeder. Fullam had graduated first in the academy class of 1877 and was already an important naval thinker. In his papers published in 1890 and 1896 in the *Proceedings* of the US Naval Institute, not only did Fullam strongly propound the removal of marines as part of the regular complement of ships' companies, but he argued that they be trained for use in expeditionary forces. Fellow board member Schroeder agreed with him, and as the decade transpired, so did many naval officers of significance during the 1890s—Robley Evans, Albert Niblack, Richard Wainwright, and Greer himself, believing that the marine detachment aboard ship remained an anachronism. Lieutenant Commander Wainwright, in charge of the Office of Naval Intelligence, went a step further, suggesting the marines be concentrated in force at advance bases. Navy Secretary Tracy, however, did not agree with them, feeling strongly that a body of sea soldiers must remain available at all times within the fleet.

Marine Commandants Charles McCawley and Charles Heywood hardly assisted in the necessary evolution of the corps, holding adamantly to the traditional role of marines as ships' guards, a view unchanged from that of Archibald Henderson, who had been commandant in the decades just before the Civil War. To offset the criticism, McCawley suggested that the shipboard marines could be trained to man certain guns aboard, in the same fashion that they had been trained decades before to pull their own rowing boats. To that argument, the proponents of getting rid of the shipboard marines countered essentially with, "Okay, then make them sailors, wearing the same uniforms as the other gunners." The marines' best argument for staying aboard ship, however, lay with their other traditional role, as the core of the landing party. On the other hand, naval officers, who simply wanted the marines off their ships, desired landings to be accomplished by sailors, as both infantrymen and artillerymen, with marines reserved for larger expeditions requiring longer periods ashore. A handful of marine officers did agree. In 1897, Capt. George Barnett, a future commandant, submitted a letter to then-current Commandant Heywood urging that the primary function of marines aboard or ashore become that of an expeditionary force, lest they be considered "merely a collection of watchmen."

The argument for the proper role of marines in the US Navy extended into the twentieth century and eventually involved none other than the president of the United States.

The First Decade of the Twentieth Century and the Mark I Field Gun

The long-held notions that the Marine Corps should be entirely subservient to the navy, such that marines going ashore required naval gun crews for the artillery landed and would be armed with lesser shoulder arms than those the sailors carried (the marine weapon remained routinely in line with the standard army rifle), began disappearing in the 1890s. The marines viewed as unacceptable the established tradition that a naval officer present ashore of whatever rank must be senior to the marine officer of whatever rank, where the disparity could be substantial. By the time the First Marine Battalion landed at Guantánamo Bay during the Spanish-American War, the marines carried the shoulder arm standard within the fleet and trained marine crews manned the machine guns and field guns landed. The signalmen, who often braved hostile fire at terrible risk, were also trained marines. The senior officer ashore, Col. Robert Huntington, answered directly to the senior officer afloat, Cmdr. Bowman McCalla, although junior to Huntington in sleeve rank. To the satisfaction of the marines on the beleaguered hilltop, however, a marine officer commanded them in essentially an all-marine landing operation.

The recent experiences of the marines at Guantánamo and in China, and further thinking by Fullam, now a lieutenant commander, led in 1901 to a trial landing in strength on Nantucket Island to secure an advance base. A paper given by Capt. Dion Williams, USMC, published the following year in the Naval Institute's *Proceedings* and widely read, advocated that the Marine Corps adopt a new role, that of defending the existing coaling stations and other naval bases beyond the continental United States. When a first lieutenant during the Spanish-American War, Williams had led the landing party taking possession of the Cavite naval base in Manila Bay. In times of war, he argued, this role would be extended to wresting such useful outposts from the enemy. Unlike the British empire, the United States did not then possess a great many such stations around the globe. Captain Williams recommended that the corps be enlarged to 15,000 men, including 515 officers schooled at the naval academy, in order to man the existing mainland and insular stations.

Fullam, a long-time critic of marines aboard American warships, agreed wholeheartedly with Williams that the marines were ideal for the new task. Moreover, he suggested that they be provided dedicated transports in order to serve as a quick-response force, with their battalion organization preserved, obviating the need for landing bluejackets. Fullam thought such vessels ought to be designed for easily landing the marines and dismounting certain guns for their use on shore. It would be even better, he submitted, if the marines posted ashore and afloat be rotated to ensure that all men receive thorough training in both roles.

Other far-thinking marine officers, exemplified by John Lejeune, realized more and more that the major mission of the Marine Corps had to change, perhaps to one involving its deployment as an expeditionary force. This thinking stood in contrast to that of the commandant, Maj. Gen. George Elliott, and other senior officers, who continued to believe that the marines were merely ships' guards against untrustworthy and unruly sailormen.

By 1908, 29 percent of the 9,300-man Marine Corps remained aboard ship. Fullam asserted that if the corps would only bring those men ashore, it would have enough men for the advance base forces. With Germany holding the Pacific islands that stood astride the supply routes between the American west coast and the Philippines, and Japan located in close vicinity to that archipelago, the US Navy became increasingly aware that in a Pacific Ocean war it would need advance bases. The very close war scare with Japan in 1907 firmly persuaded those who had earlier played down that threat. Finally tiring of the slow pace in accomplishing the conversion of the Marine Corps into an expeditionary force, Fullam and his friend, Lt. Cmdr. William Sims, naval aide to President Theodore Roosevelt, approached the president. They were supported by Rear Adm. John Pillsbury, who wrote to the secretary of the navy, Victor Metcalf, to provide additional persuasion. Roosevelt duly issued Executive Order 969 that essentially removed the marines from American warships by no longer including that duty among those assigned to the corps. Congressional pressure, however, put them back on board, where they remain today.

Not long into the new century, the Bureau of Ordnance provided field guns to the marines as standard issue, as opposed to being temporary loaners. In November 1902,

Above and below: The standard shoulder arms and uniform items issued before and after the turn of the century are disclosed in these two photographs of shipboard marine detachments. In the upper one, armored cruiser *New York*'s marines around 1899 carry Model 1895 Winchester-Lee 6mm rifles and wear 1897 Pattern forage caps; in the lower, the stacked arms aboard battleship *Georgia* in 1908 are Model 1898 Krag-Jørgensen caliber .30-40 rifles, and the forage caps are the 1904 Pattern (note their more steeply angled visors). In both instances, white cap covers are worn over the standard-issue blue caps. More significantly, despite the opposition of many naval officers, marine detachments remained aboard larger warships into the new century, routinely serving in landing parties. (*Library of Congress* and *NavSource Naval History Photo Archives*)

the ordnance inventory submitted by the Asiatic Squadron based at Cavite in Manila Bay included seventeen Mark I 3-inch field guns, sixteen of them in the possession of the 1st Marine Brigade and only one assigned to the naval station. But in October 1905, when the corps desired a 3-inch field gun for its barracks in Honolulu, the bureau replied that all Mark I guns were unserviccable owing to the defective recoil cylinders. The bureau insisted that warships had priority for the Mark I mod 1 and Mark IV landing guns, so no guns remained available for the Oahu outpost. The 1st Marine Brigade confirmed that by mid-1907, new recoil cylinders and counter-recoil springs had been installed in ten of its guns. One year later, the brigade headquarters reported that two additional guns, included on the original 1902 list and later transferred to the Peking legation guard, also had received the new recoil apparatus.

Just after the end of the first decade, the Bureau of Ordnance declared the Mark I field gun obsolete for fleet use and instructed that the guns be turned in to store. Those pieces in marine hands were for the most part relinquished at that time, although there remained exceptions that continued to see use for a year or two more.

The Marines Obtain Their Own Field Artillery: The Mark VII Landing Gun

An inquiry made in 1904 by the Bureau of Ordnance to Hans Tauscher, the North American representative of the Rheinische Metallwaaren und Maschinenfabrik, Düsseldorf, manufacturer of the Ehrhardt pattern ordnance, would ultimately be of benefit to the corps. Although the navy was initially interested in the Ehrhardt sliding block breech mechanism for its 3-inch/50-caliber shipboard guns—it eventually rejected that design because the service sought a semiautomatic mechanism—the inquiry soon focused on the sliding block and long recoil designs as potential assets for a new pattern naval landing gun. Indeed, disappointed and frustrated with the landing guns then in its possession, the navy needed a better piece to equip its new dreadnought battleships, beginning with the Michigan class (also known as the South Carolina class).

Because of the restrictions in the appropriations process that did not authorize the American armed services to purchase weapons from a foreign manufacturer, a three-party contract became necessary for the acquisition of the Ehrhardt landing guns. The three parties were the US Navy, Tauscher, holding the power of attorney for Rheinische Metallwaaren und Maschinenfabrik, and the American & British Manufacturing Company of Bridgeport, Connecticut, which Rheinische licensed under US patents to build the guns.

What became the Mark VII landing gun was a 23-caliber piece built of nickel steel, with three shrunken-on bands and a horizontal sliding wedge breech mechanism. It would be the first American naval landing gun to have a protective shield and a panoramic sight for indirect laying. The recoil generated was the longest yet of any American naval field gun. Its maximum elevation and range were improved over the previous landing gun designs, increased to 17 degrees and about 6,000 yards. Unlike the previous marks of landing guns, which the gun pointer fired when ready, the plugman fired the Mark VII gun at the direction of the pointer.

According to the details agreed upon in early 1909, for the right to manufacture the Ehrhardt 3-inch gun, field carriage, and limber, A&BMC agreed to pay royalties on the first 200 outfits it made, with a different rate on the second 100 than the first. Thereafter it would pay no royalties and have the exclusive right to manufacture the gun for the US government. Rheinische agreed that no royalties would be imposed on replacement parts for this ordnance. On February 19, the Connecticut company offered to build twenty-five Ehrhardt landing outfits for $7,500 per outfit, and in early March the Bureau of Ordnance accepted the offer for the US Navy.

Because of a pending law suit brought by Krupp Grusonwerk AG against the US Army on the basis of alleged patent infringement by Rheinische in the design of the Model 1902 3-inch field gun, the navy remained especially wary of such a predicament. Needing additional assurance, the Navy Department required a letter from Rheinische declaring that Tauscher possessed the power of attorney to act for the company. Taking a step further, the navy requested an expansion in the license for the American & British Manufacturing Company to manufacture the German ordnance now to include the explicit permission for the government to use the ordnance. A&BMC explained that the army had proceeded directly with the manufacture of their guns under the Ehrhardt patents and thereafter had assumed responsibility for their use, whereas in the contract under consideration, the Navy Department would be relieved of all responsibility from the possibility of patent infringement.

In early May, A&BMC advised the bureau of certain discrepancies in the drawings it had obtained and requested that the navy ship to Bridgeport the pilot gun and carriage that Rheinische had provided the service for testing. The bureau had assigned the register No. 1012 to the gun and No. 189 to its field carriage, respectively designated the Mark VII and Mark V. After some additional fine-tuning, the contract between the navy and the American & British Manufacturing Company was signed on September 29, 1909, for twenty-five 3-inch long-recoil landing outfits, at the previously agreed-upon price per outfit of $7,500. Crucible Steel of Pittsburgh provided the steel forgings for these twenty-five guns. Their serial numbers ran from 1055 to 1079, with those of their field carriages 221 to 245.

A&BMC advised on January 11, 1911, that the first Mark VII landing gun, No. 1055, and box-trail field carriage, No. 221, in the first production run, would be shipped to the Indian Head Proving Ground in three days. The company wished to be notified in advance when the navy planned to proof-fire them, so that it could send a representative to witness the test. The intention was to give the first production gun and carriage a full proof test, with subsequent pieces fired ten times each at various pressures (muzzle velocities) up to 11.8 tons per square inch (1,640 feet per second), and at various elevations and degrees of traverse. In response to the navy's request, A&BMC provided a brief description of the Rheinische protocol for proofing its 3-inch landing guns.

Two months later, in reference to the appropriation by the 61st Congress, A&BMC offered to manufacture a second lot of Ehrhardt-type landing guns at the same price—$7,500 per gun and carriage—but urged a signed contract soon, to obviate any increased cost of manufacture and interruption of the work between lots. Despite difficulties in achieving sufficiently high muzzle velocities in its proving ground tests, at which Rheinische expressed astonishment, the Bureau of Ordnance pursued the contract for A&BMC to build a second batch of twenty-five Mark VII landing guns, which the navy shortly authorized.

Those parties finalized that contract on September 25, 1911, within days of the two-year anniversary for the first one. The navy specified that jobs done by the Naval Gun Factory on the first order, such as cutting sight holes in the shields and providing sight pads and trail wheels, this time be done by A&BMC. The superintendent anticipated the second lot would also show an improvement in the action of the recoil mechanism, a problem that had plagued the first twenty-five guns.

Apparently, one of the drawings for the new guns specified gun steel rather than nickel steel, which the inspector of ordnance operating out of Brooklyn noticed and requested a correction of. Once again, the forgings for the guns came from Crucible Steel, although Fore River Shipbuilding provided those for the breechblocks. Not for the first time, Fore River experienced difficulties, this time in the tensile strength of the steel, although the elastic limit attained met the standard demanded. When A&BMC asked for an exemption from the standard, the navy demurred and demanded that the steel meet all specifications.

In March 1912, the new ordnance received its serial number sequences. The guns would run from Nos. 1153 to 1177, while the field carriages would occupy Nos. 246 to 270. Only for the carriages would the serial numbers be continuous with the previous run.

This time around, however, A&BMC made a number of production errors, which it wished the bureau to excuse. First, there was a problem with a poorly machined breechblock. Two months later, three guns suffered incorrect boring. Although some officers put this faulty construction down to the inexperience of the new firm, others were more hard-nosed about the performance. The ordnance inspector on the scene attributed virtually all of the errors to carelessness in the machining. In every instance, he strongly recommended against acceptance of the flawed gun, because in his opinion the remedy would result in the proposed new liner being worrisomely thin and likely weak, and therefore subject to cracking. Moreover, the inspector declared an overruling of his initial denial would create a harmful precedent and favor the continuance of bad work. A&BMC, however, convinced the chief of bureau, Rear Adm. Nathan Twining, to accept the three guns with the insertion of Bethlehem nickel steel tubes with jacket locking rings (ironically in one tube, the steel elongation came slightly below specification and the company once again had to seek an exception). The new configuration resulted in only two bands shrunken on the guns, one fewer than their near sisters, and the bureau designated the three guns Mark VII mod 1.

The Connecticut company shipped the first four completed guns of the new contract on September 9, 1912 and it completed the contract on the 26th, a day later than the deadline specified. A&BMC had had to overcome more than a few difficulties in that completion, not the least of which had been a delay in obtaining the steel forgings. In early October, the panoramic sights from the army's Ordnance Department still had not arrived.

But if A&BMC had fallen down in some aspects of building the guns, so did the navy in its turn. The company complained in mid-November that two months after shipping the guns to the Washington Navy Yard, they had not yet been sent on to the proving ground. This unexplained tardiness delayed the $60,000 owed to the company being paid, because that payment remained contingent upon all guns being successfully proofed. The company's complaint lighted a fire under the bureau, which in mid-December reported that all guns had completed proofing.

At that time, however, there appeared a new fly in the ointment, namely a weakness in the box trail of the Mark V carriage. The explanation for the damage lay in the trail bar not being in the correct position when the gun had been fired. After some discussion on the wisdom of shortening the trail bar 15 inches, the bureau decided to make no changes at that time. But four months later, during shrapnel firing tests at the army's Sandy Hook Proving Ground, the trail of the Mark V test carriage buckled. In order to simulate conditions in the field, the gun crew had fired the piece on stony ground with the trail placed against a solid backing, so causing damage. Although the pressure during the firing varied to somewhat more than the one expected in normal service, the bureau pointed out that the trail should have been rugged enough to withstand any pressure encountered in action. A&BMC queried the gun's designer about the problem, because it did not consider it should be responsible for repairing the trail. The bureau felt the trail could be strengthened by stiffening pieces or cross braces. Whatever the case, it remained clear that the trail as designed remained unsatisfactory.

There also existed concerns about the state of the test gun itself, No. 1172, because the naval ordnance inspector had urged on April 1, 1913, that the gun not be fired "any more in its present condition." Whatever the circumstances, after having been moved back to the Indian Head Proving Ground, gun No. 1172 resumed firing. In August of that year, its barrel split, and soon thereafter the Bureau of Ordnance authorized its scrapping.

However vexing these setbacks, a great deal more serious impediment proved to be a long-standing one: unreliable ammunition.

The Second Decade, the Advance Base Force, and the Issuing of the New Arms

By early 1911, the marines looked forward to replacing their Krag-Jørgensen Model 1898 caliber .30-40 infantry rifles with the army Springfield Model 1903, the latter based on the German Mauser system and more accurate than the Krag. Moreover, it fired .30-caliber rounds of a type not interchangeable with that of its predecessor. Most marines had the new rifle in hand by the end of 1912, with the Colt Model 1911 semiautomatic pistol also being issued to those authorized to carry it. Although the hand-cranked Gatling guns were, at long last, being phased out, the marines found themselves still stuck with the Colt-Browning "potato digger" machine gun. But bigger changes were in store for the marines than simply their small arms and automatic weapons.

The advance base force began to see greater development in 1911, after the Advance Base School, established the previous year at New London, Connecticut, moved to the Philadelphia Navy Yard, to be in closer proximity to the dedicated ordnance being amassed there. For starters, fifteen of the sixteen Mark III heavy carriages to mount shipboard 3-inch/50-caliber guns were moved from the Norfolk Navy Yard to the yard on Philadelphia's League Island, followed in 1912 and 1913 by a number of Mark VII 3-inch landing guns. During the course of the decade, the advance base force on the Atlantic coast increased from a battalion to a regiment, and by late 1913, the official inception of the advance base force increased to a two-regiment brigade. One regiment was to act in fixed defense, the other in a mobile role.

Although small-scale amphibious exercises had previously taken place on Culebra Island, Puerto Rico, in January 1914, the entire Advance Base Brigade carried out a major training operation there, with the two regiments acting in concert. The regimental commanders, Lt. Col. Charles Long (fixed defense) and Lt. Col. John Lejeune (mobile defense), and the brigade commander, Col. George Barnett, were all visionaries regarding the role the Marine Corps would play in decades hence. During those years, many more exercises would be carried out on Culebra Island, and by the early 1920s on nearby Vieques Island as well, because of their similarity to islands in the Pacific Ocean. During the course of those exercises, the expeditionary and advance base force missions once envisioned for the marines merged into one.

In the Pacific, because of the formidable naval threat presented by Japan and Germany, the advance base force developed from the marine formation in the Philippines (by then, marines were posted to both Cavite and Olongapo). Even before the Philippines contingent had been designated an advance base force, it had undertaken rigorous field training, including amphibious exercises, hastened by another war scare with Japan in 1913. The following year, the same one in which Col. Joseph Pendleton—who would earn his reputation in the banana wars—established a temporary advance base force barracks at San Diego, California, the fifteen obsolete, unwieldy, and unwanted Mark III 3-inch gun carriages and Mark II limbers were removed from the advance base force at Olongapo and shifted to Guam for its coast defenses. With the decision to strengthen the Guam defenses seven years later, the 3-inch carriages and limbers found their way to Mare Island, and in the 1930s were finally ordered scrapped, along with their counterparts on the east coast.

During the six-year tenure of Maj. Gen. George Barnett as commandant beginning in February 1914, the Advance Base School moved again, to Newport, Rhode Island, henceforth to be an adjunct of the Naval War College.

Influenced by farsighted marine officers such as Barnett, Lejeune, Pendleton, Ben Fuller, Dion Williams, Eli Cole, Earl (Pete) Ellis, John Russell, and Robert Dunlap, the advance base force evolved into the fleet marine force of the 1920s and 1930s, and the amphibious assault formations and defense battalions so successful during World War II. The lessons learned at Culebra and Vieques in the numerous amphibious landings

Although both marine officers on the left, identified by their leather service boots, wear the up-to-date 1912 Pattern marine field hat with its Montana peak, the enlisted marines are still adorned in the fore-and-aft-creased 1904 Pattern army campaign hat. Khakis have replaced whites as the marines' summer uniform. Their shoulder arm is the Springfield Model 1903 .30-caliber rifle, having a magazine capacity of five rounds, not interchangeable with those for the Krag-Jørgensen caliber .30-40 rifle previously arming the corps. The difference in the need to seek protective cover, just outside Veracruz, Mexico, in 1914, suggests a posed shot. Photograph by Harris & Ewing. (*Library of Congress*)

A column of marines hauling Mark VII landing guns returns shoreside after an exercise in the interior of Culebra Island, Puerto Rico, sometime before World War I. The trials and experiments done here over a period of decades, initially to train for the seizure of advance bases for the navy, provided valuable experience for the much larger scale amphibious landings of World War II. (*USMC History Division*)

led to improvements in landing craft, and revealed the necessity of combat-loading naval transports. Not a few of these lessons had to be relearned with the onset of actual war, but the development of these larger scale landings is beyond the scope of this book.

At the beginning of 1912, the bureau began the general distribution of the Mark VII landing guns. The first four, fitted with panoramic sights, were allotted to the legation guard at Peking, and six more for Cavite and Olongapo in the Philippines. Other stations so favored included the Philadelphia, Norfolk, Mare Island, and Puget Sound Navy Yards and the US Naval Academy, with the balance to the quartermaster depot in Philadelphia, in reserve for the advance base force. Those intended for Panama also went to Philadelphia, to be doled out as needed. Because the marines had recently withdrawn from Sitka, Alaska, no such guns would be required there. The instructions to the various stations included that to return any Mark I and Mark I mod 1 field guns. A year later, Mark VII landing guns went out to additional marine barracks and posts: the Charleston, Philadelphia, New York, Boston, and Portsmouth Navy Yards, the Washington Marine Barracks, the Guantánamo and Guam Naval Stations, and Port Royal (Parris Island), South Carolina, among others.

Maj. Dion Williams, widely recognized as the marines' authority on reconnaissance procedures and amphibious operations, and by 1913 commanding the legation detachment in Peking, urged that until the ammunition supply for the Mark VII landing guns at Peking was increased and the internal situation there stabilized, the two old Mark I field guns should be retained so as not to give the impression that the American defenses were being weakened. After all, the siege of the international settlements at Peking had happened a little more than a decade before. It took over a year to resolve the issue. The Bureau of Ordnance considered the Mark I guns likely to be dangerous and was determined to have them withdrawn. The commandant suggested that two Mark VII guns originally intended for Guam might be sent instead to Peking, so the Mark I pieces could be returned to the Washington Navy Yard. The bureau concurred and in early 1915 sent 700 rounds of shrapnel to Peking, along with the two additional Mark VII guns.

There remained another issue: the weight of the field ordnance to be moved upon landing ashore. The sheer weight of the Mark VII gun and Mark V field carriage was about 1 short ton, give or take a few hundred pounds (official sources differ appreciably), and that of the Mark IV limber was in excess of 1,700 pounds without ammunition. When fully loaded with ammunition and the necessary accessories, the weight of the gun and limber to be managed well exceeded 2 tons. Although the traditional hauling of landing guns by men in harness could be done over relatively short distances, such as those encountered when directly in action, marching greater distances in order to confront hostile forces required animal or vehicle transport. Putting ashore supporting artillery from ships' boats as in the days of yore now presented a weighty challenge.

The bureau expressed concern in October about the unsatisfactory 3-inch shrapnel rounds, with the hope that reliable shrapnel previously ordered from the army would reach the outposts by January 1913. As reported by Capt. Robert Underwood, the

A gun crew, wearing summer khakis with the 1904 Pattern forage cap, trains on the new Ehrhardt-designed Mark VII 3-inch landing gun in 1913 at the Marine Barracks, New York Navy Yard. Clearly apparent is the characteristic straight shield with a hinged bottom portion. The navy model ammunition limber, carrying seventy-two rounds—eight each in nine removable boxes (one sits on the ground)—was disproportionally heavy for such a limited number of rounds, and more than one officer urged its replacement. (*National Archives and Records Administration*)

On the parade ground of the Marine Barracks, New York Navy Yard, in 1913, thirty-odd men execute a gun run with the Mark VII 3-inch landing gun and its heavy navy Mark IV limber. The marines doing the heavy pulling have slung rifles, while those responsible for steering (in front of the limber) or restraining when descending a hill (behind the gun) wear pistols. Such hauling teams did not endure this speed for very long if the ground was overly rough or steep, or the temperature terribly high. (*National Archives and Records Administration*)

shrapnel used in battle in Nicaragua set for 2,000 to 2,500 yards had exploded only a few hundred yards from the gun muzzles, and thus most of the artillery fire intended to support the infantry assault had been ineffective. Clearly, a precise and reliable fuse was needed. Underwood also strongly recommended an effective common shell to knock down masonry walls.

In addition to seeking dependable ammunition, the marines grappled with the best means of delivering those rounds accurately—that is, proper fire control. One choice involved the battery commander's telescope versus the compass aiming circle, both mounted on a tripod. Each had its advantages and disadvantages. The BC scope had graduations on the reticule, enabling readings over small arcs without moving the eye from the instrument, and it could be positioned like a periscope behind a sheltering wall or parapet, which the exposed aiming circle could not do. On the other hand, the aiming circle had the advantages of greater compactness and lightness, and possessed convenient crosshairs, illuminated within a window, not a feature shared by the BC scope. The aiming circle could be used without its compass when both the gun battery and the target were visible from the instrument, but it needed the compass when only the target could be seen. The BC scope had greater power but a lesser field of view than the aiming circle: the former 10 power, with a field of 4 degrees; the latter 4 power, with a field of 10 degrees. In the end, both saw use.

Not very long after the Bureau of Ordnance had issued the marines the first Mark VII 3-inch landing guns, four of these pieces saw combat. The action in Nicaragua proved more significant as an experimental laboratory than it did as a memorable battle.

At the Philadelphia Navy Yard in October 1913, one marine loads a fixed 3-inch round through the horizontal sliding wedge breechblock, while the other members of the gun crew assume their correct positions during firing drill on the Mark VII landing gun. The sergeant in command is posted on the farthest left. Each marine is armed with the Colt Model 1911 .45-caliber semiautomatic pistol. Note that the trail wheel has been unshipped and placed aside. (*National Archives and Records Administration*)

Coyotepe and La Barranca

US intervention in Central America began on a larger scale than previously as a result of President Theodore Roosevelt's expanded interpretation of the Monroe Doctrine and successor President William Howard Taft's policy of "dollar diplomacy." The intention of both men was to achieve two ends: To prevent political and financial instability in a Latin American nation from creating a vacuum into which a European power could move. And to permit safe and stable conditions in which large US commercial firms—whether in the business of oil, mining, sugar, or fruit, nicely termed "American interests"—could carry out their enterprise unhindered by local unrest. Needless to write, these policies were not intended to benefit the ordinary nationals of those countries and, indeed, all too often guaranteed their exploitation.

The Washington Convention of 1907 strove to promote greater stability in the frequently volatile nations of Central America through diplomacy directed at averting revolutionary upheaval and war. Within a short time, a number of countries neighboring Nicaragua complained to the United States that President José Santos Zelaya had violated the convention. A dictator by any definition, Zelaya ordered the execution in 1909 of several hundred insurgents, including two Americans. A marine battalion commanded by Maj. Smedley Butler landed at Bluefields in late May 1910 and remained there until shortly before Zelaya resigned and left Nicaragua. His departure, however, did not end the troubles, and the Zelaya and Emiliano Chamorro factions continued to clash, often violently. José Estrada came and went as president, replaced by his vice president, Adolfo Díaz; the minister of war, Luis Mena, was arrested for plotting to overthrow the presidency and was released; revolutionaries blew up two military magazines in late May 1911; and fourteen months later, a full revolution broke out in Nicaragua, and the US Navy moved warships to that nation's coastal waters.

In August 1912, in response to a request for American assistance from President Díaz, gunboat *Annapolis* (six 4-inch guns) put ashore a landing party at Corinto on the west coast of Nicaragua and protected cruiser *Tacoma* (ten 5-inch/50-caliber guns) did so at Bluefields on the east coast, landing mostly bluejackets with some marines. A few days later, Major Butler returned from Panama with a force exceeding 350 marines. Before the month ended, additional marines and sailors landed from armored cruiser *California* (two twin 8-inch gun turrets) and protected cruiser *Denver* (ten 5-inch/50-caliber guns) at Corinto and entrained for the capital city of Managua, nearly 100 miles to the southeast. *California* returned to Panama for the 1st Provisional Marine Regiment, commanded by Col. Joseph Pendleton, which had come down from Philadelphia on the transport *Prairie*. The armored cruiser thereupon transported the formation to Corinto. Maj. William McKelvy commanded 1st Battalion, while E Company, equipped with two Mark VII 3-inch landing guns and under the command of Capt. Robert Underwood, constituted the regiment's 2nd Battalion. Not long after Pendleton's arrival, Butler's command, with its own pair of Mark VII landing guns, became 3rd Battalion.

The regiment carried out drills and marches until September 15, at which time Butler's battalion and a company of bluejackets off *Annapolis*, originally intended to bolster the legation guard in Managua, boarded a train for Granada. Nearing the way station at Nindiri, the train came under artillery fire from the forces of insurgent Gen. Benjamín Zeledón, entrenched on Coyotepe Hill overlooking the station. Once the train had pulled back out of range, Butler sent a message to Zeledón, requesting a conference. On the 17th, the discussion essentially failed and the emissaries of Zeledón and Mena informed Rear Adm. William Southerland, commander of the US naval expedition, that their forces would resist the American advance to Granada. Pendleton's orders, however, were to push on and capture Granada, which he and Butler accomplished at the end of September. Mena surrendered his force and its arms, and went into exile in Panama.

Doubting that the loyalist forces could capture the entrenchments atop Coyotepe and the adjoining hill, La Barranca, from which Zeledón's force still threatened the railroad, Pendleton moved his expeditionary force back there from both Managua and Granada. His battalions began arriving in the afternoon of October 2. Their four 3-inch landing guns constituted the primary artillery support, and a two-company sailor battalion off *California*, commanded by Lt. Cmdr. George Steele, reinforced the marines. The bluejackets brought along their own Mark I mod 1 3-inch field gun. Pendleton at once established communication with Zeledón, informing him that it was his intention to reopen the railroad and telegraph line between Granada and Corinto, and to keep them operating. The colonel warned him that unless his forces evacuated their positions and surrendered by the next morning, laying down their arms, the Americans would attack and clear those positions.

When, before the morning deadline, Zeledón refused, all five American guns opened fire on the positions on Coyotepe, at ranges between 1,500 and 2,500 yards, three from the southeast and two (Underwood) from the northwest. The marine gun crews, at least, had had no prior training or experience firing fieldpieces, and both officers and enlisted gunners faced great difficulty in providing effective fire. The absolute lack of communication between the assaulting infantry and supporting artillery only made things worse.

Pendleton ordered Butler to have his battalion move into assault position during the night of October 3–4, ready to jump off at 5:15 a.m. The infantry attack began only three minutes late, with 1st Battalion in the center, 3rd Battalion on the left, and the two sailor companies on the right. The expected participation of Nicaraguan federal troops did not take place. While the rest of the force provided rifle, machine gun, and shrapnel fire, the 1st Battalion's C Company made the final assault, suffering most of its casualties—four dead, several wounded—when it encountered a barbed wire entanglement not far from the summit trenches. The dead marines included Pvt. Charles Durham, who sacrificed his life cutting the barbed wire for his mates. A little more than a half-hour after the assault began, the defenders broke and ran. Soon afterwards, the forces defending La Barranca also withdrew. Zeledón died during the battle, likely at the hands of his own men.

Four months later, Captain Underwood submitted a ten-page critique of the Mark VII landing gun, its ammunition, and range-finding. After firing about 100 rounds, Underwood found the guns, breech mechanisms, recoil systems, carriages, wheels, sights, and firing attachments sound and reliable, with no jams, misfires, or other failures. On the other hand, he declared the ammunition deficient both in quality and quantity. With 100 rounds per gun, there were too few supplied for the fire support mission at hand, and the ammunition fired—shrapnel only—performed poorly in too many instances. In some cases, the rounds exploded prematurely—one just after clearing the muzzle—and many rounds failed to explode at all. The fault was found with the army's fifteen-second combination (time and percussion) fuse manufactured at the Frankford Arsenal. When the guns switched to purely percussion fuses, the results were better, but problems were still encountered, most particularly premature detonations. At that time, Frankford Arsenal was working to improve its combination fuse, and Underwood's recommendation was to adopt it when available.

Equally problematic, the gun, carriage, and limber weighed too much for easy hauling by the gun crew, obviating its value as a landing piece. Because draft animals could not readily be brought ashore, they had to be sought among the local population. If that population remained hostile, draft animals of choice such as horses and mules would be hidden and difficult to acquire, so the navy or marines would have to settle for less desirable animals. In Nicaragua, the insurgents had previously snatched up all the usable horses, forcing the marines to use oxen, with their considerable strength but slow speed. Underwood iterated a problem already known to the navy, indeed the reason for its consigning the Mark VII landing gun almost entirely to the marines: the navy-designed Mark IV limber weighed too much and carried too few rounds (seventy-two) for its substantial weight. Underwood urged a better-designed limber, lighter but carrying more rounds in more easily accessible horizontal compartments, and offering better protection to the members of gun crew serving the limber. The Model 1902 army caisson for its 3-inch field gun offered a considerable improvement: it was 500 pounds lighter and although carrying fewer rounds, it provided better storage and easier service of these rounds. Also, the army limber gave greater protection to the men carrying out ammunition service.

The Need for a Curved-Trajectory or Mountain Gun

In his 1913 report, Captain Underwood conceded the limitations of the flat trajectory of the Mark VII 3-inch landing gun, and acknowledged that a curved trajectory piece, such as a mountain howitzer, would be useful. On the other hand, he questioned such a battery on three counts. Given the low muzzle velocity common in such guns, their value in battering masonry works would be limited. Also, mountain guns and howitzers needed to be broken down and mule-packed to get them where needed in the mountains, so the difficulty in collecting a sufficient number of such animals once the landing party had gotten ashore remained an unpredictable or even doubtful factor. Training of both men and animals would also have to be taken into consideration for this special application of artillery.

Two years later, however, an apparent change of heart had taken place. In August 1915, the artillery battalion headquartered at the Marine Barracks, Annapolis, reminded the commandant that the standard 3-inch artillery piece of the corps had excessive weight, a flat trajectory, and narrow wheel treads, making it less than ideal either as a landing gun or a mountain gun. The stipulation was made that while the army's 2.95-inch mountain gun, an elderly Vickers-Maxim design from the 1890s, would not be adequate, the new army mountain howitzer, Model 1911, might be quite acceptable to equip two navy mountain batteries. Assistant Commandant Lejeune agreed with that sentiment and recommended the acquisition of two batteries of the army mountain howitzers if the funds could be obtained.

Replying to the navy's inquiry, the army Ordnance Department declared that while it had not lost interest in acquiring mountain guns, it had not adopted either of the two experimental mountain howitzer types recently considered. Neither type had been found to be satisfactory because of an overly complicated design. It informed the Bureau of Ordnance that one battery of four guns would cost about $40,000, with the ammunition for same about $10 per round. Alternatively, the Ordnance Department suggested that the navy might borrow the army mountain battery for its own testing. The bureau decided to put the decision to provide the Marine Corps with mountain artillery in abeyance until a satisfactory design became available.

Las Trencheras and Guayacanas

Beginning in the late nineteenth century, the Dominican Republic suffered a number of attempted or successful coups, on some occasions descending to armed revolution and leaving the country in a political vacuum. This dangerous void pulled in the United States, which became the island nation's financial minder. At first content with collecting customs revenues in order to pay the Dominican Republic's foreign debts and fund its government, in 1907 the United States forced it to enter into a treaty that permitted the American government to exert greater fiscal control. Initially, things went well, with the national debt of the Dominican Republic progressively reduced, but the assassination of President Ramon Caceres in November 1911 and the consequent political instability returned the nation to its former profligacy. Increasingly tactless and peremptory demands by the United States hardly improved relations between the two countries. More crises followed, and in April 1916 President Juan Jimenez faced an armed rebellion led by his minister of war, Gen. Desiderio Arias. Predictably, the Americans sent in the marines, backing Jimenez.

USS *Prairie* landed two companies of marines, commanded by Capt. Frederic Wise, in early May at Santo Domingo on the south coast. One of them, under Capt. Eugene Fortson, brought four Mark VII landing guns in the event that artillery support became necessary. Wise secured safe conduct from Arias, whose forces controlled the capital city, to move foreign residents out and rations for his men in. General Perez, the commander of the loyalist forces, requested rifles and ammunition and artillery support from Wise, who refused the small arms but agreed to support the loyalists. On the following day, however, President Jimenez resigned and the impending battle became a moot point.

In due time, the US Navy gathered a force of 400 marines and sailors under the command of Rear Adm. William Caperton. The admiral demanded that General Arias

Upon the arrival of naval transport *Prairie* at Santo Domingo, Dominican Republic, in early May 1916, the marines aboard landed four Mark VII 3-inch guns. One of them is shown emplaced near this port city on the south coast of the island, while a signalman wigwags a message. (*USMC History Division*)

disband his army and lay down his arms. Arias refused, but did assent to evacuate the capital. More marines arrived, including the 4th Marine Regiment in early June, commanded by Col. Joseph Pendleton. By that time, Arias's army had withdrawn to Santiago in the north, and it was clear that the marines would have to drive him out and defeat his forces if peace was to return to the Dominican Republic.

Pendleton's plan involved two columns marching on the city. His regiment, supported by artillery, would proceed east by road from Monte Cristi in the northwest, and the other, composed of two companies and two ships' detachments, under the overall command of Captain Fortson, would advance south on the rail line from Puerto Plata. The intention was for the two columns to converge as soon as possible and attack Santiago as a joint force. Pendleton's column of nearly 840 men, motor vehicles, and animal-drawn carts, would cut loose from its base and subsist on the rations carried.

On June 27, 1916, Pendleton's men deployed to assault the insurgents, dug in on two hills arranged in tandem at Las Trencheras. The rebel-occupied positions lay astride the route to Santiago, with the ground in front flat and brush-covered. Early in the morning, the Mark VII landing guns, commanded by Capt. Chandler Campbell and supported by the machine gun platoon, opened fire from a hill overlooking the rebel positions. The marine infantry came under heavy fire, fortunately for them not highly accurate, about 1,000 yards from those positions. The infantry, with the artillery fire lifted at the last minute, launched a bayonet attack and drove the rebels from their trenches. The latter rallied briefly on the second hill until chased off again. One marine was killed and four were wounded.

During an early encounter with the insurgents between Monte Cristi and Santiago, Dominican Republic, in mid-June 1916, the marine skirmish line fires from behind brush cover. The men are armed with the Model 1903 .30-caliber rifle and are adorned with the marine 1912 Pattern field hat. (*USMC History Division*)

Firing against the earthworks at Las Trencheras in late June 1916, a Mark VII 3-inch landing gun displays its long recoil on the Mark V field carriage. Beyond the gun, a heavy navy Mark IV limber holds the rounds of ammunition. The five members of the gun crew in view wear the army 1904 Pattern field hat, though by then replaced in most units. (*USMC History Division*)

A five-man army gun crew poses with a Hotchkiss-designed Benét-Mercié Model 1909 .30-caliber machine rifle. The gunner's left hand holds down the wrist of the stock, as light automatic weapons mounted on bipods tend to rise as they recoil. The loader (assistant gunner) is about to insert a thirty-round strip, the most common manner of ammunition feed for this machine rifle. The marines and navy used the weapon as well, in Mexico and Central America. US Army photograph.

Six days later, the marines confronted the insurgents again, in the trenches at Guayacanas. Here the vegetation grew so thickly that supporting artillery fire was not possible. But the undergrowth worked for as well as against the marines, who dragged up their Colt-Browning and Benét-Mercié machine guns to within 200 yards of the rebel trenches before they opened fire. While the main assault drove forward, one company was shifted to thwart a flank attack on the marines' supply train. Again the rebels fled the field, with the cost to the marines of one man killed and ten wounded.

Meanwhile, the second column advanced south along the railroad as intended, some riding in a makeshift military train, with a 3-inch landing gun mounted on a flatcar pushed ahead of the locomotive. At Llanos de Perez, the fire from the landing gun broke up a would-be rebel ambush. On June 29, the day after Maj. Hiram Bearss had taken over command, the marine column ran into a rebel entrenchment blocking the railroad at Altamira. Sending a company up a mountain trail to flank the opposing force, Bearss led the remainder of his men in a frontal attack, which included a dash through a 900-foot-long railroad tunnel to prevent its destruction.

The two marine columns joined on the 4th of July, a peace commission arrived the following day, assuring Pendleton that Arias would capitulate, and on the 6th the marines marched into Santiago. Although the full-scale military campaign had ended, operations against small rebel bands and outright bandits would continue for months.

References Consulted

Cole, E. K., "The Necessity to the Naval Service of an Adequate Marine Corps," US Naval Institute (hereafter USNI), *Proceedings*, 1914, 40(5): 1395–1400.

Davis, H. C., "Advance Base Training," USNI, *Proceedings*, 1911, 37(1): 95–100.

———, "Some Notes on the Training of Marines for Advance Base Work," USNI, *Proceedings*, 1911, 37(3): 837–44.

Dictionary of American Naval Fighting Ships, Naval History and Heritage Command (online). Consulted for each US naval warship and auxiliary mentioned in this chapter.

Ellsworth, H. A., *One Hundred Eighty Landings of United States Marines 1800–1934*, History and Museums Division, HQ, US Marine Corps (Washington, DC: Govt. Print. Office, 1974 [originally printed 1934]), pp. 69–70, 125–27.

Fullam, W. F., "The System of Naval Training and Discipline Required to Promote Efficiency and Attract Americans," USNI, *Proceedings*, 1890, 16(4): 473–95. Discussion following, 16(4): 495.

———, "The Organization, Training, and Discipline of the Navy Personnel as Viewed from the Ship," USNI, *Proceedings*, 1896, 22(1): 83–116. Discussion following, 22(1): 116–19. 156–57.

Fuller, B. H., "The Mission of the Marine Corps," *Marine Corps Gazette*, November 1930, 15(3): 7–8.

Fuller, S. M., and Cosmas, G. A., *Marines in the Dominican Republic 1916–1924*, History and Museums Division, HQ, US Marine Corps (Washington, DC: Govt. Print. Office, 1974), pp. 1–24.

Lejeune, J. A., "The Mobile Defense of Advance Bases by the Marine Corps," *Marine Corps Gazette*, March 1916, 1(1): 1–18.

McAulay, J. D., *Rifles of the United States Navy & Marine Corps 1866–1917* (Woonsocket, RI: Mowbray Publishing, 2017), pp. 164–65, 169, 214–19, 225–26.

McClellan, E. M., "American Marines in Nicaragua (continued)," *Marine Corps Gazette*, June 1921, 6(2): 164–75.

Millett, A. R., *Semper Fidelis: The History of the United States Marine Corps*, Chapter 10, "The Creation of the Advanced Base Force," pp. 267–86 (New York: The Free Press, rev. edit. 1991).

Niblack, A. P., "The Enlistment, Training, and Organization of Crews for Our New Ships," USNI, *Proceedings*, 1891, 17(1), 3–29.

Shetter, R. M., "Assaulting the Littorals: The Development and Evolution of a Dedicated American Amphibious Assault Force," master's thesis, Department of History, California State University, Long Beach, 2012, pp. 40–109.

Smith, J. E., *Small Arms of the World*, 10th rev. edit. (Harrisburg: Stackpole, 1973), pp. 63–65, 130-31, 181–82, 616–17.

Underwood, R. O., "Report on 3-inch gun Mark VII Landing Gun," Marine Barracks, Philadelphia, Pa., February 16, 1913, 10 pp., *in* US National Archives, RG 74, as given below.

———, "United States Marine Corps Field Artillery," *Field Artillery J.*, 1915, 5(2): 296–310.

US Militaria Forum, "Army Issued USMC Field Hats 1899 to 1917," 2020 (online).

US National Archives, Record Group 74, Records of the Bureau of Ordnance, Washington, DC.

Williams, D., "The Defense of Our New Naval Stations," USNI, *Proceedings*, 1902, 28(2): 181–94. Discussion following, 28(3): 593–95.

———, "The Fleet Landing Force," *Marine Corps Gazette*, June 1926, 11(2): 116–27.

Williford, G. M., "The American Defenses of Guam 1909–1932," *Coast Defense J.*, 2017, 31(3): 1–23.

Wingfield, T. C., and Meyen, J. E., (eds.), "Lillich on the Forcible Protection of Nationals Abroad," Appendix I. "A Chronological List of Cases Involving the Landing of United States Forces to Protect the Lives and Property of Nationals Abroad Prior to World War II," *International Law Studies* 77: 156–57, 161–62.

10

Veracruz, April 1914

Between the battle at Coyotepe/La Barranca, Nicaragua, in October 1912, and those at Las Trencheras and Guayacanas, Dominican Republic, in June 1916, a rather bloodier affray took place in April 1914 in the city of Veracruz (also spelled Vera Cruz), state of Veracruz, on the east coast of Mexico. This event would have a profound and lasting effect on the US Navy, and its willingness thereafter to put ashore landing parties of any great size.

Background

Of considerable concern to the government of the United States were the stability and tenure of Mexico's presidents and their governments. Porfirio Díaz, first ascending to the presidency in 1876, had ruled continuously from 1884 to 1911, when after a decisive military defeat, he resigned from office and went into exile in France. His elected successor, Francisco Madero, fared less well, being assassinated in 1913 along with his vice president, José Pino Suárez, in a bloody coup led by the ruthless, corrupt, and alcohol-besotted Gen. Victoriano Huerta. In 1914, the Huerta government faced opposition from the northern constitutional forces led by Venustiano Carranza. Moreover, that government had displeased the new American president, Woodrow Wilson, who refused to recognize its legitimacy.

For its part, the United States hardly remained just a concerned neighbor. Its aggressive capitalism had ensured the control or outright ownership of an enormous slice of Mexico's mineral and petroleum resources, and of its national railroad. The US ambassador, Henry Lane Wilson, had conspired in the downfall of Madero, almost certainly with the knowledge of previous President William H. Taft. Thus the ordinary Mexican had no reason to love the *Norte-Americanos*, almost certainly agreeing with Díaz's sentiment, "Poor Mexico! So far from God, so near to the United States!"

But Mexico's problems did not all stem from its northern neighbor. Chaos reigned in its national government, which successive Mexican presidents either tried to solve or took advantage of, some becoming outright dictators. Whichever, the disarray and internecine war continued, and the US Navy hovered never far away, ready to intervene in order to assist the many American citizens living as expatriates in the troubled land.

That is how the situation stood in the spring of 1914.

The Tampico Incident

The difficulty began in Tampico, a port city in the state of Tamaulipas, 240 miles north of Veracruz, and at the time under siege by the insurrectionist force commanded by Gen. Pablo González. In response, the Huertista garrison of the city, commanded by Gen. Ignacio Morelos Zaragoza, put up a stiff defense. Shortly before the fighting began in earnest, the American gunboat *Dolphin* (six 6-pounders), commanded by Lt. Cmdr. Ralph Earle, had negotiated the narrow Pánuco River between the Gulf of Mexico and the city, and had fired the requisite salute in doing so. The warship was there to show the flag and provide protection for the sizable number of American citizens who either resided or had taken refuge in Tampico or adjacent Madero. Many of them were involved in the petroleum extraction industry that flourished in this oil-rich region.

Earle, a future chief of the Bureau of Ordnance, duly arranged with a petroleum dealer to purchase a supply of canned gasoline for his ship's motor launches, and on April 9, 1914, he sent a pulling whaleboat to pick it up. The warehouse that stored the gasoline stood near Iturbide Bridge, a railroad trestle spanning the Carpentero Canal, where fighting had recently taken place and now threatened to break out again. After landing near the bridge, Asst. Paymaster Charles Copp, the officer commanding the boat detail, ordered six men ashore and left the coxswain and another crew member aboard to load the cans most efficiently. None of the nine Americans bore arms, and national ensigns flew plainly fore and aft on the whaleboat.

Shortly after they had begun loading, a Mexican officer confronted the sailors, informed them they were in an off-limits zone, and insisted they accompany him and his armed men. The situation was still salvageable, but the Mexican officer ordered the two seamen in the boat to come ashore as well. International law recognized that large or small, a vessel flying the colors of its nation remained inviolate while peacefully engaged in the waters of another nation and not in contravention of that nation's laws. The circumstances only got worse when the soldiers pointed their rifles at the men in the boat.

After an exchange at the local command post, Col. Ramón Hinojosa ordered the junior officer to return to the docks with the Americans and await developments. In a short time, the colonel ordered the US naval personnel to be released, and conveyed apologies from General Zaragoza. After he had learned of the incident, General Huerta did likewise to the US government.

Now it was the Americans' turn to err. Instead of conceding the reality that its sailors had blundered into a zone where the military situation remained undecided and edgy nerves persisted, and thereupon accepting the apologies tendered by the Mexicans, the United States treated the matter as a violation of its national honor.

For Rear Adm. Henry Mayo, commanding the US Atlantic Fleet's 4th Division, mostly anchored beyond the Tampico Bar, and perceiving that past insults had been made disproportionately against Americans, the incident constituted the last straw. On his own volition, the admiral ordered a more formal apology for the offense and a demand for the Mexicans at Tampico to render honors to the flag of the United States. Although Huerta had accepted the need for an apology, he rejected the 21-gun salute on Mexican soil as an affront, in no small measure because the United States had refused to recognize the Huerta government. Instead, he suggested a mutual honoring of each nation's flags in order to put an end to the matter. Mayo rejected the suggestion.

The US government had been fully informed of the incident, and although there were high-ranking military officers and government officeholders, including Sen. Henry Cabot Lodge, who counseled a less bellicose response, President Wilson remained a man convinced of his own rightness. He despised Huerta and that animus proved decidedly unhelpful. Without sufficient reflection, Wilson decided to back the senior naval officers on the scene and insisted that the Mexicans at Tampico comply unconditionally with the American demands. The April 18 deadline imposed by the US government came and went. Two days later, Wilson asked for congressional approval to use the American armed forces to enforce its demand that the Mexican government recognize "the rights and dignity of the United States." The resolution passed overwhelmingly in the House, but after some debate, got deferred in the Senate.

Six days before his address, the president had ordered the Atlantic Fleet, commanded by Rear Adm. Charles Badger, to Mexico's east coast in order to bolster the naval force already there. In addition to the gunboat *Dolphin*, the American scout cruiser

Protected cruiser *Des Moines*, seen here in peacetime livery, was one of the units of the US Atlantic Fleet anchored off Tampico in April 1914, doing little other than maintaining their presence. Five years later, *Des Moines* sailed to Arkhangelsk, northern Russia, to take part in the evacuation of the American naval landing force put ashore a little more than a year previous. (*Naval History and Heritage Command*)

Chester (two 5-inch/50cal guns) and the small protected cruiser *Des Moines* (ten 5-inch/50cal guns) remained anchored in the Pánuco. Other warships of the US Navy stood off the Tampico Bar, including the battleships *Connecticut* (flagship; two twin 12-inch gun turrets) and *Minnesota* (the same armament), and the minelayer (former protected cruiser) *San Francisco* (eight 5-inch/40cal guns). The transport *Hancock* (six 3-inch guns) had just arrived with the 800-man 1st Marine Regiment, Advance Base Brigade, under the command of Col. John Lejeune. Presently aboard *Dolphin* and faced with a delay in landing, the intractable Admiral Mayo chafed at the bit to put his battalions ashore.

Meanwhile, the pot began bubbling to the southeast.

Down Veracruz Way

The stern gaze of the United States had shifted indelibly from Tampico to Veracruz, a port city edged by sand hills and areas of lush greenery. The explanation for that change in focus lay in one word: *Ypiranga*. She was a 8,100-ton steamship owned by the Hamburg-America Line, and diplomatic intelligence had confirmed she was loaded with arms and munitions intended for the Huerta government. Moreover, reliable information placed her arrival at Veracruz on April 21. President Wilson resolved that the ordnance aboard *Ypiranga* would never reach the Huertistas. If an act of retribution was required because of the Mexicans' affront to American honor, then it was reasoned that the US Navy should administer a two-birds-with-one-stone stroke at the intended destination for the arms, preventing them from being landed. That the destination, Veracruz, had been invaded by Gen. Winfield Scott's army in March 1847 during the Mexican-American War—the largest amphibious landing undertaken by United States armed forces until that time (or indeed until World War II)—furnished a symbolism missed by few.

Initially, the battleships *Florida* (flagship; five twin 12-inch gun turrets) and *Utah* (the same armament) of the Atlantic Fleet's 1st Division, commanded by Rear Adm. Frank F. Fletcher—clearly forgiven for his unfortunate landing gun design—and the transport *Prairie* (3-inch guns) lay off Veracruz. The two battleships each could put ashore a battalion of bluejackets, typically three rifle companies and a heavy weapons company equipped with .30-caliber machine guns and a 3-inch landing gun. *Prairie* had aboard a battalion of the 2nd Marine Regiment, Advance Base Brigade. Although the designations of battalion and company are illusory, being considerably smaller than like-termed units of the US Army, still, a landing battalion amounted to more than 300 men, suitably armed and equipped.

The navy and marine automatic weapons in use included Colt-Brownings and Benét-Mercié Model 1909 machine rifles; the landing guns comprised mostly the Bethlehem Mark IV, with a lesser number of the Mark I mod 1, with which some battleships and other vessels were still equipped. As related in the previous chapter, because their Ehrhardt Mark VII 3-inch fieldpieces were not considered to be suitable landing guns, when the marines ashore needed field artillery support, navy gun crews provided it. Many photos exist of marine Mark VII pieces at Veracruz, but they arrived largely during the occupation period following the active fighting in the city.

At Fletcher's request, on April 20 *San Francisco* proceeded from Tampico to Veracruz, to add her own landing force to that already available. Late that day, *Utah*, in command of her executive officer, Cmdr. Hutchinson Cone, put to sea to warn *Ypiranga* of the disturbed conditions existing at Veracruz. Because a state of war did not exist between the United States and Mexico, the cargo of munitions aboard the merchantman could not legally be taken as contraband, any more than the US Navy could establish a blockade of the port. Fletcher could hope at best to delay the landing of the arms she carried.

After a flurry of radio transmissions during the night of April 20–21—enhanced in these early, range-limited wireless sets by the nocturnal changes in the ionosphere—Secretary of the Navy Josephus Daniels directed Fletcher to send a landing force ashore to secure the customs house and keep the arms carried aboard *Ypiranga* out of Huertista hands. Fletcher received this message at around 8 a.m. and immediately radioed Mayo to dispatch the scout cruiser *Chester* south to Veracruz, and then Cone to return *Utah* without delay, despite not having sighted *Ypiranga*. After consulting with Capt. Henry Huse, his chief of staff, Capt. William Rush, commanding *Florida* as well as the naval brigade at hand, Cmdr. Herman Stickney, *Prairie*'s captain, and Lt. Col. Wendell Neville, commanding the 2nd Marine Regiment, Fletcher decided to order the brigade ashore. The meteorological threat of an oncoming norther only made the timing of the landing more urgent.

Captain Huse visited the US consulate in Veracruz at once, conveying the information that the naval brigade would land and seize the customs house, cable station, and related buildings on or very near the waterfront. He requested the American consul, William Canada, to inform the other foreign consulates of the US Navy's intentions, as well as the willingness of the service to accept all foreign residents who wished to seek refuge on board the Ward liners *Mexico* and *Esperanza*, which the navy had chartered for the purpose. After the landing had commenced, the consul telephoned Gen. Gustavo Maass, the commander of the Veracruz Military District, to inform him of Fletcher's intentions and to urge him strongly not to interfere with the landing. Indeed, Maass and most of the Veracruz garrison soon withdrew to Tejeria to the west, taking all but one locomotive and some of the rolling stock, and ripping up the track behind them. Before departing, however, Maass ensured that a number of soldiers, as well as prisoners released from the military prison and provided arms, would remain behind to contest the Americans.

Consul Canada also telephoned to inform the collector of customs and the chief of police of the American intentions. They were astonished at the news he conveyed, but although they promised to cooperate, in the end they failed to do so. For his part, Fletcher dispatched junior naval officers to notify the commanding officers of the British and Spanish warships anchored at Veracruz that US forces were about to effect a landing. Such a messenger cautioned the military and naval commanding officers at insular Castle San Juan de Ulúa, a late sixteenth-century bastioned fort connected by redoubts and bridges to the mainland, against any aggressive move. The commander at hand agreed, but declared that if the Americans fired on the fort, the harbor defense guns there would reply.

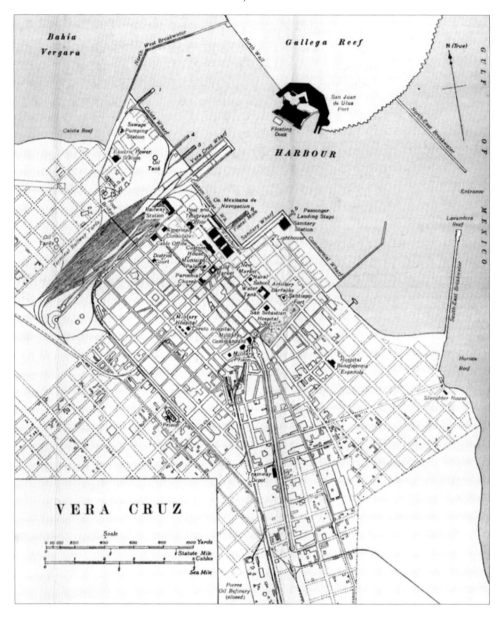

The 1918 Ordnance Survey map of Veracruz shows the harbor entrance in the eastern breakwater. US Navy warships entered here and landed their battalions at the commercial wharves to the west. Bluejackets seized the essential buildings near the docks, such as the customs house and post and telegraph office, and assisted the marine battalion in taking possession of the railroad yard along the north side of the dock area. After that part of the city had been secured, the naval landing force moved south, and within two days, resistance in the city had largely been quelled. *A Handbook of Mexico*, 1919. (*University of Texas Libraries*)

The Occupation of Veracruz, First Day

Unloading from transport *Prairie*, Maj. Randolph Berkeley's 1st Battalion, 2nd Marine Regiment, landed at the Ward Line Pier No. 4 at 11:20 a.m., April 21, and moved quickly to secure the cable office, the railroad yard, and the remaining motive power (one locomotive needing repair had been left behind). Supplementing this battalion, the fleet marine companies from *Florida* and *Utah* came ashore, and the former seized the city powerhouse to the northwest of the railroad terminal. These two marine companies landed at the same time as *Florida*'s naval battalion, commanded by Lt. Richard Wainwright III, the son and namesake of the officer commanding *Gloucester* during the Spanish-American War. The bluejackets thereupon moved southwest into the buildings near the waterfront.

Ens. Theodore Wilkinson's 2nd Company took possession of the post office, while Ens. George Lowry's 1st Company, encountering increasingly stiff resistance from municipal policemen and other armed persons, moved to seize the customs house. Lt. (j.g.) Leland Jordan's 3rd Company initially remained in reserve in the dock area, then reinforced the marines at the powerhouse. Ens. James Cresap's smaller ordnance company brought its single 3-inch landing gun to shore in a motor launch and then hauled it to the Terminal Plaza fronting the US consulate. Of the nearly 800 men in this initial stage of the landing, the marines accounted for 64 percent.

During the approach of Lowry's men to the customs house, Mexicans fired at them from the streets, rooftops, and upper-story windows. The riflemen included federal troops, men recently released from the military prison, and ordinary civilians. The bluejackets also came under machine gun fire from various places, including the east yard of the naval academy (Escuela Naval Militar). Resolute nineteen-year-old Lt. José Azueta manned a Hotchkiss machine gun at the academy. Not only did he teach at the academy, Azueta had organized a machine gun school at the artillery barracks across the street (Calle Estéban Morales). Moreover, the lieutenant was the son of Commo. Manuel Azueta, commandant of the academy. At least one American sharpshooter, perhaps two, hit the young Azueta three times while he fired his machine gun. He died in hospital on May 10, after repeatedly refusing American medical aid offered by Admiral Fletcher.

Before the noon hour, Captain Rush, who had also come ashore with the *Florida* battalion, established his headquarters in the Terminal Hotel, facing the waterfront opposite Pier 4. His brigade signal officer, Ens. Edward McDonnell, immediately set up a semaphore station on the hotel roof. That station stood in an exposed position, attracting sniper fire of an ever-increasing volume. Given the amount of firing and the mounting number of wounded naval personnel as the day wore on, it was fortunate that a field hospital had been promptly established on Pier 4. The medical officers of the British armored cruiser *Essex* (fourteen 6-inch guns) and the Spanish armored cruiser *Emperador Carlos V* (two single 280mm gun turrets), anchored in the inner harbor, subsequently volunteered their services, which were accepted with gratitude. Fletcher radioed Mayo at Tampico, informing him that he needed the hospital ship *Solace* to depart for Veracruz at once.

Her stern seen beyond the Spanish armored cruiser *Emperador Carlos V*, the US naval transport *Prairie* disgorges strings of pulling boats towed by steam launches. The boats are carrying the 1st Battalion, 2nd Marine Regiment, under the command of Maj. Randolph Berkeley, which landed on the Veracruz waterfront not long after 11 a.m. on April 21, 1914. Later in the day, *Prairie* exchanged fire with Mexican pockets of resistance. (*USMC History Division*)

The American naval landing gun seeing the most common use during the fight for Veracruz was the Bethlehem Steel Mark IV 3-inch piece, issued exclusively to battleships. The bluejackets hauling this one and its Mark III limber reveal the second most common function of such landing guns: the conveyance of blanket rolls, food, and other gear not necessarily related to the ordnance at hand. (*Naval History and Heritage Command*)

Veracruz police and Mexican federal personnel fire their Mauser rifles at American bluejackets during the early phase of the US landing, April 21, 1914. One of the municipal policemen, Aurelio Monffort, would be the first Mexican to die at the hands of the Americans. Photograph by Walter P. Hadsell. (*Library of Congress*)

As the action developed, lookouts on the American battleships spotted smoke on the horizon, which turned out to be that of *Ypiranga*. Upon arrival in the outer harbor a little past noon, her captain was informed that he could not leave the port with the arms still on board. With little choice, he decided to anchor in the outer harbor in sight of the battleships.

With the reports of large numbers of federal troops firing on his men, at 12:30 p.m. Fletcher ordered in *Utah*'s four-company bluejacket battalion, commanded by Lt. Cmdr. Guy Castle. It arrived, however, without the ship's landing gun. The marines under Lt. Col. Neville had also begun to take increased firing, and had deployed their Colt machine guns accordingly. Skirmishers on the ground, covered by sharpshooters on the rooftops, cleared the houses from which the worst firing originated. Following this effective sortie, Rush ordered the marines back to their positions near the warehouses on the waterfront.

But in fact the Mexican snipers had dispersed seemingly everywhere, and continued to inflict casualties on the Americans. McDonnell's signal section on the roof of the Terminal Hotel remained particularly exposed, with the intensity of the fire worsening as the afternoon progressed. Rush, himself wounded, sent a squad of marines to the roof in order to provide counterfire. Pvt. Daniel Haggerty, the first marine to reach the signal station, immediately received a mortal wound; Elec3c Edward Gisburne, who attempted to drag Haggerty to safety, had his knee shattered for his bravery. Rush had no choice but to remove the marines, although the signalmen remained steadfastly at their post, courageously wig-wagging messages to Admiral Fletcher. Another signalman suffered a wound before the long day was done.

To the southeast, Ensign Lowry's rifle company continued the fierce battle with the snipers. His machine gun, commanded by a gunner's mate named Wertman and entirely in the open, attempted to seek out the hostile riflemen in upper rooms and on roofs, but one by one, Wertman's crew suffered wounds. Two of the men were shot by riflemen in the tower of the Faro Benito Juárez. Lowry dispatched his runner to let Rush know he needed support, and the brigade commander at once ordered Ensign Cresap to shift his Mark IV 3-inch landing gun and shell the lighthouse. The ensign moved his gun to the front of the American consulate, and with ease, his gun crew put six shells into the tower, blowing away large chunks and ending the firing from that eminence.

Meanwhile, *Utah*'s 4th Company, commanded by Ens. Paul Foster and put ashore an hour after the order had reached the battleship, double-timed forward to relieve academy classmate Lowry's company. The *Florida* company had by now been reduced in effectiveness because of its numerous casualties. Rush thereupon directed Cresap to move his gun to the northwest, near the railroad warehouses, in order to support the marines. At that juncture, it remained the only landing gun brought on shore.

The landing area at Pier 4, including the field hospital set up there, proved not immune to hostile fire. Captain Rush instructed the beachmaster, Chief Boatswain John McCloy, to determine the main source of this annoying fire and do something about it. McCloy, who had won the Medal of Honor for his gallantry at Tientsin in 1900, thereupon led three steam launches belonging to the battleships at their best speed toward the waterfront opposite the naval academy. At that point, he ordered a salvo from the long 1-pounder RF guns mounted singly on the bows. Although their 37mm rounds could do little against the academy's thick masonry walls, reinforced within by mattresses and furniture placed by the Mexican naval cadets, good shooting could put the projectiles through the windows.

McCloy's shells, however, only triggered a response in kind from Mexican 1-pounders, which the cadets and gunners from the barracks nearby had dragged into place to fire out of those same windows. Rifle fire from numerous points of resistance, including vessels at dockside, struck the steam launches. Three successive 1-pound shots from the academy disabled McCloy's launch. The hostile fire seriously wounded the chief boatswain and his gun loader, and fatally wounded his gun pointer. As the launches withdrew, *Prairie*'s 3-inch deck guns found the range and pounded the academy. Within its walls, sixteen-year-old Naval Cadet Virgilio Uribe was killed, and his classmates were warned henceforth to stay away from the windows. *Prairie*'s fire silenced the Mexican 1-pounders, for the moment at least, and the cadets withdrew that night to Tejeria. For his part, after receiving medical attention, Chief Boatswain McCloy remained on duty until at last ordered to evacuate. For his heroism, the chief boatswain won a second Medal of Honor and subsequently a commission.

Ensign Foster's rifle company, which had relieved that of Ensign Lowry, cleared the new customs warehouse of its considerable number of Mexican riflemen. Once inside, they discovered an enormous trove of goods, which then gave Foster an inspiration. Quickly loading barrels, crates, and other stores onto the numerous hand trucks found, they soon fashioned a mobile barricade used to effect in clearing the streets nearby.

Early in the fight for the city, a sailor infantry unit, armed with Model 1903 .30-caliber rifles and two tripod-mounted Model 1895 .30-caliber Colt-Browning machine guns, pauses at a street intersection, keeping wary eyes peeled for snipers. At least one hot bluejacket has needed to take a breather. Photograph by Walter P. Hadsell. (*Naval History and Heritage Command*)

While a good many riflemen are resting in the shelter of a building, a 3-inch landing gun section moves at the double to some hotspot farther along the street. At this stage of the fighting, the vast majority of sailors are still wearing their blues, with a few clad in coffee-stained whites. Note the fixed bayonets. Photograph by Walter P. Hadsell. (*Library of Congress*)

A member of this Mark IV 3-inch landing gun crew, evident from the pistol he carries (the holster indicates a Model 1911 .45-caliber piece), draws a bead with a Model 1903 .30-caliber bolt-action rifle. The trail wheel has been removed from the landing gun for firing. Sailor uniforms evident are a mixture of blues and dyed whites. Photograph by Walter P. Hadsell. (*Courtesy of Karl Schmidt*)

Three steam launches armed with long 1-pounder guns engaged guns of similar size firing from the windows of the naval academy. The hostile fire seriously wounded Chief Boatswain John McCloy, the flotilla commander, and two of his gunners, one fatally. His vessel disabled, McCloy ordered the launches to withdraw. In the launch shown, note the Model 1903 rifles lying adjacent to the junior officer, nicely immaculate in his white uniform. (*National Archives and Records Administration*)

Although Lowry's casualty-depleted company was to have gone into reserve, the exigency of the moment required that this unit be employed to clear the waterfront of its numerous snipers. These men were taking potshots at the bluejackets from within shacks and behind boxes and bales at dockside, and from the deck of the Mexican Navigation Company's steamship *Sonora*, moored to the seawall near the Fiscal Wharf. As skirmishers, the company quickly cleared the dock area of these pesky riflemen. When he reached the dock office, Lowry encountered *Sonora*'s captain and her first and second mates. The young ensign pulled his sidearm and flatly told the captain that if he did not order his crew to cease firing, he would shoot him out of hand. The captain complied with alacrity, his crew disembarked, and Lowry's men thoroughly searched the vessel. After the removal of all the arms found, the bluejackets permitted the crew to reboard their ship. Thereafter, *Florida*'s 1st Company returned to the Terminal Plaza and rightfully went into reserve.

At around 2 p.m., Admiral Fletcher shifted his flag from *Florida* in the outer harbor to *Prairie* inside the breakwater. The admiral's being closer to the action, however, had little effect on the first day's fighting. It remained as it had been: a confusing array of independent fights, in which streets were cleared block by block, house by house, and floor by floor. Rooftop shooting matches between men at odds were not uncommon. Such battles had ever been the realm of the junior officer, proving (or not) his mettle, physical courage, initiative, and leadership ability. In those fights on or near the waterfront of Veracruz, the junior officers of the US Navy's Atlantic Fleet shone, as did virtually every bluejacket and marine in the landing force. Their Mexican opponents largely fought with courage and tenaciousness, defending their city against a foreign invader who, they fervently believed, had no right to be there under arms.

The firing diminished by mid-afternoon, finally becoming desultory, and the senior American naval officers took stock of the situation. In addition to the customs house and post office, and the American consulate, the bluejackets and marines had taken possession of the telegraph station, cable office, power and light plant, water pumping station, and railroad terminal and trackage extending as far southwest as the locomotive roundhouse. They had secured most of the wharf area. Sporadic fire still came, however, from vessels at dockside and from buildings fronting or close to the waterfront: the naval academy, New Market, Municipal Palace, the steeple of Nuestra Señora de la Asunción, and the roofs of various hotels and other premises surrounding the Plaza Constitución, which lay directly behind the Municipal Palace. Moreover, *Ypiranga*'s arms had been kept out of the hands of the Huertistas, at least for the moment.

The navy had no wish for the bloody fight necessary to take the entire city, or to shell the city in achieving that end, which would inflict terrible casualties on the city's inhabitants. Fletcher knew that General Maass had withdrawn with most of his garrison and offered no immediate threat. The admiral sent his chief of staff, Capt. Henry Huse, ashore to confer again with US Consul William Canada. After some conversation, they agreed that an emissary under a flag of truce should attempt to seek out either a senior military officer or civil official who could arrange for an armistice, with that to be accomplished as soon as possible. The first man dispatched under a white flag never returned and later was reported to have been shot. A second messenger went out, and though he frequently attracted sniper fire, he returned safely. The Americans never found an official to make the arrangements they wished, likely because Mexican law prohibited any degree of collusion with an invading foreign power.

Canada and Huse also concurred that the time was right for the refugees sheltering in the consulate to go to the Ward steamer *Esperanza*, which offered a great deal more safety. Under constant fire and audibly protesting, they scampered to the Sanitary Wharf, where boat crews took them out to *Esperanza*, herself subjected to occasional sniper rounds. The steamer lay anchored under the temporary command of Admiral Fletcher's stalwart nephew, Lt. Frank J. Fletcher. Nicknamed "Fighting Jack," he would become a notable flag officer during World War II. Of the 350 refugees eventually taking shelter aboard *Esperanza*, none suffered harm of any kind.

That remained the status quo as the landing force settled into defense positions for the night, the marines even loopholing the railroad warehouses they had previously taken. During the hours of darkness, sporadic shots could be heard, sometimes spiking to a crescendo, and occasionally including the clatter of one or two machine guns. Shipboard searchlights periodically swept the streets near the waterfront, but detected no more than small bands of men moving furtively from one place to another. Fletcher remained apprehensive, however, and he radioed Admiral Badger, heading toward Veracruz with the main division of the Atlantic Fleet, to be prepared to land his own battalions.

The Occupation of Veracruz, Second Day

San Francisco arrived from Tampico at 8:30 p.m. Following a conference between Fletcher and her captain, Cmdr. William Harrison, and a light meal of sandwiches aboard, the minelayer's landing force shoved off just past midnight. Joining the defensive line to the left of the *Utah* battalion, they began to build a barricade for themselves. Also proceeding south from Tampico, but at higher speed, the scout cruiser *Chester* arrived off the breakwater at midnight. Her commanding officer, Cmdr. William Moffett, a brash and opinionated aristocrat hailing from the American South, found the harbor lights of Veracruz switched off. Fletcher signaled Moffett to use his own discretion, including the option to remain in the outer harbor until dawn. Instead, in a daring show of his excellent seamanship, Moffett ran *Chester* straight through the narrow opening of the breakwater and deftly dropped his hook off the Sanitary Wharf. Even Fletcher was impressed.

Chester had aboard two seaman companies and a marine company; she also carried a marine battalion commander, the inimitable Maj. Smedley Butler. They came ashore at 3 a.m., with their detachments sent to the powerhouse and locomotive roundhouse. A little before the *Chester* force landed, the sailors on the dock belatedly unloaded *Utah*'s 3-inch landing gun.

Not long after these events transpired, Admiral Badger's five battleships, flagship *Arkansas* (six twin 12-inch gun turrets), along with *South Carolina* (four twin 12-inch gun turrets), *Vermont, New Hampshire,* and *New Jersey* (each having two twin 12-inch gun turrets), made their appearance. Each battleship fielded a rifle battalion and a 3-inch landing gun. Because Fletcher remained more familiar with the situation at hand, Badger directed him to retain command of the naval brigade on shore. After discussion, the two admirals agreed to reinforce Lt. Cmdr. Allen Buchanan's 1st Seaman Regiment, initially the rifle battalions off *Florida* and *Utah*, with that from *Arkansas*. Her landing gun would pair up with the piece just disembarked from *Utah*. The remaining four battalions, some 1,200 riflemen, constituted the 2nd Seaman Regiment, led by Capt. Edwin Anderson, the commanding officer of *New Hampshire*. The battleships' four landing guns, plus six machine guns, were organized into an independent artillery battalion, commanded by Lt. John Grady. The fleet marines from the four warships formed a battalion, with Maj. Albertus Catlin in command. Although later, when additional marines became available, it formed the basis of the 3rd Marine Regiment, for the moment it reinforced Neville's 2nd Marine Regiment.

Fletcher continued his efforts to contact senior Mexican authorities during the morning of April 22, in order to arrange a cessation of fire, but he and Badger agreed that if it hadn't happened by 7:45 a.m., the naval brigade would be forced to take the city. After that deadline, the 1st Seaman Regiment—now consisting of the battalions off *Florida, Utah, Arkansas, San Francisco,* and *Chester*, with the new battalions landing at Pier 4—would advance southwest, then southeast along the axis of Avenida Zaragoza, clearing the entire area between Avenida Independencia and the waterfront. This advance was to be supported by the landing guns from the four battleships named.

Although the circumstances of this posed shot remain uncertain, note that the sailors flanking the Mark I mod 1 3-inch field gun are dressed as bluejackets, typical of the uniforms worn on the first day. From his naval uniform and leggings, one may assume that the man in the saddle is one of the landing force. Associated Press photograph. (*NavSource Naval History Photo Archives*)

The 2nd Seaman Regiment forming up on the Fiscal Wharf would advance on the 1st Regiment's left, to its southeast, and then bear south, to the sand hills. The 2nd Marine Regiment was to clear the southwest part of the city, from Calle Morelos all the way to the sand hills.

Because the heavy blue uniforms worn by the traditionally attired bluejackets coming ashore in the first landing had turned out to be overly hot, the seamen in the second landing put on whites colored by a variety of dyes—coffee, iodine, or iron oxide (rust)—to make themselves less conspicuous to hostile riflemen. Needless to write, such uniforms were anything but smart and military in appearance. The notion of providing seamen in landing parties with khaki uniforms similar to those worn by the marines arose periodically during those years, but little ever came of it.

As the sun rose over the Gulf of Mexico, additional ships arrived at different times: the battleships *Michigan* (four twin 12-inch gun turrets), *Minnesota*, and *Louisiana* (each having two twin 12-inch gun turrets); the transport *Hancock* carrying Col. John Lejeune's 1st Marine Regiment, Advance Base Brigade; the collier *Orion* (four 3-inch guns); and last of all, the badly needed hospital ship *Solace*. At about that time, the firing from the Mexican riflemen resumed in volume, and in response, the machine guns of the *Utah* and *San Francisco* battalions sheltered behind their barricades opened up on the hotels and other structures near the waterfront. As a final measure, the bluejackets resorted to volley fire to clear those buildings.

In the vicinity of the Fiscal Wharf and the Faro Venustiano Carranza—commonly called the New Lighthouse—the 2nd Seaman Regiment formed up. Its commanding officer, Capt. Edwin Anderson, was about to lead the units at the head of his formation into disaster. Although Captain Rush had cautioned the senior officers newly on the scene about the strong possibility of ambush and the need for advance guards and flankers, for some reason Anderson ignored those injunctions. Instead, he marched his men in close order as if they were on parade. Lt. T. Gordon Ellyson, an officer on *South Carolina*—he was also the first US naval aviator—pleaded with the regimental commander to push out a point element. When Anderson refused, Ellyson urged him at the very least to open up the ranks into sensible skirmish order. Again, Anderson's response was a decided no. This obstinate stupidity would unnecessarily cost the lives of a number of the regiment.

The leading battalion, from Anderson's own *New Hampshire*, passed in front of the New Market, naval academy, and artillery barracks. In his best parade ground voice from his days at the US Naval Academy in the early 1880s, and duly echoed by his unit commanders, Anderson ordered column right onto Calle Francisco Canal, running from northeast to southwest. With predictable certainty, intense fire from rifles, machine guns, and 37mm (1-pounder) guns hit the leading ranks of the column. Stunned and confused, and amidst the screams of the wounded and dying, those elements froze, came apart, and then broke, their seamen streaming to the rear.

Although most the men ran mindlessly back toward the waterfront, others kept their heads and with great courage ran into the fire zone to drag wounded men to safety. Some of these heroes had to suspend saving a wounded man to deliver return fire. Junior officers pulled their pistols and attempted with varying degrees of success to restore

Above and below: Capt. Edwin Anderson (1860–1933), commanding both the Connecticut-class battleship *New Hampshire* and the 2nd Seaman Regiment ashore, foolishly brought his men forward into a hot fire zone in marching formation, thus causing them to suffer heavy casualties. Instead of being given a justifiable court martial, he was awarded a Medal of Honor for the physical bravery he displayed in rallying his men in their shattered ranks. Anderson subsequently rose to the rank of admiral and commanded the US Asiatic Fleet. (*Naval History and Heritage Command* and *National Archives and Records Administration*)

order. Ens. George Dale managed to corral a machine gun crew and ordered the men to set up their automatic gun and return the hostile fire. Other sailors were firing wildly without effect, so the commissioned and petty officers closest at hand had first to effect a ceasefire, before they could initiate a more directed fire against their adversaries.

From *Prairie*'s bridge, Fletcher could clearly see the awful debacle that had developed. At once he conveyed permission to provide supporting fire from the warships in the inner harbor, all of which had their gun crews standing at the ready. *San Francisco* and *Chester* began firing their 5-inch and 3-inch deck guns at the naval academy and the New Market. Most of the rounds were on target, inflicting great damage, and in short order, the firing from these points evaporated.

Meanwhile, the 2nd Regiment began to rally and regroup. Although Captain Anderson had been guilty of impetuous and irresponsible leadership on the march in, his impressive history of physical courage asserted itself. He and Lt. Cmdr. Rufus Johnston, his regimental adjutant—also his executive officer aboard *New Hampshire* and a man with steely nerves equal to Anderson's—pulled the formation together and got it started again clearing the city blocks assigned, this time in skirmish order and supported by the landing guns.

On the other side of the waterfront, Lieutenant Castle's *Utah* battalion began its advance southwest from the barricades at the customs house and railroad warehouses, with their Colt-Browning machine guns and Benét-Mercié machine rifles in support. Suffering no casualties, various sections of the battalion secured the Plaza Constitución and the buildings surrounding it: the Municipal Palace, the Hotel Diligencias, and the church of Nuestra Señora de la Asunción. Following those seizures, the *Utah* battalion went into reserve in the Terminal Plaza, while the *Arkansas* battalion continued the advance southeast on Avenida Zaragoza, encountering increasing resistance and needing the support of its machine guns and 3-inch landing guns, the latter more often than not firing over open sights. Farther south, the seamen met even more intense fire from federal troops in the military barracks on Calle Ocampo, and had to pull back while *Chester* shelled the area, including the Instituto Veracruzano.

Several vessels moored at the waterfront harbored snipers, including once again *Sonora*, and had to be dealt with by clearing parties. Farther northwest, Neville's marines were heavily involved in house-to-house fighting. As usual, Maj. Smedley Butler found himself "fighting like hell." By mid-morning, the leathernecks had passed Calle Benito Juárez, and moved on to clear the block to Calle Miguel Lerdo. Butler would win the first of his two Medals of Honor in the street fighting there (his second was awarded a mere nineteen months later, for valor during an action in the mountains of Haiti).

To the south, the 2nd Seaman Regiment found itself in identical fighting, now supported by six landing guns, as those from *Florida* and *Utah* had been added to Grady's battery of four such guns. These guns shelled several houses where resistance proved particularly fierce. The good shooting of the gun crew directed by Ens. R. B. Hammes, off *South Carolina*, drew praise from Lieutenant Ellyson, the ship's field artillery commander. One of their shells hit the ammonia pipes in the ice-making plant, and the released gas drove out the Mexican riflemen more effectively than bullets. The regiment thereafter occupied the military headquarters and adjacent barracks, located on Calle Ocampo. Most of the regiment, including part of its field artillery battery, advanced south along the railroad to the Waters-Pierce oil refinery, with one company continuing on to the sand hills. As the afternoon shadows lengthened, the sailors who remained at the southernmost end of the palm-tree-lined Paseo de los Cocos halted there for the day. The long, curved concrete benches encircling the alameda's terminal plaza offered protection for the riflemen, landing guns and tripod-mounted machine guns. Watching over them was the Estatua de la Libertad, Veracruz's Statue of Liberty.

More unrest and uncontrolled firing occurred that night than the previous, but this time largely by the Americans. A lively exchange of fire took place between an outpost occupied by *New Hampshire* riflemen and the newly landed battalion from *Minnesota* (lacking its 3-inch landing gun because of an otherwise-engaged ship's crane). At the time, ironically, the *Minnesota* battalion was marching to support Grady's artillery. Other than this friendly fire incident, *Minnesota*'s battalion and those off *Michigan* and *Louisiana*—the eighth, ninth, and tenth battleships sent to Veracruz—as well as the 1st Marine Regiment, Advance Base Brigade, aboard the transport *Hancock* and commanded by Col. John Lejeune, arrived too late to see action.

When Mexican officials of sufficient rank still failed to materialize, on April 26 Admiral Fletcher declared martial law. Two days later, the transports carrying the US Army's 5th Brigade, reinforced, arrived at Veracruz, and Gen. Frederick Funston, hero of the fighting in the Philippines, came ashore. On April 30, the soldiers landed, and following a handing-over ceremony and pass in review, the bluejackets and fleet marines

Battleship *Vermont*'s sailor riflemen, Mark IV 3-inch landing gun crewmen, and a single marine (on the far right) peer intently at the sand hills to their outpost front. The difference in shades of the sailors' dyed white uniforms is a function of the type of dye used and the length of time of immersion. One crewman wields a Model 1911 .45-caliber semiautomatic pistol, while another stands by with a 3-inch round to pass on to the loader. Two spent cartridge cases indicate that the landing gun has previously been fired. (*Naval History and Heritage Command*)

Although they will not see action, members of the battleship *Michigan*'s seamen battalion engage in practice firing while still at sea. Most of the sailors fire their Model 1903 .30-caliber bolt-action rifles, with a Mark IV 3-inch landing gun, an ammunition limber, and a tripod-mounted Colt-Browning .30-caliber machine gun also in view. (*Library of Congress*)

With the addition of Col. Littleton Waller's marine provisional brigade to the US Army's occupation force remaining in Veracruz, the leathernecks brought ashore their own supporting field artillery. A battery of four Ehrhardt-designed Mark VII 3-inch landing guns stands in an artillery park, watched over by a marine sentry. An observation and/or signal post has been established in the adjacent tower. (*USMC History Division*)

began their departure. After much discussion and not a little argument, a three-regiment provisional brigade of marines commanded by Col. Littleton Waller remained behind under army command as part of the occupation force. In November 1914, the Americans finally cleared out, no more beloved by the Mexicans than when they had arrived.

Aftermath

By the time the first landing force of marines and bluejackets shoved off from their respective warships, the Tampico Incident had been all but forgotten. *Ypiranga* departed Veracruz and in due course steamed to Puerto México (in 1936, the name of the port reverted to Coatzacoalcos). She arrived there on May 27, 1914, and with little fanfare unloaded the arms and munitions, which made their way directly to the Huertistas. One is thus understandably puzzled as to what precisely the American occupation of Veracruz managed to achieve, above and beyond the substantial bloodshed and destruction, and decades of ill will.

The United States had overreacted badly, imposed its will arbitrarily, used force excessively and unnecessarily, and taken Mexican lives—too many lives—needlessly. Moreover, it had appeared petulant in the eyes of the international community. There would be no more landings of this scale. Veracruz was at once the high- and low-water marks of American naval landings in force.

Nonetheless, changes to landing ordnance were made as a result of the Veracruz incursion. The experience reminded the marines of the excessive combined weight of their Mark VII 3-inch landing gun, Mark V field carriage, and Mark IV ammunition limber. Using the army limber as a template, the marines made modifications to its design and the Bureau of Ordnance began the gradual replacement of the overly heavy navy Mark IV limber, with its awkward ammunition service. The bureau soon designated the army type limbers the Marks V and VI. To make possible indirect laying with a forward aiming point, the marines strongly urged that the top shield on the Mark V carriage be cut through horizontally below the tops of the wheels—in order to protect the gun if it turned over—and then hinged for dropping. After a little delay, the navy thought it could make a dozen such conversions with the money at hand.

References Consulted

Action reports by the brigade adjutant, and various battalion and detachment commanders, US naval forces ashore at Veracruz, April 21–30, 1914, *in* US National Archives, RG 45, as given below.

Author not specified, "Veracruz: Estatua a la Libertad en la alameda Díaz Mirón," *Veracruz Antigua*, Sept. 7, 2011 (online).

Canada, W. W., US Consul at Veracruz, "Occupation of the Port of Veracruz by the American Forces, April 21st and 22nd, 1914," dated August 11, 1914, 15 pp., *in* US National Archives, RG 45, as given below.

Dictionary of American Naval Fighting Ships, Naval History and Heritage Command (online). Consulted for each US naval warship and auxiliary mentioned in this chapter.

Ellsworth, H. A., *One Hundred Eighty Landings by United States Marines 1800–1934*, History and Museums Division, HQ, US Marine Corps (Washington, DC: Govt. Print. Off., 1974 [originally printed 1934]), pp. 116–18.

Fletcher, F. F., (Rear Adm., USN), Comm. US Naval Forces on Shore, "Seizure and Occupation of Vera Cruz, Mexico, April 21st–April 30th, 1914," dated May 13, 1914, pp. 1–30, *in* US National Archives, RG 45, as given below.

McAulay, J. D., *Rifles of the United States Navy & Marine Corps 1866–1917* (Woonsocket, RI: Mowbray Publishing, 2017), pp. 228–31.

McCandlish, B. N., "A Khaki Uniform Needed for the Bluejacket Landing Force," US Naval Institute, *Proceedings*, 1913, 39(2): 146.

Quirk, R. E., *An Affair of Honor: Woodrow Wilson and the Occupation of Vera Cruz* (New York: McGraw-Hill, 1964).

Sweetman, J., *The Landing at Veracruz: 1914* (Annapolis: Naval Institute Press, 1968).

US National Archives, Record Group 45, Naval Records Collection of the Office of Naval Records and Library, Washington, DC.

11

World War I and Intervention in Russia

The Fighting US Navy

On July 19, 1918, south of Long Island, New York, the American armored cruiser *San Diego* (previously named *California*), with twin 8-inch guns in a single turret fore and aft as main armament, and a displacement of around 15,000 long tons when fully loaded, steamed westward at a moderate speed. She was zig-zagging, with lookouts in place, including one aloft in the crow's nest, as she headed for the Port of New York to meet a convoy bound for France. In late morning, a large explosion rent her port side, stopping her dead in the water and causing great flooding. The wireless proved inoperative, so Capt. Harley Christy sent his gunnery officer and a boat crew ashore to notify the nearest naval station and summon aid. Less than a half-hour after the explosion and long before rescue vessels arrived, however, *San Diego* capsized and sank. Mercifully, only six men died in the sinking, including the lookout trapped in the enclosed crow's nest. Hers was the only loss of a major warship suffered by the US Navy during its nineteen-month involvement in World War I. The explosion may be attributed in all likelihood to a contact mine laid by German submarine *U-156*, which had been operating in those waters, and as she returned home, would be lost to a similar device in the North Sea [Mine] Barrage laid by the US Navy.

Given the rapidity in which *San Diego* went down, she took most of everything stowed aboard with her. Among that gear remained Mark I mod 1 3-inch landing gun No. 381, all but certainly still on the ship despite her having been involved in rigorous convoy duty for the past several months. Ironically, the loss of this piece was the only role a US 3-inch naval landing gun would play during the war, at least in American hands.

Near war's end, however, the 530-man Naval Railway Batteries went ashore to man five 14-inch railway guns (originally intended for battleships). A smoothly working collaboration between the Naval Gun Factory and the Baldwin Locomotive Works of Eddystone, Pennsylvania, got the railway carriages designed, the units built, and the completed guns sent Over There in record time. The personnel consisted mostly of reservists and naval militiamen; except for the commanding officer, Rear Adm. Charles

Plunkett, USN, the officers had recently been commissioned into the US Navy Reserve Force (USNRF). They wore army uniforms displaying commissioned rank devices or naval sleeve ratings. Between September and November 1918 on the Western Front, the guns effectively shelled German troop concentrations and railway marshaling yards.

The war had begun in August 1914, more than two and a half years before the United States became a belligerent. During the unsettled conditions in prewar Europe, Greece decided it needed a couple of new battleships to counter the naval threat by its long and bitter rival, Turkey (Greece had become independent of the repressive Ottoman empire eighty years before). As it happened, the United States had two battleships that it could afford to part with. The last eight American pre-dreadnoughts consisted of six 16,000-ton battleships of the Connecticut class (two twin 12-inch gun turrets), all of which would see service in Mexican waters in 1914, and two 13,000-ton battleships of the Mississippi class (same main armament). Because of the several reductions in their design, these two warships, *Mississippi* and *Idaho*, possessed less power and carried a smaller amount of coal, and had a shorter length, than the contemporary Connecticuts. As a result, they were slower, shorter-legged, and had indifferent seakeeping qualities in the open ocean, reminiscent of the thoroughly outmoded coastal battleships of the 1890s. They were, however, rather better suited for service in the Aegean, Ionian, and Mediterranean Seas surrounding Greece. Through an intermediary, Greece purchased the two warships in July 1914, just before World War I began, renaming them *Kilkis* and *Lemnos* respectively.

As part of the sale, their 3-inch landing guns, Mark I mod 1 Nos. 397 and 412, mounted on Mark II field carriages Nos. 154 and 152, also went to the Greek Navy. Almost certainly, the Greeks faced the same reality the Americans had some years before with the Elswick 12-pounder landing guns aboard the New Orleans-class protected cruisers, i.e. the need to provide an entirely alien family of ammunition for a very small number of artillery pieces. The American landing guns thus may have experienced a short history in Greek service aboard these ex-Mississippi-class battleships. The ships themselves would be lost to German dive bombers in April 1941.

Despite the strong intention otherwise on the part of the national government, by 1916, the concern in both the army and the navy was that the United States would be drawn unprepared into the world conflict. Attention in the navy was thus concentrated on the state of readiness of the fleet. Landing parties were not normally a wartime function, and the navy remained all the more determined that the number of men detached ashore must never cause the number kept aboard to fall below that sufficient to handle the ship and man the guns. The Atlantic Fleet did, however, give some thought to its landing force, suggesting that entrenching tools and pioneer equipment be furnished for that function. On the weaponry side, there existed widespread agreement that the number of machine guns attached to landing parties needed increasing.

Also in 1916, the navy effectively terminated support for the state naval militias, although they were given a stay of execution for the duration of the war in order to facilitate the transfer of their unit personnel into federal service. By the teen years, many of the obsolete but still useful Mark I mod 1 3-inch landing guns had been issued to those militias. With war's end, the naval militia was no more, and the navy would thereafter depend upon the newly created fleet reserve.

In August 1918, apprentice seamen assigned to the battleships *Virginia* and *Georgia*, but still in training ashore, struggle through the manual of arms. Most of the recruits have failed to heed their drill instructor's admonition, "Don't move your head!" The lieutenant (junior grade) in the foreground appears to be greatly displeased with the results so far. International Film Service. (*National Archives and Records Administration*)

Outside its Seattle Armory in May 1915, the 1st Artillery Section (of two such sections), 1st Division, Washington Naval Militia, forms up astride its Mark I mod 1 3-inch field gun. Although the end of World War I also saw the termination of the states' naval militias, such units enjoyed ample support in the years just preceding the war. The Washington Naval Militia was not created until 1910 and thus had a short life. (*Washington National Guard Museum*)

Following the occupation of Veracruz, the US Marine Corps had requested the replacement of the navy Mark IV ammunition limber with the army type, to be designated the Marks V and VI, and the cutting through and hinging of the top shield on the Mark V carriage to permit indirect laying with a forward aiming point. Some progress had been made with the former intention, but all three types of limber were seen well into the postwar period. Although an estimate of twelve hinged shields had been made in 1915 with the money at hand, perhaps only ten such conversions were completed before American engagement in the war.

Landing Gun Cartridges for Firing Y-Gun Projectors

Upon the entrance of the United States into the war in April 1917, its navy fully realized that the warships of the Imperial German Navy it would face in greatest number were submarines—the enemy navy's deadly U-boats. In developing an antisubmarine weapon to arm destroyers, submarine chasers, and converted yachts for use against submerged submarines, the US Navy followed the lead of the Royal Navy and opted for the depth-charge projector. In American naval service it became the Y-gun from its pair of Y-shaped firing barrels. In that development, the Bureau of Ordnance appreciated that a ready means of projecting both depth charges simultaneously lay in the 3-inch landing gun cartridge case filled with discrete measures of black powder to achieve different ranges: 1 pound for 50 yards; 1.25 pounds for 66 yards; and 1.5 pounds for 80 yards. Thus the depth charges could be fired in a pattern covering a wider danger zone to the enemy submersible.

With most of the wartime US Navy in the Atlantic Ocean, a number of its east coast ammunition depots were enlarged and improved (relatively little was done to those on the Pacific coast). In that period of time, nearly 18,000 depth-charge cartridges were prepared at the Iona Island Naval Ammunition Depot on the Hudson River, 13,000 of them holding the standard charge. Downriver in New York harbor, the ammunition depot in old Fort Lafayette delivered almost 9,400 Y-gun charges to US fleet units. The Iona Island facility also produced nearly 31,000 fixed 3-inch rounds to be fired from its 3-inch/23-caliber landing guns and 3-inch/23-caliber deck guns.

The Navy's Final 3-inch Landing Gun: The Mark XI

Long conscious of the navy's need for a modern landing gun, in October 1916 the Bureau of Ordnance put forth the specifications, in general, of the 3-inch gun it wanted. The memorandum addressed to the Washington Navy Yard approximated twenty-five landing guns to arm both recently constructed battleships and those to be built in the near future. First, it included a reminder of the limitations of the two most recent landing guns. The Bethlehem-made Mark IV lacked a bullet-proof shield, mounted sights of insufficient precision, had a restricted traverse and inadequate elevation, and possessed inconveniently narrow tires on its wheels. Nothing was

The weight increase in field carriages and ammunition limbers persisted as an unfortunate trend in navy and marine landing guns. The army type limber, carrying fewer rounds and weighing less than the navy Mark IV design, was adopted in limited number by the marines as the Mark V limber. The horizontal storage of rounds in this limber, as well as the panoramic sight on the Mark VII 3-inch landing gun, are evident in this photo taken during a 1920s artillery practice in Haiti. (*USMC History Division*)

Experience showed that the top shield on the Mark V field carriage needed to be hinged and turned down for indirect laying with a forward aiming point. Such a modified shield appears on this Mark VII 3-inch landing gun/Mark V field carriage during firing practice at the Marine Barracks, Quantico, VA, sometime around World War I. (*National Archives and Records Administration*)

mentioned about its dubious breech mechanism. The Ehrhardt-designed Mark VII needed an extended shield in order to provide greater protection, weighed too much, necessitating animal hauling, was equipped with awkward elevating and traversing gear, and had an unsuitably low trail, unhandily narrow tires, and unsatisfactory ammunition stowage.

What the bureau desired in the new 3-inch/23cal landing gun, which would fire the same ammunition already in use, included a bullet-proof shield, a panoramic sight permitting indirect fire, elevating gear that provided for high-angle fire, a split trail that accommodated a substantial gun traverse on its carriage, smooth recoil and counter-recoil, wide wheel rims, and a trail wheel usual for a naval landing gun. As important, the total weight of gun, field carriage, and limber carrying at least forty rounds was not to exceed 3,500 pounds. The navy remained only too acutely aware that its landing guns had been getting progressively heavier. The task at hand involved the design of a new landing gun outfit that somehow balanced a modern light fieldpiece with a total weight that could reasonably be manhandled by bluejackets ashore.

The Bureau of Ordnance did not make a public advertisement for the new landing gun, but rather sent a confidential invitation to a small number of ordnance makers. The letters went out in December 1916, estimating the need for between twenty and forty pieces and desiring estimated price, time of delivery, and a specific design based on the navy's general description of needs. Because it had been burned in a previous effort to develop a special 3-inch gun for the navy, American & British Manufacturing

Contrary to the original intention of having light guns landed with relative ease, the weights of successive US naval landing guns escalated, none more than that of the Mark VII 3-inch piece consigned to the marines. The combined weights of the gun, carriage, and fully loaded limber exceeded 2 tons and required hauling by animals. The navy could not readily land such draft animals, thus they had to be seized once the force came ashore. This artillery unit is training in Mexico during the occupation that lasted until November 1914. (*USMC History Division*)

Company declined to submit a bid. The Bethlehem and Midvale Steel companies did proffer bids, and in February 1917, Rear Adm. Ralph Earle, chief of bureau, appointed a board of three officers to consider the merits of each design. By that time, the number of guns required had settled on fifty.

Midvale Steel contemplated a licensed version of the proven Schneider hydropneumatic recuperator, whereas Bethlehem Steel proposed to design its own mechanism. The Bethlehem design provided for a gun elevation of 40 degrees, a traverse of 20 degrees on each side, the first gun in five months, and the contract completed in eighteen months, for a price per gun of $15,250. Against that, the Midvale design embodied a gun elevation of 28.5 degrees and a paltry traverse of 2.5 degrees on each side. It estimated its first gun in ten months, the second gun one month later, and the completion of contract in twenty-one months, at $15,450 per gun. On top of those more favorable factors in Bethlehem's case, many of the patent rights were already controlled by the United States and the company promised to protect the navy from additional patent claims. The March 12, 1917 award to Bethlehem, summarized by the lawyerly acting secretary of the navy, Franklin D. Roosevelt, contained the proviso that if the company's hydropneumatic recuperator did not operate properly, it would have to be replaced by the older (but less robust) hydrospring type. In response to the last, Bethlehem promised one gun and field carriage ahead of the schedule specified and strongly requested their earliest testing, so the company could improve the recuperator as needed.

In the months following, the Bureau of Ordnance designated the marks and serial numbers for the new pieces: Mark XI 3-inch landing gun and vertical sliding wedge breech mechanism, Nos. 2104–2153; Mark VI field carriage, Nos. 273–322; Mark VII limber, Nos. 172–221; Mark XIX panoramic sight, Nos. 9162–9211. The navy put forward in June the need for an additional fifteen such landing guns and carriages, for which Bethlehem Steel agreed to build for $15,000 each; the delivery date thereupon extended an additional two months to November 1918. At the end of the month, Acting Secretary Roosevelt knocked back that added number to six, now specified for the naval reserve. Bethlehem understandably balked at the price previously agreed upon, and that amount was now increased to that of the original contract, $15,250 per gun set. The added serial sequences were guns Nos. 4975–4980, carriages Nos. 327–332, and limbers Nos. 222–227.

The navy also expressed its interest in acquiring two batteries (eight pieces) of 3-inch mountain guns. Having one in its gun catalog, Bethlehem made the sole response, with a price of $15,000 per gun set, which included nine mule packs, four of them to carry the gun, carriage, accessories, and tools, and the other five packs to carry ten rounds of ammunition each. The final contract, dated June 18, 1917, halved the number of such guns, now fixed at $15,500 each. As off-the-shelf items of an older Bethlehem design, they needed modification only to chamber and fire the existing navy 3-inch field gun round. Bethlehem's short 15-caliber gun, composed of nickel steel, elevated to 27 degrees and traversed a total of 7 degrees on its carriage (one manual claims a trifle less than 5 degrees). The gun, carriage, wheels, and steel shield weighed a little more than 820 pounds. The bureau numbered the four Mark XII guns 5482 to 5485 and the Mark VII mountain carriages 323 to 326.

One of the primary goals sought for both the landing and mountain guns, that the pieces be easily manhandled and thus readily brought up for action, proved difficult to achieve. Moreover, the products would be long delayed. That story continues in the following chapter.

Machine Guns in American Service During the War and Postwar

During the war in Europe and elsewhere, American military and naval forces used a variety of machine guns, many of foreign origin, but others the design of native sons or of companies founded by native sons. They included the Lewis light machine gun, created by Dr. Samuel McClean and Col. Isaac Lewis, US Army; the Hotchkiss Model 1914 heavy machine gun (pioneer Benjamin Hotchkiss had died in 1885); and a British improvement on Hiram Maxim's mankiller of the late nineteenth century, the Vickers Mark I water-cooled machine gun, with its American counterpart, the Colt-Vickers Model 1915, almost identical, but chambered for the US Model 1906 .30-caliber cartridge. Both variants bore the distinguishing muzzle booster. The new automatic weapons devised by the inventive John Moses Browning arrived too late to see much combat, with the result that the Colt-Vickers .30-caliber machine gun armed thirteen US divisions sent to Europe.

At Marine Barracks, Quantico (VA), Lewis guns, boxes of ammunition pans, and boxes containing other components, all carried on Cole carts purpose-designed by marine officer Edward Cole (KIA 1918), have been set aside under guard. Although the guns are identified as belonging to the 6th Machine Gun Battalion, that unit was not so designated until January 1918, after reaching France. Once there, the entire 4th Marine Brigade would be divested of its Lewis light machine guns. (*USMC History Division*)

The American 4th Marine Brigade, which would earn distinction and sustain losses in the fighting on the Western Front, exchanged its Lewis guns shortly after arriving in France for either the Hotchkiss heavy machine gun (6th Machine Gun Battalion) or the Chauchat Model 1915 machine rifle (infantry units). Much vilification has been heaped upon the French automatic rifle, but to be fair (only slightly), it was 10 pounds lighter than the Lewis gun and not encumbered by the large and awkward ammunition pan atop the receiver. Thus the Chauchat could readily be fired from the hip, providing walking fire in an infantry assault, or at least so until it jammed. The US Navy, including its landing parties, continued to use the Lewis gun during the war, the Russian interventions, and the interwar decades.

The United States Sends Naval and Military Forces to Russia

A. *The Naval Landing in North Russia*
Much to the distress of the remainder of the Allied powers and the United States, in early March 1918, Russia withdrew from the Great War. That vast nation had heretofore pulled a considerable number of troops of the Central Powers away from the Western Front, troops that were now freed up to move west against the Allied armies battling in France and elsewhere. Those nations in league with Russia had supplied it with considerable amounts of ordnance and other useful matériel to enable it to continue the fight. The Allies at once became determined that these military supplies would not fall into German hands or those of their Finnish allies. Of course, there was more to it than that. The European allies particularly feared the spread of Bolshevism. During the course of events, thousands of Allied troops, as well as American sailors, soldiers, and marines, would go ashore in North Russia (Murmansk and Arkhangelsk) and Vladivostok in the east.

In response to British (mostly) and French pressure, the United States ordered Capt. Bion Bierer to take his ship, the 1890s protected cruiser *Olympia*—she had been Commodore Dewey's flagship during the battle of Manila Bay—to the port of Murmansk, ice-free year-round, and the northern terminus of a railway from Petrograd (St. Petersburg). She also carried Maj. Gen. Frederick Poole, British Army, and his staff, as well as a French civil mission. Poole would be the ranking officer to be landed in North Russia. Sailing from the British fleet anchorage at Scapa Flow, *Olympia* (as modified and rearmed the year before, ten 5-inch/51-caliber guns) arrived at Kola Inlet on May 24, 1918, finding major warships of the Allied nations already present. Those warships included the pre-dreadnought HMS *Glory* (two twin 12-inch gun turrets), the flagship of the senior naval commander, Rear Adm. Thomas Kemp, Royal Navy, and the French armored cruiser *Amiral Aube* (two single 194mm gun turrets). Two hundred British marines and 100 French marines had so far been landed in order to thwart any incursion by German forces, or White Finnish ones not far to the west.

Fifteen days later, pursuant to Admiral Kemp's order to Captain Bierer, *Olympia* put ashore a two-company landing party, 100 bluejackets and eight officers, under the command of Lt. Henry Floyd. The shore party amounted to about one quarter of the

The protected cruiser *Olympia* is viewed on her port side in 1920, anchored in the main basin of Venice harbor, with six of her replacement 5-inch/51cal guns in view. She steamed to many places during her period of service, most famously as Commo. George Dewey's flagship during the battle of Manila Bay in May 1898. In late World War I, she traveled to North Russia in order to protect weapons and matériel provided by the Allied nations. Photograph by S. Marco.

ship's complement (the cruiser did not possess shipboard marines at that time). The sailors moved into timber barracks ashore, which they shared with the British marines. The American officers found more comfortable quarters in a Pullman car. At the time, the weapons equipping the US force were drawn from the ship's armory: Model 1911 .45-caliber pistols, Model 1903 .30-caliber rifles, and Colt-Browning Model 1895 .30-caliber machine guns. The clumsy automatic gun constituted the heaviest weapon in the sailors' hands and its reputation had preceded it, as one seaman grumbled in his diary. The American landing force functioned in much the same way as the port garrison, and effectively formed a part of it, with duties commensurate with such service.

Although the initial reception by the mostly anti-Bolshevik Russians had been friendly, all too soon trouble in the port arose, and in July, the British sent search parties through the town and uncovered a substantial number of arms in private homes. The Americans were favored with a number of these weapons, including Lewis light machine guns, far preferable to the sailors' Colt-Brownings. The sailors had been experiencing difficulty with weak ejectors in their Springfield rifles, so they exchanged them for Mosin-Nagant 7.62mm infantry rifles that had been made under contract by Remington and New England Westinghouse in the United States. After the usual grousing by men accustomed to their own rifles having to turn them in for those of foreign design, the American bluejackets decided they could put up with the rugged Mosin-Nagant, which was accurate enough for that terrain. The Russian rifle also armed subsequent American naval and military forces landed in North Russia for the sensible logistical reason that ammunition supply would not be a problem. Inside the country of the rifle's origin, the correct rounds could be found in plentiful supply.

At month's end, Floyd moved south with one company of bluejackets—two additional officers and about fifty men—and joined with Allied troops to occupy Arkhangelsk, located 430 miles south across the Barents and White Seas. The city, ice-bound in winter, lay 16 miles up the Dvina River delta (that is, directionally south). Substantial Allied war matériel had been unloaded and stockpiled there previously, which the coalition wished to safeguard. Because the would-be repossessors had tarried at such length up in Murmansk, however, the supplies had long been removed by the Bolsheviks, called "Bolos" by the British and Americans.

Soon following their arrival in Arkhangelsk on August 2, the coalition forces moved against Bolshevik forces in the marshy land to the south. A twenty-five-man detail commanded by Ens. Donald Hicks, USNRF, ferried across the Dvina to the Bakharitza railway yard, where they got their hands on a wood-burning locomotive and several flatcars. After fortifying the cars with sandbags and machine guns, the party headed south for a bit of derring-do. They soon found all they could handle.

After engaging in a stern chase with a train of Bolsheviks, which involved a spirited exchange of fire, Hicks and his men settled into a passive defense, and then, under orders, joined mixed-nationality Force B commanded by Lieutenant Colonel Haselden, British Army. The force headed up the Dvina by barge, disembarking at Siskoe. Thereafter, it moved by land 30 miles south to attack Seletskoe on August 15, defended by 250 Bolshevik sailors, equipped with several machine guns, a pair of 18-pounder field guns, and an armored car, originally British gifts to the tsar's Imperial Russian Army. The Russian sailors offered stiff resistance, and during the ultimately successful attack on the town—it required five hours—one British officer and five men

This Russian 0-4-0 shunting locomotive may be similar to the one commandeered in August 1918 by Ens. Donald Hicks's party of bluejackets to chase Bolsheviks down the line south of Arkhangelsk. US Army photograph.

were killed. American Seaman George Perschke was wounded in the arm, the first US casualty of this remote war, but hardly to be the last.

At month's end, Force B was ordered west and north to join the coordinated assault on Obozerskaya on the railway, with Force A attacking from the north. During the march, however, Haselden received information that a large Bolshevik force was moving on Seletskoe, which his men had just captured. By this time, the British officer thought highly of the American naval detachment. Learning that Seaman Corbin Hardaway could ride, Haselden mounted the sailor on a sturdy horse and sent him back to warn his headquarters at Tiagra, not far north of Seletskoe. Hardaway rode alone more than 30 miles through the remote and desolate and by no means friendly territory, arriving safely at Tiagra with the news of the Bolshevik advance. He was subsequently awarded the Navy Cross for his hazardous achievement.

Force B meanwhile attacked Obozerakaya, but the assault stalled due to fierce resistance. Worse news came on September 1, with the Bolsheviks having taken not only Seletskoe but Tiagra. Force B's situation now grew perilous; it had become squeezed between two enemy formations that heavily outnumbered it. Accordingly the force pulled back more than a dozen miles into a defensive position. Haselden posted the Americans and French to an advanced trench line, where they repulsed two enemy bayonet charges, but thereafter things began to unravel. The French departed, leaving the Americans alone to defend the advance line. When the latter at last pulled out, they found the main camp abandoned, with the supply wagons raided by their putative allies. The American sailors unharnessed and released the horses, and destroyed what remaining supplies they could not carry. That night Hicks and his men took to the bog forest and swamps of the surrounding taiga, soon running out of rations. Fortunately, on the day following, they reached the railway and obtained food from a French contingent. The Americans were thereafter detailed as guards aboard a train carrying prisoners of war, and so returned to Arkhangelsk. They had been in the field fourteen weeks and had suffered no deaths. In their absence, deadly influenza had begun to spread among the ships on the North Russia station.

During the period in which Hicks and his men were making their way through the marshy wilderness, the US Army's 339th Infantry Regiment, American North Russia Expeditionary Force, arrived at Arkhangelsk on September 4 and immediately sent its 3rd Battalion south by train to reinforce Obozerakaya. There, two platoons of K Company moved farther south in an attempt to link up with Force B. The detachment found equipment and personal items abandoned or lost by the force, including Ensign Hicks's diary. Force B had of course already reached safety on the railway, so it may be assumed that communications had largely broken down. Subsequently, the 339th Infantry assumed the cognomen of "Polar Bears," a name shared with the 31st Infantry in Vladivostok.

The other twenty-five men from *Olympia* landing at Arkhangelsk in early August, under the command of Ens. James Williamson, proceeded up the Dvina River and eventually joined with a French detachment manning an armored train on the railway. Their service is a deal less well documented than that of the Hicks detachment, but Boatswain's Mate Emil Keranen won the Navy Cross for gallantry during a sharp action.

Above and below: After a month either battling the Bolsheviks or slogging through Russian marshes, the Hicks detail has returned somewhat the worse for wear to the railway yard at Bakharitza. They are armed with the Russian Mosin-Nagant Model 1891 7.62mm infantry rifle. Ens. Donald Hicks (1894–1953) appears in the lower photograph wearing a patrol jacket and, like his men, a woolen watch cap. Surrounding the bluejackets are members of the 339th Infantry, North Russia Expeditionary Force, doubtless impatient to hear of the sailors' adventures—tales that will very likely be much embellished. US Army Signal Corps photographs. (*Naval History and Heritage Command*)

Rear Adm. Newton McCully and thirty-six men arrived at Murmansk on October 24, 1918, constituting the command element of the US Naval Forces Northern Russia. The admiral immediately hoisted his flag on *Olympia*. McCully was a flag officer fittingly qualified for his assignment because of his previous duty as naval attaché; not only did he speak Russian, he well understood Russia. Although the actions in which the US Navy had already been engaged and the US Army would be engaged may argue otherwise, the firm instructions to McCully, Bierer, and the army commander were not to interfere in Russian internal affairs. The admiral retained his flagship for less than a month; *Olympia* departed Arkhangelsk on November 8 and Murmansk three days later, never to return to Russia. She left behind Ens. Donald Hicks with eleven men for service with the international naval brigade, and Ens. C. S. Bishop with fifty men for manning a Russian destroyer. These men were pulled out in December 1918 and January 1919. But the US Navy's landing efforts in North Russia were not finished.

The steel steam schooner *Yankton* (six 3-pounders) reached Murmansk in early February 1919 and the protected cruiser *Galveston* (ten 5-inch/50cal guns) two months later. Two motorsailers were landed at Murmansk, *Yankton*'s 34-footer and *Galveston*'s 30-footer, each armed with a 1-pounder rapid-fire gun and Lewis machine guns, and taken the final few miles to Lake Onega by a hastily improvised railway. Lt. Douglas Woodward off *Yankton* commanded the small motor gunboats, naming them "Atlanta" and "Georgia." They experienced their first battle not long after arrival, firing 136 1-pounder shells and 1,600 rounds from their Lewis guns during a night action in support of White Russian troops, destroying any vestige of American neutrality in that nation's internal affairs. The two motorsailers soon found themselves not only outnumbered but outgunned by Bolshevik vessels on the lake, and thereafter were able to carry out only opportunistic raids. They were ordered out in early July 1919, with Woodward and his men going aboard protected cruiser *Des Moines* (ten 5-inch/50cal guns), *Yankton* regaining her motorsailer, and the British taking possession of the sailer formerly belonging to *Galveston* for continued service on the lake.

At the end of the month, Woodward went ashore again, commanding a fifty-man landing party to Solombola, on the north bank of the Kusnetchikha River (one of the major channels within the Dvina delta), just across from Arkhangelsk. The situation to the south had been rapidly deteriorating, with the Bolsheviks, now infused with tough, experienced units of the Red Army, increasingly ascendant. The American bluejackets continued to patrol the streets of Solombola against any outbreak of trouble. The landing party withdrew in early September, and a week later, *Des Moines* stood out from that port and departed Russian waters. As it did Ens. Donald Hicks for his actions earlier in North Russia, the service awarded Lt. Douglas Woodward the Navy Cross for gallant leadership.

B. *The Landing at Vladivostok*

The landings and operations executed by a coalition of nations, including the United States, at the primary Russian eastern port of Vladivostok owed a nearly identical rationalization to the landings carried out in North Russia. With one notable exception among them, those nations wanted mainly to protect their military investments sent to Russia. Much of the ordnance and other matériel, intended to move west on the Trans-Siberian and Trans-Baikal Railways to the Eastern Front, still sat in Vladivostok, in the open and unguarded. In 1917, the United States alone had invested close to a billion dollars in military aid to keep Russia in the war. During the brief Kerensky government, America had provided that nation with 300 steam locomotives of preferred Russian types and more than 10,000 railway cars (or wagons) of European style. That motive power and rolling stock were stretched out in yards and depots along the Trans-Siberian Railway as far as Lake Baikal and Irkutsk, as well as along the Chinese Eastern Railway beyond Harbin in Manchuria.

To manage those railways, the United States established the Russian Railway Service Corps (RRSC), composed of experienced civilian railroad men, but uniformed and organized along military lines. During March 1918, the corps entered Siberia through Harbin. Because of the experience of the United States in transcontinental railroading, this arrangement proved acceptable to the Allies, or at least it did at first.

The one exception alluded to among the coalition nations was Japan, which had rapacious designs on eastern Asia, particularly on the natural and commercial resources of Siberia and Manchuria. The Japanese made the cynical decision to destabilize local government in Siberia and to thwart the efforts to restore peaceful conditions to eastern Russia by its putative allies, in particular those of the United States. Toward that end, the Japanese landed 72,000 troops—by far the largest military contingent in Siberia and much larger than agreed upon by the Allies—and at least initially, allied themselves with mendacious and murderous Cossack atamans, such as Grigori Semyonov and Ivan Kalmikoff. These warlords controlled lengthy stretches of the Trans-Siberian Railway, and would prove to be sharp thorns in the American side when it came to bringing order to the railway. The Japanese further ingratiated themselves with them by the bestowal of opulent gifts. These problems, however, were those more directly of the US Army and the RRSC, and less those of the US Navy and Marine Corps.

One other relevant group deserves mention, the 40,000-man Czechoslovak Legion, constituted in 1918 from former Czech and Slovak prisoners of war in Russia, Russian Slavs, and Austro-Hungarian army deserters. As a result of the fighting in the west, and the generally now victorious Red Army, the legion moved east along the Trans-Siberian Railway, with the hope of evacuation from Vladivostok. That goal was wholeheartedly supported by the government of the United States. The legion would eventually have contact with the American navy and marines, and would obtain a most curious gift from them.

By May 1918, warships representing several Allied nations had anchored in Zolotoy Rog (Golden Horn Bay), along with the armored cruiser USS *Brooklyn* (four twin 8-inch gun turrets). Some of these ships already had marines or bluejackets in the city to protect their consulates and commercial assets. When the disorder there descended into chaos, Adm. Austin Knight, commander-in-chief of the US Asiatic Fleet, directed that marines be sent ashore. On the evening of June 29, 2nd Lt. Conrad Grove took thirty-one marines of the *Brooklyn* detachment to the American consulate, set up defensive posts in a one-square-block perimeter, and allocated men to participate in the Allied patrols around the city. Later that night, Capt. Archie Howard, commanding officer of the ship's marine detachment, arrived to assume command at the consulate. He and Grove alternated in leading patrols to ensure that Americans and other foreigners were safe on Vladivostok's streets. When sporadic skirmishes erupted, however, the American marines—under orders not to engage—refused to become involved in the fighting, much to the astonishment and displeasure of the naval detachments of the other nations.

The army's 27th and 31st Infantry Regiments, reinforced to nearly 8,000 men and constituting the American Expeditionary Force Siberia (AEFS), arrived at Vladivostok in August 1918 and began moving west along the Trans-Siberian Railway. The commander of the AEFS, Maj. Gen. William Graves, came in early September aboard USAT *Thomas*, with a restrictive aide-mémoire originating from President Woodrow Wilson, although issued through the State Department, prohibiting him from interfering in the internal affairs of Russia. This constraint proved as difficult to enforce in Siberia as it did for the American forces fighting in North Russia. Although the army arrived in strength, the participation of American marines and bluejackets hardly ended in Siberia. In conjunction with the detachments representing other nations, marine patrols from whatever American warships lay in the harbor continued to patrol the streets of Vladivostok. A marine detail was also maintained to guard the US naval hospital and radio station on Russky [Russian] Island, just southwest of Zolotoy Rog.

Sometime after her return to the Vladivostok station in the fall of 1918, *Brooklyn* unloaded two Mark VII 3-inch landing guns, Nos. 1072 and 1073, that she had picked up at the marine base at Olongapo, Luzon, in September. Thereupon, at the order of Admiral Knight, *Brooklyn*'s supply officer transferred them, with ancillary gear and spares, and an amount of 3-inch ammunition, to the Czechoslovak Legion. Knight also directed that four army 2.95-inch mountain guns, also originating from the Philippines, be turned over to the Czech force. As late as 1922, the navy's Bureau of Ordnance was still trying to puzzle out this seemingly gratuitous gift, for which few records survived and no one in the chain of command wished to take responsibility. If the pair of former marine pieces saw action in Czech hands, they were very likely the only American naval landing guns of that size to do so during the Russian campaigns.

In July 1919, several companies of the army's 31st Infantry Regiment began operations to clear the Suchan Valley, an area directly east of Vladivostok containing numerous bituminous coal mines that served for both domestic heating and firing locomotives on the railway. The fighting was intense, often vicious, and both the Americans and Red partisans sustained serious casualties. The history of the 31st Infantry indicates that American sailors and marines occupied several villages reclaimed beforehand from the partisans by the American soldiers. For much of July, protected cruiser *Albany* (ten 5-inch/50cal guns) remained active in the vicinity of Amerika Bay and put ashore landing parties of short duration. This bay, more than 50 miles east of Vladivostok, has been renamed Zaliv Nakhodka, at whose head stands the village of Amerikanka, its name originating from the Russian corvette *Amerika* that took refuge there from a fierce storm during the mid-nineteenth century.

At month's end, *New Orleans*, which had relieved her Elswick-built sister *Albany* a few days previous, landed thirty-one marines commanded by 1st Lt. Leland Swindler at Tyutuke Bay near Vladivostok in order to protect American interests. Perhaps for

Above and below: Docked on the far right in Vladivostok harbor, the armored cruiser USS *Brooklyn* clearly displays her early 1890s design. The large warship moored nearby is HMS *Suffolk*, an armored cruiser built in the early twentieth century. Judging from the snow on the ground and the cold-weather attire of the marines disembarking from *Brooklyn*, the images date from late 1918 or early 1919. US Army Signal Corps photographs. (*Naval History and Heritage Command*)

the Red partisans in the area, such distinction was too thin, given that the landing coincided with a White Russian operation in the area. In the event, nothing much transpired there, and the marines were back on board within two days.

With the White Russian forces badly disintegrating, the Red Army arrived in the vicinity of Vladivostok in January 1920, and began preparing artillery positions to shell the port city. Three companies of bluejackets and a company of marines were landed from the armored cruiser *South Dakota* (two twin 8-inch gun turrets; soon to be renamed *Huron* to free up her name for new construction). The marine company, led by Capt. David Barry and Lt. Edgar Allan Poe, Jr., pushed the Reds beyond the line established by the Allied powers, and that foray ended the problem for a time. Three months later, the United States pulled its forces out of Siberia, although the naval radio station remained on Russky Island, guarded by a detachment of marines. The navy dismantled that facility in November 1922, one month after Red forces occupied Vladivostok, and took off the men and equipment on the gunboat *Sacramento*. Thereafter the United States and the Soviet Union eschewed full diplomatic relations for another eleven years. Worse yet, the antagonism between the United States and Japan only intensified until the latter country brought the former into World War II in the Pacific on December 7, 1941.

A large parade down the main street of Vladivostok in autumn 1919—likely on Armistice Day—includes both US Army and US Navy contingents. The army contingent, passing from view on the right, is almost certainly from the 31st Infantry Regiment, and the naval party following is off the Elswick protected cruiser *Albany*, which had recently relieved her sister *New Orleans*. As most everywhere, a group of boys and a small dog are marching along beside. US Army Signal Corps photograph. (*Naval History and Heritage Command*)

References Consulted

Author unspecified, "Sideshow in the Soviet Union. Siberia Expedition," *VFW*, November 1993 (World War I Commemorative Issue), p. 27.

Beers, H. P., *U.S. Naval Forces in Northern Russia (Archangel and Murmansk), 1918–1919*, Administrative Reference Service Report No. 5, November 1943, Office of Records Administration, US Navy Department, Washington, DC.

Bethlehem Steel Company, *Mobile Artillery Material*, undated, pp. 16–17, and drawing.

Colbourn, C., "Far-off Northern Lands: Marines Ashore in Siberia," *Leatherneck*, March 2007, 90(3): 30–33.

Cooling, B. F., *USS Olympia: Herald of Empire* (Annapolis: Naval Institute Press, 2000), pp. 182, 189–99.

Daugherty, L. J. III, "'Bluejackets and Bolsheviks': The U.S. Navy's Landings at Murmansk April 1918–December 1919," *J. Slavic Mil. Studies*, 2005, 18(1): 109–52.

——————, "'... In Snows of Far Off Northern Lands': The U.S. Marines and Revolutionary Russia, 1917–1922," *J. Slavic Mil. Studies*, 2005, 18(2): 227–303.

Dictionary of American Naval Fighting Ships, Naval History and Heritage Command (online). Consulted for each US naval warship and auxiliary mentioned in this chapter.

Ellsworth, H. A., *One Hundred Eighty Landings by United States Marines 1800–1934*, History and Museums Division, HQ, US Marine Corps (Washington, DC: Govt. Print. Office, 1974 [originally printed 1934]), pp. 141–43.

History Committee, 4th Bn. (Mech.), 31st Inf., *History of the 31st Infantry and 73rd Yearbook*, Chapter 2, "Siberia 1918–1920," Fort Sill, OK, Winter 1988–1989, pp. 5–7, 13–14.

Jackson, C. V., "Mission to Murmansk," US Naval Institute (hereafter USNI), *Proceedings*, 1969, 95(2): 82–89.

McClellan, E. N., "American Marines in Siberia during the World War," *Marine Corps Gazette,* June 1920, 5(1): 173–81.

Smith, G. B., "Guarding the Railroad, Taming the Cossacks: The U.S. Army in Russia, 1918–1920," *Prologue*, Winter 2002, 34(4): no pagination.

Smith, J. E., *Small Arms of the World*, 10th rev. edit. (Harrisburg: Stackpole, 1973), pp. 74–75, 111–14, 131–33. 270–71, 578–79, 681–83.

Staff writers, "The U.S. Colt Vickers Model of 1915 Water-Cooled Machine Gun," *Small Arms Review*, July 2000, 3(10) (also online).

Tolley, K., "Our Russian War of 1918–1919," USNI, *Proceedings*, 1969, 95(2): 58–72.

US National Archives, Record Group 74, Records of the Bureau of Ordnance; Washington, DC.

US Navy, "Ammunition Depots," *Navy Ordnance Activities, World War 1917–1918* (Washington, DC: Govt. Print Office, 1920), pp. 86–89.

——————, *Depth Charge Projector Mark I (or Y-Gun)*, Ordnance Pamphlet No. 63, January 1918 (rev. October 1918), pp. 3–7.

——————, *The United States Naval Railway Batteries in France*, Publication No. 6, Historical Section, Office of Naval Records and Library (Washington, DC: Govt. Print. Office, 1922).

Willett, R. L., "Bluejackets vs. Bolsheviks," *Naval History,* August 2016, 30(4): 16–21.

12

US Naval Landing Guns in the 1920s

The Mark XI 3-inch Landing Gun and the Mark XII 3-inch Mountain Gun

Bethlehem Steel Company delivered the pilot Mark XI 3-inch/23-caliber landing gun and Mark VI field carriage on March 14, 1919, two years after the letting of the contract, and nineteen months late according to the terms. The final guns of the total of fifty-six would be delivered in 1922, more than three years late. And in March 1921, precisely three years late, the navy at last got its hands on the Mark XII 3-inch/15-caliber mountain guns, a mere four pieces of Bethlehem design available from its ordnance catalog. Needless to write, the navy voiced its great displeasure at these turns of events. After all, one persuasive reason for awarding the major contract to Bethlehem Steel had been its promised delivery time, earlier than that of competitor Midvale Steel. The ensuing tale is fraught with incompetence, sloth, poor workmanship, and lame alibis.

It began early in 1918 with a letter from the secretary of war, Newton Baker, to the secretary of the navy, Josephus Daniels, explaining that the Bethlehem Steel ordnance works were choked with the production of two different army field gun models, having a total of 704 units—none of which would be completed in time to see combat in Europe—and requesting the navy to have its fifty-six 3-inch landing guns built elsewhere. The navy did not wish to repeat the whole process of advertising for this construction and awarding a new contract in the face of maximum war production in virtually every other plant, and so refused, reminding the army of the small number of naval guns involved. Despite Bethlehem having done work on a single piece to date, the navy expressed confidence that the company could complete the naval contract without undue interference with the War Department work. That confidence would prove badly misplaced.

Bethlehem resumed work on the navy contract, but within months, Rear Adm. Ralph Earle, chief of the Bureau of Ordnance, complained that Bethlehem had made an unwarranted number of requests to waive defects in machining. Earle concluded that the numerous defects resulted from poor shop supervision or outright carelessness, and demanded that Bethlehem Steel take the necessary steps to prevent a repetition of such errors. In late April, Earle complained again about the lack of progress in the landing gun contract. Bethlehem promised to look into the matter. But three weeks later, the company reported yet another machining error and made the inevitable request for its waiver.

Some of the squabbles were simply needless because of long understood norms. One of them lasting half the year (1918), in which Republic Iron & Steel Company, Cleveland Welding & Manufacturing Company, Archibald Wheel Company, Bethlehem Steel Company, and the US Navy haggled about the sulfur and phosphorus content of the steel to be used in the metal tires of the wheels for the field carriages and limbers, should have been settled at the outset. The welding company strenuously pointed out that the content specified would have made the welding of those tires impossible. That it took six months for all parties to agree to the same standards set for the steel tires of army field guns was just one example of the obstacles that just kept coming.

At the end of October 1918, Admiral Earle urged that the pilot set, expected in one month's time, be sent to the Naval Proving Ground as soon as available in order to examine the component designs. At that point another big problem arose: the unavailability of the panoramic sights for the fifty-six landing guns. The company that had subcontracted to produce these optical sights, the Recording & Computing Machines Company of Dayton, Ohio, engaged in a lengthy charade, in which it refused to ship the sights to the prime contractor, Bethlehem Steel, without permission from the Bureau of Ordnance. In February 1919, Earle informed the Dayton company that it did not need such a release and directed it to send the sights forthwith to the prime contractor. Despite persistent efforts, Bethlehem still had not obtained the sights by late May. Two months thereafter, the Recording & Computing Company informed Bethlehem that it had not been able to fill the order for the fifty-six panoramic sights. Moreover, the army had already cancelled its larger contract with the Dayton company. The navy thereafter agreed to adopt the army panoramic sight, as well as its methodology, for its Mark XI landing gun and Mark XII mountain gun. That potential setback turned out to be not a bad thing, because the army's panoramic sight was a later model, and its wartime experience brought superior practices to the navy's use of this optical device.

Some apprehension on Bethlehem Steel's part accompanied the delivery of the pilot landing gun, No. 2107, and field carriage, No. 273, in mid-March 1919. The company unnecessarily reminded the navy that the gun represented a special design, largely seeing fruition during its building, and thus needed more that the usual examination. Bethlehem strongly insisted to the navy that its representative be present to explain the gun's unique features even before any preliminary tests were performed. Moreover, it urged that the proof tests be completed as soon as possible in order to incorporate any changes that the navy demanded in the subsequent production run.

The problems discovered at the Naval Gun Factory were hardly minor or expected. The hydropneumatic recuperator developed by Bethlehem operated on a pressure that could not be maintained, necessitating that a compressed air bottle accompany each battery or section of guns. In a pinch, a two-stage hand pump could serve, but this necessity amounted to an Achilles heel. The gearing beneath the gun proved to be overly small, exposed to sand and grit, and difficult to disassemble for cleaning. Perhaps most damning was the lost or empty motion in elevating or depressing the gun, a factor the navy attributed to poor workmanship. The navy noted that no obvious means of taking up the lost motion existed and the company representative did not suggest one when confronted with the problem. The inspectors also found objectionable the make-do leather packing used in the recoil system. And those problems had been discovered *before* the gun and carriage went to the proving ground. Taken together, such blunders were considered inexcusable on the part of a seasoned ordnance maker.

Although the gun and field carriage passed their proof tests, two features that exceeded specifications, and at first glance implying unexpected gifts, turned out to be liabilities or at least challenging problems. The specifications called for a gun shield made of 0.15-inch steel plate, but the one Bethlehem delivered was one-tenth inch thicker. The navy likely would have found 0.19-inch plate acceptable, as that thickness proved resistant to a conventional infantry bullet fired from as close as 50 yards. Quarter-inch plate, however, added another, unnecessary 60 pounds to a landing gun outfit already taxingly heavy.

In addition to being the navy's first landing gun to have a split-trail field carriage and the first not to have a trail wheel, the Mark XI was specified for a maximum gun elevation of 40 degrees. The pilot carriage, however, permitted an elevation of 50 degrees, which should have provided a greater flexibility in the trajectory of the piece. As were most then-contemporary field guns, the Mark XI landing gun was meant to have variable recoil, such that the length of the recoil progressively shortened as the gun was elevated. This feature prevented the recoiling breech from striking the ground. At 40 degrees, however, the gun recoiled 18 inches, clearing level ground by only half an inch, and at 50 degrees it recoiled 19 inches (measured either by raising up the entire gun or placing the breech over an excavation). While it remained apparent that the variable recoil feature had not been perfected in the pilot carriage, the naval inspector of ordnance at the Bethlehem plant strongly urged that the recoil be constrained to no more than 18 inches.

The Bureau of Ordnance declared its unhappiness that the proofing of the pilot gun and carriage had been done in the absence of a panoramic sight—the failure of the Recording & Computing Machines Company to produce this essential item would not become generally known for another two months—and insisted that the proof of the second gun set should include such a sight. The bureau desired a modification of the shield design, mostly to ensure no gap remained between the upper and lower halves when the top shield was folded upward. It also approved the provision of dust covers for these pieces.

By June 1922, Bethlehem Steel Company had still not completed the contract let more than five years before. Its litany of lame excuses beggared the collective imaginations of the senior members of the Navy Department, including Theodore Roosevelt, Jr., the assistant/acting secretary and the distant cousin of the previous

US Naval Landing Guns in the 1920s

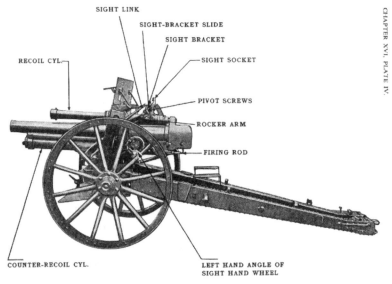

Above and below: The Mark XI was the US Navy's final purpose-built landing gun, the first to have split trails, and the last in the line of 3-inch pieces deployed for such service. Although the navy let the contract for these fifty-six guns with Bethlehem Steel Company before America's entry into World War I, they did not see completion until mid-1922, years late. Viewed both in left-side and breech-end views, the gun's various parts are identified, although perhaps as a reflection of its ill-starred history, the panoramic sight for indirect fire is not mounted. *Naval Ordnance* (USNA textbook), 1939 edition. (*HathiTrust Digital Library*)

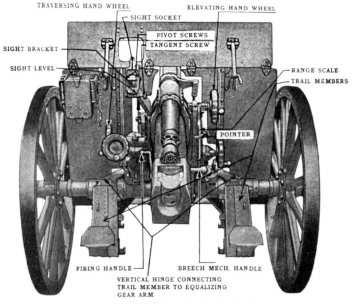

assistant/acting secretary and future president, Franklin D. Roosevelt. First, the company offered up as reasons the bedlam arising from wartime conditions and the general concession by the navy that the army production should take precedence (the navy at first disputed that concession, but eventually decided not to pursue the matter vigorously). Next, the company explained that despite the war having ended more than three years before, it had moved the navy work from the machine shop properly set up for ordnance production to one far less ideal, as if such a self-fulfilling interference with the progress of the work offered a proper excuse. Finally, and least credible of all, Bethlehem claimed that the latest deadline, May 1, 1921, which the company had put forth and promised as inviolate, was not binding because the navy had not formally conveyed the date as final. In July 1922, the contract was at last completed and closed out. As the younger Roosevelt wrote that August, the causes of the delay lay "within the control of the contractors."

Much the same complaints attended the Mark XII mountain gun, a pre-existing Bethlehem design. It had been modified only to chamber the old navy 3-inch field gun round, providing a muzzle velocity of 1,100 feet per second. In May 1920, the navy agreed to the company's request to extend the delivery date from March 18, 1918, to October 31, 1920. Bethlehem missed that deadline, too, finally delivering the four mountain guns on March 28, 1921, another 148 days added to its tardiness. Part of the blame rested with the subcontractor not delivering the panoramic sights on time. As well, Bethlehem's wheel subcontractor had delivered the wrong size wheels—the specifications enumerated those of 29-inch diameter. Not for the first time, Bethlehem claimed that the navy had not formally conveyed to it the final delivery date, even though the company had stipulated that date. For the next two years, numerous letters were exchanged about the money owed by the navy to Bethlehem for the guns, versus the assessment of liquidated damages against the company for its substantial lateness in delivery.

It remained eminently clear even to the most biased observer that the Bethlehem Steel Company had badly botched at least two naval gun contracts. But March 1921 had brought in a Republican administration, that of Warren G. Harding. With the new president came a new secretary of the navy, Edwin Denby, who at the end of October 1923 pronounced the navy responsible for the delay in delivery. This early skullduggery within the military-industrial complex permitted Bethlehem to skate free. Although the company did not face retribution for its poor performance, Denby would. During that same October 1923, the US Senate began investigations on the transfer of control of naval oil reserves in Teapot Dome, Wyoming, and Elk Hills, California, to the Department of the Interior, with subsequent corruption and graft that resulted in the gain in personal loans—read, bribes—to Interior Secretary Albert Fall. Although Fall went to jail for one year, Denby was not similarly charged with fraud, but with neglect of his duties in thus ramrodding the transfer. Having done substantial damage to the US Navy, the secretary resigned his office in March 1924 and returned to private life.

Above and below: The Mark XII 3-inch mountain guns were the first and last pieces of this type purchased by the navy, because of a perceived need by the marines. Although only four guns were ordered from Bethlehem Steel Company and despite being an existing company design, their completion in 1921 was years late. Their service seems to have been brief. The expansive steel shield seen in both left and right side views is typical of mountain guns, which anticipated close combat in such high and rugged terrain. A panoramic sight is mounted for indirect fire. *Assemblies, 3-inch Mountain Gun, Mark XII*, Ordnance Pamphlet No. 124, May 1923. (*National Archives and Records Administration*)

Postwar Doldrums and Adjustments in the Landing Force

In May 1919, whereas the navy disapproved any landing force aboard colliers and oilers, from army and marine experience in the war, it did approve a revision of the *Landing Force Manual* to require Browning automatic rifles (BARs) and hand grenades in landing parties from the expected types of warships. Its supply base for landing force equipment moved a year later from the New York Navy Yard to the Navy Operating Base, Hampton Roads.

The marines were also busy consolidating their assets. They requested the transfer of 200 machine gun carts to the Philadelphia Navy Yard for issue to warships as part of the landing force equipment. The light hand cart as devised by Maj. Edward Cole, USMC, during the early years of World War I (he died in action at Belleau Wood, June 1918), came in a variety of shapes and sizes as a function of what each type was intended to carry. The loads commonly included the Lewis gun, the Browning water-cooled .30-caliber machine gun, the 3-inch Stokes mortar (later the 81mm mortar), ammunition to accompany the weapon at hand, range finders of different types, tools, spares, etc. Although the handcart was most closely identified with the interwar marines, in actuality it saw use by all three services into World War II. The Cole cart used by marines and sailors generally had light 19-inch wheels with motorcycle-type wire spokes. The army developed its own handcart with smaller wheels and heavier spokes that saw extensive use in the European theater.

Beginning in the early 1920s and ending a decade later, the navy retired twenty-six of its cruisers that had been built during the late nineteenth and early twentieth centuries. They included the armored cruisers *Brooklyn*, *Pittsburgh* (ex-*Pennsylvania*), *Rochester* (ex-*Saratoga*, ex-*New York*), and *Huron* (ex-*South Dakota*), the Denver-class protected cruisers *Des Moines*, *Cleveland*, *Galveston*, and *Denver*, and the scout cruiser *Birmingham*, all of which had some type of 3-inch landing gun aboard. In the case of *Birmingham*, the Bureau of Ordnance ordered her landing gun set retained with the ship at the Philadelphia Navy Yard during the lengthy hiatus between her decommissioning in 1923 and scrapping at decade's end. She had been the flagship of the Special Service Squadron and perhaps the navy had thoughts of bringing her back into service. In January 1930, the bureau ordered *Birmingham*'s Mark XI landing gun No. 2112, Mark VI field carriage No. 281, and Mark VII limber No. 188 sent to the Naval Gun Factory in Washington to be overhauled for reissue to the fleet.

At mid-decade, the new gunboat *Tulsa*, the only vessel of this type then carrying a Mark XI 3-inch landing gun (her sister *Asheville* had an aging Mark I mod 1 field gun on board), reported lost motion in that gun. Although her gunnery officer ascribed the problem to the wear of the training and elevating gears, almost certainly it remained the same trouble discovered during the trials on the pilot gun and carriage. Rear Adm. Claude Bloch, chief of bureau, queried the Naval Gun Factory about this imputed weakness. Apparently unfamiliar with the report made a mere six years before, the superintendent reported difficulty in discovering the origin of the lost motion within the several bevel gears and worm. It is equally clear that this major problem had not been addressed and fixed. The superintendent took refuge in Ordnance Pamphlet No. 131, which suggested

A detachment of bluejackets somewhere on the Asiatic station passes in review, hauling Cole carts loaded with guns and ammunition boxes. Such two-wheeled carts carried weapons, ammunition, and other gear when ashore. Although they used the same 19-inch-diameter wheels with wire spokes, their shapes and sizes varied considerably. (*Naval History and Heritage Command*)

The artillery section of armored cruiser *Brooklyn*'s landing force hauls a Mark I mod 1 3-inch field gun somewhere in the Far East, perhaps Russia, around the time of World War I. The warship spent almost all of the last few years of her service life with the Asiatic Fleet, including a stint as its flagship. The navy decommissioned her in March 1921 and subsequently sold her for scrapping. (*Naval History and Heritage Command*)

manipulating an adjusting nut and washer. One assumes that *Tulsa*'s gunnery officer had already taken that step, and someone along the line should have realized that the problem remained more serious—and widespread—than one correctable by a nut and washer.

Much of the evolution of the landing force during the 1920s was owed to the Asiatic Fleet. The frequent sending of destroyers to trouble spots on the China station led to a request to increase the allowance of landing force gear on these small warships to forty-five fully equipped outfits. The bureau acceded to that request, made possible by the redistribution of such equipment aboard *Huron*, the flagship of the Asiatic Fleet, with orders to return stateside by year's end. Oddly, the equipment transferred to the destroyer tender *Black Hawk* included no ordnance. *Huron*'s field gun and all her machine guns, rifles, and pistols were to be landed at Cavite Navy Yard, except those the commanding officer thought necessary to see his ship safely home.

As regards those small arms, another useful piece would soon be added to landing parties: the (in)famous Tommy gun. As originally developed by retired Col. John T. Thompson, US Army, the Model 1921 Thompson .45-caliber submachine gun, which operated by blowback, fired 800 rounds per minute from a fifty-round drum. This rate of fire made accuracy difficult except at the shortest ranges. The Navy Model 1928 submachine gun fired 600 rounds per minute and was normally fitted with a Cutts compensator to reduce muzzle rise during automatic firing. This gun used both the fifty-round drum and a twenty-round box magazine. Because this model was not entirely interchangeable with the Model 1921, the navy held back the thirty-four Thompson guns originally intended for Yangtze River gunboats *Panay*, *Oahu*, *Luzon*, and *Mindanao*, until the more satisfactory guns became available. After that need had been satisfied, the bureau undertook their distribution to the landing parties of other warships. The marines in Central America had been among the gun's earliest recipients, and soon one man in every ten in Nicaragua wielded a Tommy gun.

The Naval Gun Factory also had been keeping its eye on the larger landing guns deployed. In early 1928, the superintendent called attention to the practice of cannibalizing stored Mark XI 3-inch landing gun sets because of the urgent need aboard warships in service. He pointed out that such missing parts needed replacement as soon as they could be provided, a request the Bureau of Ordnance quickly agreed to. Shortly thereafter, the gun factory set aside eight landing gun sets for the new heavy cruiser construction. It is quite probable that the assignment of non-consecutively numbered Mark XI 3-inch guns (the serial numbers ranged from Nos. 2106 to 2147), Mark VI field carriages (Nos. 275 through 316), and Mark VII limbers (Nos. 176 through 214) was based upon the intactness of those components. The new cruisers—all too soon to be derisively nicknamed "treaty tinclads" because of their meager armor protection to satisfy the weight limitations imposed by the Washington Naval Treaty—were *Pensacola*, *Salt Lake City*, *Northampton*, *Chester*, *Louisville*, *Chicago*, *Houston*, and *Augusta* (three or four triple or dual 8-inch gun turrets).

At the end of 1928, the Bureau of Ordnance took an inventory of the Mark XI 3-inch landing gun sets already assigned to the fleet and its training establishments, with an eye on the number of compressed air bottles needed to maintain the efficiency of those guns. Along with the eight reserved for the new cruiser construction, the navy had such landing guns aboard eighteen battleships, five old cruisers, and one oceangoing gunboat. An unknown number of Mark XI landing guns had also been assigned to the Naval

Training Centers at Newport, Rhode Island, and Hampton Roads, Virginia, and to the US Naval Academy. The December 17, 1928 letter described the air bottle for charging that particular landing gun's recoil cylinders as holding about 168 cubic inches of air at a pressure of 2,000 psi at 70 degrees Fahrenheit, with a safety disk protecting the bottle to 2,800–3,000 psi. The volume of compressed air was estimated as one and a half to two charges of air for each recoil cylinder, hardly sufficient for an entire battery of these guns as stipulated by Bethlehem Steel Company during the proof of the pilot gun in 1919.

Landing Actions and the Loss of Landing Guns

By the end of the 1920s, the little wars in the Americas had been reduced to sporadic political upheavals in Nicaragua and Honduras, with the odd shoot-out with bandits there and in Haiti. During that decade, marines had trained constabulary or civil guard units to keep the outlaw gangs at bay. A proposal made in 1925 to place the Mark XII 3-inch mountain guns on four small World War I-era warships of the Asiatic Fleet—gunboats *Palos* and *Monocacy* and minesweepers *Penguin* and *Pigeon*—had come to nothing, so those pieces went to their originally intended recipients. The mountain guns that the marines had waited for for so long got shaken out in the mountains of Nicaragua, with more than a little help from some stalwart mules. Following decade's end, the foursome disappeared into the whimsies of leatherneck lore, their departure almost certainly hastened by the development of the army's superior 75mm pack howitzer.

January 1924 saw gunboat *Sacramento* landing members of the Philippine Constabulary, supported by American marines equipped with machine guns, at Socorro, Surigao del Norte, Philippines. At the time, the bitter enemies of the constabulary were embodied in the growing insurgency known as the Colorum, with their stronghold at Socorro. After a sharp fight, the mixed force landed by the gunboat chased the insurgents out and took control of the town.

That same year, facing the reality that the Spanish-built gunboats serving in the Asiatic Fleet for a quarter-century were nearly worn out, the navy asked congress for funding to build six new gunboats for the China station, mostly to patrol the Yangtze River. On that river, landings by sailors continued routinely, although the necessity to bring heavy weapons on shore remained rare. In what could have been a scene written by old China hand Richard McKenna for his novel, *The Sand Pebbles*, in November 1927, gunboat *Asheville* sent bluejackets up the Makyoung River to the American Maryknoll mission at Yeung Kong, only to find that the missionaries had not been threatened. A few weeks later, during the Red Guard uprising in Canton, the seemingly ubiquitous *Sacramento* put ashore a landing party bolstered by her field gun to protect the American consulate. With the easing of the conditions the following day, the warship withdrew her detachment. Also on that day, December 13, a motor launch carrying a second *Sacramento* detachment accompanied gunboat *Pampanga* in order to escort forty-six Americans and other foreign nationals from Canton's native city to safety on the island of Shameen. By coincidence, *Asheville* and *Sacramento* were among the few US warships still carrying the Mark I mod 1 3-inch field gun.

In January 1927, soon after being deployed as the flagship of the Asiatic Fleet, armored cruiser *Pittsburgh* (ex-*Pennsylvania*) sends ashore a landing party to protect American and European citizens in Shanghai during the civil war then raging. The helmeted bluejackets include riflemen and stretcher bearers, as well as several Chinese crewmen assigned to duty ashore. The warship survived until 1931 before being decommissioned and scrapped. (*Naval History and Heritage Command*)

Year's end of 1927 saw ocean-going gunboat *Sacramento* put ashore a landing party at Canton, China, accompanied by her Mark I mod 1 3-inch field gun, in order to protect the US consulate and other American interests. She is seen at Tsingtao during the late 1920s or early 1930s, with at least four S-boats of the Asiatic Fleet's submarine force appearing in the foreground, and sampans and junks sailing farther out in the bay. (*Naval History and Heritage Command*)

Two notable gun losses due to misadventure occurred during the 1920s. On October 24, 1922, tug *Kewaydin* commenced towing *Lighter No. 91*, laden with both government freight and dependents' personal goods, from Hampton Roads to Washington. The vessels soon encountered heavy weather in Chesapeake Bay and the lighter capsized in what the navy determined to be the fault of naval officers in responsible positions, including the tug's captain. In addition to the forfeiture of property by the various naval officers or their wives, the marines lost Mark VII 3-inch landing gun No. 1055, Mark V field carriage No. 221, and Mark IV limber No. 121.

The mishap nearly fifteen months later turned out to be the worse by far, because loss of life accompanied it. Assigned to the navy's Special Service Squadron, Denver-class protected cruiser *Tacoma* (eight 5-inch/50cal guns) patrolled the waters near Veracruz, Mexico, in January 1924. During rough weather on the 16th, she ran aground on Blanquilla reef. For the following week, the crew labored arduously but unsuccessfully to free her. Somehow in that attempt, Capt. Herbert Sparrow and three enlisted men drowned. Mark XI 3-inch landing gun No. 2118, mounted on Mark VI field carriage No. 287, also did not survive the disaster.

References Consulted

Dictionary of American Naval Fighting Ships, Naval History and Heritage Command (online). Consulted for each US naval warship and auxiliary mentioned in this chapter.

Officers of the US Navy, *Naval Ordnance: A Textbook Prepared for the Use of the Midshipmen of the United States Naval Academy*, Chapter XVI, "The Naval Landing Gun" (Annapolis: US Naval Institute, 1939).

Smith, J. E., *Small Arms of the World*, 10th rev. edit. (Harrisburg: Stackpole, 1973), pp. 117–19, 151–52, 663–66, 683.

US Congress, *Public Act No. 375*, 67th Cong., 4th sess., Dec. 28, 1922.

US House Comm. on Appropriations, Subcommittee on Navy Department, *Navy Department Appropriation Bill, 1940 (Bureau of Ordnance): Hearing on H.R. 6149*, 76th Cong., 1st sess., Apr. 3, 1939, p. 534

US House Comm. on Claims, in regard to the relief of Ann Margaret Mann, H.R. 8423, 68th Cong., 2nd sess., Jan. 19, 1925, and *Report No. 1286 to accompany H.R. 5264*, 69th Cong., 1st sess., May 21, 1926, pp. 1–2.

US National Archives, Record Group 74, Records of the Bureau of Ordnance; Washington, DC.

US Navy, *Assemblies, 3-inch Mountain Gun, Mark XII, General Description*, Ordnance Pamphlet No. 124, May 1923.

———, *Assemblies, 3-inch Naval Landing Gun, Mark XI, with Carriage and Limber*, Ordnance Pamphlet No. 131, July 1921.

———, *Landing Force Manual, United States Navy 1920* (Washington, DC: Govt. Print. Office, 1920).

US Navy Dept., *Annual Reports*, Secretary of the Navy, *1924*, p. 7; *1928*, p. 5.

———, *Annual Reports, 1927*, Marine Corps, p. 1193.

13

US Naval Landing Guns in the 1930s

Replacement of the 3-inch Landing Gun by Other Ordnance

The 1930s would see not only the retirement of three types of US 3-inch naval landing guns previously in service, but by mid-decade the removal of all 3-inch landing guns from warships of the fleet and the brief use of smaller pieces to fill that role. The changes in policy and doctrine would evolve against the backdrop of the aggressive wars of acquisition undertaken by the militaristic powers in Asia and Europe. The trend for the US Navy during this prewar decade continued to be the withdrawal from intervention in those parts of the world it had hitherto kept a close watch on. With the need thus lessened for landing parties to maintain the peace, show the flag, or protect American interests, the weaponry carried aboard ships of the fleet also changed in composition, with the heaviest disappearing first.

During the summer of 1930, the Philadelphia Navy Yard shipped four partially overhauled Mark VII 3-inch landing guns to the Naval Gun Factory, serial numbers 1075, 1153, 1169, and 1174, along with their field carriages and Mark II limbers. Although all of them had been initially been allocated to the Marine Corps, the last two had seen their final service aboard the Connecticut-class pre-dreadnoughts *Minnesota* and *Kansas*, a rare thing for this mark of landing gun. The Philadelphia yard declared that additional work and spare parts were required to put these pieces into proper condition for reissue. Well into the following year, the Washington yard estimated a total cost of $3,925 to place the four pieces into a serviceable condition, including $200 to modify the upper shield on one of the carriages. Subsequently, the bureau instructed the gun factory to hold the work in abeyance pending any decision to reissue them to service (likely never done). The navy thus signaled that the Mark VII landing gun had nearly come to the end of its useful life.

In March 1931, the Naval Gun Factory was confronted with a Mark IV and a Mark VII landing gun, each in poor condition with an enlarged bore and both chamber and bore pitted throughout. The Mark VII moreover proved to be badly eroded. Despite the newer gun having fired more than 1,500 rounds versus the older one having fired fewer than 900 rounds, the gun factory recommended the Mark IV be disposed of and the Mark VII be retubed to become a Mark VII mod 1. It is not clear if the facility did in fact refurbish the Mark VII landing gun, but the preliminary decision reflected the navy's long unhappiness with the Mark IV landing gun, which the service deemed the weakest of its landing guns, particularly in regard to the gun's breech mechanism.

But the Mark VII's final days were not long in coming either. At the end of 1932, the Marine Barracks at Pearl Harbor turned in its pair of Mark VII mod 1 guns, sundry accessories, and the battery commander's fire control equipment. Those return shipments continued at various marine posts, with B Company (formerly the 39th Company), Peking Legation Guard, being the last unit equipped with the Mark VII landing gun. The four pieces, Nos. 1056–1059, remained in the hands of the Peking garrison in 1936. By the following year, it had shipped them to Cavite Navy Yard, where they joined Nos. 1157 and 1158 and where all six lingered on December 8, 1941.

Also in December 1932, the Philadelphia Navy Yard reminded the Bureau of Ordnance that it still had stored, since 1911, fifteen each Mark III 3-inch/50-caliber field carriages, carriage slides, and Mark II limbers, and eleven Mark VIII sights for same. There had been some thought of this equipment being used by the advance base force, but those days had long flown, and the gear was old and sorely obsolete. Accordingly, the Philadelphia yard requested permission to dispose of it. The yard would wait some while to get an answer.

The major reductions in the types of landing guns serving in the fleet began in late August 1934, with the request from the bureau to the secretary of the navy to declare the Bethlehem Mark IV 3-inch landing gun obsolete, and to scrap all such guns and associated material then surviving. In the following month, the Bureau of Ordnance added the Mark I and Mark I mod 1 field guns to those landing pieces deemed obsolete, but wished to retain them as drill pieces. Rear Adm. Edgar Larimer, chief of bureau, wrote, "The bureau's current war plans do not contemplate the wartime use of any of these guns, which are considered unsuitable for the purpose due to an insufficient margin of safety. These old guns should be removed from otherwise valuable storage space in navy yards and depots." The bureau ordered their ammunition entirely disposed of.

Although the twenty-two Mark IV guns that survived eventually received a reprieve, and instead were to be rendered unserviceable and donated for decoration and exhibition, they would no longer be used by the navy for any purpose, even gun drill. In fact, ten Mark IV landing guns then in naval training establishments were scheduled for replacement by the older Fletcher-type field guns, of which 118 remained. At first glance this seems to have been an illogical, retrograde choice because of the Fletcher guns' lack of modernity and relevance, but it must be kept in mind that the navy had never been happy with the Bethlehem Mark IV. Its design in many aspects remained more the exception than the rule. Moreover, the navy would have had to maintain spare parts to keep these twenty-two guns in operation as drill pieces. If those spare parts had already been largely depleted, that also would explain the choice away from the Mark IV as a training gun.

Marines in Peking wearing 1922 Pattern service caps pose with their four-legged buddy and a Mark VII 3-inch landing gun. An original six-gun battery in Peking is implied by the number stenciled on the shield. The Ehrhardt-designed piece was the only US Navy landing gun to have a box trail and the last to be equipped with a trail wheel and trail stave. The Thompson submachine gun ammunition drum slung from one man's shoulder and the rifles propped against the wall close at hand testify to the unrest in China during the late 1920s. (*USMC History Division*)

By the mid-1930s, these Mark VII 3-inch landing guns assigned to B Company of the Peking garrison remained the final four such guns in Marine Corps service. The inspection party led by Lt. Col. Lowry Stephenson (1889–1939), fleet marine officer of the Asiatic Fleet, also includes Lt. Col. Alexander Vandegrift (1887–1973), detachment commanding officer, and 1st Lt. David Shoup (1904–1983), both future commandants of the US Marine Corps. (*USMC History Division*)

The equipment authorized in the early 1930s for the 6th Battalion, US Naval Reserve, assigned to the national guard armory on Randolph Street in Chicago, included two Fletcher Mark I 3-inch field guns modified with Mark I mod 2 recoil sleeves. Thorough training was done on these old landing pieces at that time, in the event the reserve unit was called to active duty and issued more modern landing guns. The guns arrayed on the upper level are training pieces for an army national guard coast artillery unit. (*Naval History and Heritage Command*)

Larimer's letter also noted that two warships in the fleet, the gunboats *Asheville* and *Sacramento*, each had a Mark I mod 1 3-inch field gun aboard, the last two pieces of this type in service. By year's end, the bureau stipulated their replacement by Mark XI 3-inch landing guns and associated equipment, and in March 1935, it ordered the exchange to be carried out. But while these large oceangoing gunboats were gaining Mark XI landing guns, the navy's heavy cruisers were being divested of theirs.

When *Augusta* returned to the Far East in the fall of 1933 to relieve her sister *Houston* as flagship of the Asiatic Fleet, she carried an augmented landing force, including one or two each Model 1916 37mm infantry guns and 3-inch Stokes mortars. A June 1933 letter anticipating her departure declared the need for Cole carts to carry these landing weapons and their ammunition, along with fifteen Thompson submachine guns and 200 steel helmets (of the Kelly type). Penciled notations, however, indicate that none of these latter-named items were then available. Some of them may have been obtainable by the time of her leaving in October, but if so, the Cole carts would have had to be acquired from the Marine Corps. The marines had been equipped with the 37mm guns and 3-inch mortars as standard infantry-support weapons since the 1920s. Completely unmentioned in the June letter is *Augusta*'s Mark XI 3-inch landing gun, casting strong doubt that she still carried it.

Two years later, however, it became unequivocal that the Mark XI 3-inch landing gun was on the way out. Twelve cruisers under construction—heavy cruisers *Quincy*, *Vincennes*, and *Wichita*, and nine light cruisers of the Brooklyn class—and two large oceangoing gunboats, *Charleston* and *Erie*, would be assigned the M1916 37mm infantry gun and the new M1 81mm mortar. The latter was an American modification

Above and below: Two long lines of riflemen, supported by a battery of Mark XI 3-inch landing guns, fire blank rounds against an imaginary foe during a training exercise at the naval academy. By 1935, the navy had begun removing 3-inch landing guns from its large fleet units, but the midshipmen at the academy still underwent training on them. Such guns would disappear from the academy training schedule by the beginning of US involvement in World War II. (*US Naval Academy Archives*)

of the French Brandt mortar, itself inspired by the Stokes mortar. Much smaller than the 3-inch landing gun, the mortar delivered explosive rounds of about the same size and effect, required much less effort to get ashore, and in a protected emplacement presented a rather smaller target against counterbattery fire. Not for nothing did it soon become a "stovepipe" in military slang.

The two gunboats in addition were slated to carry a 75mm pack howitzer. Four of these pieces, first designated the Model 1923-E2 and then the M1, and readily broken down into six pack loads for mules, had found their way into marine hands in 1930 and proved far superior to the older mountain guns. The recipients, the 10th Marines, had come into being in early 1918 along standard field artillery lines, but had not evolved beyond its bare bones stateside when the war ended. Impressed with the pack howitzer's performance, in the mid-1930s the navy managed to obtain two 75mm howitzers to use as landing pieces aboard its two newly built gunboats.

At the same time, the exchange of the 3-inch landing gun for the 37mm gun and 81mm mortar commenced on the somewhat older treaty cruisers. *Louisville* will serve as the exemplar. The four-year-old heavy cruiser took aboard the light gun in June 1935, with the mortar to be obtained when available. The Bureau of Construction and Repair authorized the magazine space then allocated for stowing the 3-inch landing gun ammunition be converted for the new rounds. The Pearl Harbor Navy Yard carried out this alteration during the ship's overhaul at the end of 1936, with the Mark XI 3-inch landing gun—likely No. 2139, mounted on Mark VI field carriage No. 277—turned in at the yard and its ammunition offloaded to the Naval Ammunition Depot on Oahu in January 1937.

The navy convened a board of officers aboard the battleship *Texas* in early January 1936 in order to identify possible weight reductions on these capital ships, such to permit the mounting of additional light antiaircraft guns and stowing their ammunition. The long list of changes included the removal of the pair of saluting guns, as well as the 3-inch landing gun, caisson (limber), and ammunition. This decision represented the final knell for the larger landing guns aboard American battleships.

While these events were happening, tests were underway at the Naval Proving Ground, by then at Dahlgren, Virginia, to correct another problem in the Mark VI field carriage. Even though the decision had been made to remove the 3-inch landing guns from fleet units, the navy likely kept an eye on maintaining these fifty-odd artillery pieces against some future need. During the proofing of the pilot carriage and its Mark XI landing gun seventeen years before, the Bureau of Ordnance had identified the cup leather packing in the recuperator, or counter-recoil cylinder, as a likely weakness. That flaw had been confirmed in the years since, with the substantial leakage of fluid around the piston and consequent reduction in pressure during firing. Moreover, the air-charging valve in the cylinder with the leather packing failed to hold its pressure. The proving ground substituted horn and fiber packing, but with no substantial improvement. When it tried Chevron packing, however, the recuperator leakage diminished appreciably and the air-charging valve held its pressure. In October 1937, following a one-on-one comparative test, the bureau ordered that the superior packing supplant the original material when each carriage came up for overhaul. Given the

The Model 1916 37mm infantry gun originated with the French army. Adopted by the US Army and Marine Corps for combat on the Western Front, these guns saw frequent use against enemy machine gun nests. Increasingly dissatisfied with its heavy 3-inch landing guns, during the early 1930s the US Navy adopted the 37mm infantry gun for landing purposes. In order to deliver heavier fire, the navy selected the Stokes 3-inch mortar, succeeded by the M1 81mm mortar, which were far more readily brought ashore, concealed, and protected. These soldiers at Fort Douglas, Utah, train on the Model 1916 37mm gun mounted on a heavy tripod. US Army photograph. (*Fort Douglas Museum*)

The Model 1916 37mm infantry gun was mounted as well on a wheeled carriage for landing purposes. The wheels seen here are the same 19-inch diameter ones on the Cole carts of various kinds used by the marines in China, but in other cases they were more traditional wooden-spoked ones (see picture on page 308). A 3-inch Stokes mortar is also in the photographer's frame. (*USMC History Division*)

The army developed its M1 75mm howitzer both to break down for transport by pack mules over rough terrain and as a light fieldpiece to accompany cavalry. The navy acquired a small number of 75mm pack howitzers for brief use as landing guns on the large oceangoing gunboats *Erie* and *Charleston*. The acute need by the marines for such pieces soon resulted in their transfer to the corps. This army M1A1 pack howitzer shown in the 1930s is mounted on the original M1 box-trail carriage, identical to that issued to the gunboats. US Army photograph.

The end of the 1930s also spelled the end of the useful service life for the Mark XI 3-inch landing gun. Such a gun appears with the landing party mustered aboard the Tennessee-class battleship *California* late in the decade. Sister *Tennessee*, also assigned to the US Pacific Fleet, looms astern. Both battleships would be present during the air attack on Pearl Harbor. (*Naval History and Heritage Command*)

late date, the wide replacement of the recuperator packing seems unlikely. With the progressive removal of the Mark XI 3-inch landing guns from fleet units, the major users of this ordnance remained the midshipmen at the naval academy, and they too would relinquish their guns sometime around decade's end.

Mid-decade also saw the tying up of other loose ends. Not having gotten any response for almost three years, the Philadelphia Navy Yard tried again in August 1935 to obtain permission to scrap the old Mark III 3-inch/50-caliber siege carriages, slides, limbers, and accessories, stored there for nearly a quarter-century. This time the yard met with success, with the chief of bureau conceding that no possible use could be foreseen for this old matériel. Some of it, such as the slides, spare parts, and tools, went to the Washington Navy Yard for stock, while the scrapper got the rest.

New small arms arrived with the 1930s. The marines and navy adopted redesigned Cole carts to transport their machine guns and ammunition ashore. The Garand semiautomatic .30-caliber infantry rifle, designated the M1, found mention in a February 1934 letter from the secretary of war to the secretary of the navy. The former emphasized the need to keep the weapon's details classified and the technical notes out of unauthorized hands. In his response, the latter wondered if the navy would be included in the first limited issue of the rifle. The M1s were obtained eventually, but a general issuance to the Marine Corps took place only after its assault infantry regiments had learned a hard lesson at Guadalcanal, equipped with their trusty bolt-action M1903 Springfields while facing Japanese mass frontal attacks. Contrary to legend, M1 rifles had been issued to the marine artillery and supporting units

The Mark VII limber accompanying the Mark XI 3-inch landing gun carried forty rounds, fewer than in previous marks, in an attempt to reduce the combined weights of gun, carriage, and limber. The empty slots in the limber and the litter of spent cartridge cases on the ground in this naval academy training exercise in 1935 reveal that about half of the rounds carried have been fired. Note the split trails on the Mark VI field carriage. (*US Naval Academy Archives*)

present on Guadalcanal, but the infantry had to wait until the semiautomatic rifle passed its acid test in combat.

To defend the costs estimated by the Bureau of Ordnance, part of the naval appropriation bill for 1940, its chief, Rear Adm. William Furlong, went before the House subcommittee on the Navy Department in April 1939. Among many other requests, the bureau had asked for more than $81,000 to purchase both sizes of the new landing pieces for battleships. In response to the questions directed by the subcommittee chairman, Congressman James Scrugham from Nevada, the admiral explained that the 81mm mortar had replaced the 3-inch landing gun on those capital ships. All too soon, however, the mortar and its 37mm sidekick would be gone as well.

Lessons Advocated by the Special Service Squadron

The Special Service Squadron, composed of gunboats *Dolphin*, *Sacramento*, *Asheville*, *Tulsa*, *Charleston*, and *Erie*, older destroyers *Jacob Jones*, *J. Fred Talbott*, *Zane*, and *Perry*, and small cruisers *Birmingham*, *Des Moines*, *Galveston*, *Tacoma*, *Richmond*, *Omaha*, and *Trenton*, serving at one time or another, had patrolled the Caribbean Sea during the early decades of the twentieth century. Later, such warships kept their vigil on both isthmian coasts of Central America, constantly alert for trouble and ready to put a landing force of bluejackets and/or marines ashore to protect Americans and American interests in that volatile region. The squadron ceased operation in 1940.

In late March 1934, Maj. Pedro del Valle, squadron marine officer, sent a long report to the commander of the Special Service Squadron, Rear Adm. Charles Freeman, with various recommendations. Based on del Valle's report, Freeman prepared his own assessment for the chief of naval operations, Adm. William Standley (subsequently copied to several naval bureaus), conveying these ideas based on the firsthand experience of both officers in landing parties.

The old dilemma persisted of furnishing a sufficiently strong force ashore versus having an adequate number of crewmen left aboard to maneuver the ship and fire the guns. Any ship's commanding officer was warned, however, that sending a weak force ashore, which suffered heavy casualties, got rebuffed, and had to withdraw, did more harm than good. A major logistical objection to maintaining sufficient strength for a prolonged period, in order to deal with a large threat ashore, remained the slim likelihood of such a naval landing arising. On the other hand, if by chance the squadron did encounter such a hostile force, the landing attack as conceived decades before had become unthinkable, because the opposition would very likely by then be armed with modern weapons, including machine guns and light- or medium-caliber artillery.

Because, more often than not, a destroyer would be providing the landing force, in the opinion of these officers, the number of men sent ashore could be increased as long as the ship remained secure from naval attack or bad weather. Thus a destroyer with a complement of 106 men and six officers could safely land seventy men and three officers, whereas a Special Service Squadron destroyer with a complement of 118 men, including thirty-seven marines, could safely land eighty men and three officers. The present small arms allocation of twenty-five M1903 rifles and M1905 bayonets, twenty-five M1911 pistols, three M1918 Browning automatic rifles, three Mark VI Lewis guns—and compellingly, twenty-five canteens—were clearly insufficient for a landing force for a destroyer of either complement. Admiral Freeman urged that the *Landing Force Manual* be changed to reflect these amplified needs, and it was done.

Despite their rigorous shipboard duties, the members of the landing force had to keep in good physical shape and be thoroughly trained in order to march long distances, utilize cover, maneuver effectively—with the knowledge of hand signals—and above all, use weapons efficiently and shoot straight. Line officers had an even greater responsibility, both to their mission and their men, in keeping informed of the most current revision of the *Landing Force Manual* and other publications, and effectively deploying and using all weapons issued to the landing force.

Drawing on experience garnered in World War I, the officers making the reports recommended additional equipment as essential. In their firm view, all landing personnel needed khaki uniforms and sturdy boots, and the bluejackets required more than just ponchos against driving rain, but also shelter halves and ground sheets for their health, comfort, and adequate sleep if they were to remain an effective fighting force. In the event of injuries or battle wounds, sufficient medical personnel needed to be present, ideally one corpsman per fighting platoon, along with sufficient first aid material and stretchers (with men assigned as stretcher bearers). If the landing force remained on shore for a longer period of time, other items such as pioneer tools, sand bags, barbed wire, demolition sets, water containers, Lister bags to ensure potable water, and mobile or pack kitchens would be needed.

Two items strongly recommended from their origin in the recent war were gas masks and steel helmets. The American Kelly helmet, derived from the British Brodie model quickly devised because of the immediately urgent need, became so hot in the tropics that a hand could not be laid upon it. Cloth, leaves, or other insulating material placed between the head and the helmet proved self-defeating because such padding lifted the helmet too far above the head and face to offer protection against grenades and shell fragments. At the time, the navy possessed few trench helmets, as the service termed them, and had to turn to the marines in order to obtain such items. All three armed services acknowledged the need for a better-designed helmet.

The strong recommendation was made that all boats aboard be motorized and most be made available for landing operations, but the navy adamantly resisted this suggestion, with the reminder that its vessels needed pulling boats to carry out man-overboard and lifesaving attempts, whatever the prevailing weather conditions.

As previously set forth, naval landing guns had by then become so heavy that animal hauling over a considerable distance was a necessity. The impossibility of maintaining draft animals aboard ship adequate for towing guns and the great difficulty of acquiring ones on shore of sufficient number and sturdiness were widely acknowledged. The navy and marines had landed draft animals on an Oahu beach during the 1932 fleet problem, with unpromising results. This long-standing obstacle offered another reason to replace the heavy 3-inch landing gun with something a good deal lighter.

Of course, as always, money loomed large over these desired additions and modifications, with funds remaining tight during the Depression years of the 1930s. Some of the sensible suggestions originating from Major del Valle and Admiral Freeman were carried out and some not. At decade's end, as gunfire muttered from both Asia and Europe, del Valle's notions became even more relevant. His thinking had gone far beyond traditional landing parties, but rather incorporated the landing of substantial bodies of infantry, larger field artillery pieces, and motor vehicles from specialized landing boats—all the stuff of the amphibious assaults that the marines, navy, and army would undertake in the coming war.

References Consulted

Biggs, C. M., Jr., *The United States Marines in North China, 1894–1942* (Jefferson, NC: McFarland & Co., 2003), pp. 141–42.

Buckner, D. N., *A Brief History of the 10th Marines*, History and Museums Division, HQ, US Marine Corps (Washington, DC: US Marine Corps, 1981), pp. 18, 40.

Dictionary of American Naval Fighting Ships, Naval History and Heritage Command (online). Consulted for each US naval warship mentioned in this chapter.

Hiatt, R. C., "Goodbye 'Little Dynamite'," *Marine Corps Gazette*, August 1949, pp. 44–47.

Plowman, T., "Pre-war/WWII Era USMC M1 Garands," undated (online).

Smith, J. E., *Small Arms of the World*, 10th rev. edit. (Harrisburg: Stackpole, 1973), pp. 85–86.

US House Committee on Appropriations, Subcommittee on Navy Department, Navy Department Appropriation Bill, 1940 (Bureau of Ordnance): Hearing on H.R. 6149, 76th Cong., 1st sess., April 3, 1939, p. 534.

US National Archives, Record Group 74, Records of the Bureau of Ordnance, Washington, DC, and College Park, MD.

14

The Beginning of World War II and the End of US Navy Landing Guns

1940–1941

The progression in the character of the US Navy's shipboard landing forces and landing guns evident in the 1930s continued into the 1940s. In late July 1940, the chief of the Bureau of Ordnance pointed out to the superintendent that the US Naval Academy retained one 75mm pack howitzer for training purposes, while only two such landing pieces served in the entire fleet. The chief inquired if this howitzer was considered essential, and if not, he intended to transfer the piece to the marines, who needed every such howitzer they could lay hands on. Two months later, the superintendent replied that after due consideration, the retention of the pack howitzer could not be justified and he approved its removal.

During September 1940, the same month that the superintendent agreed to the transfer of the academy's single light howitzer, large gunboat *Charleston* sailed for Seattle and began patrols between that port city and the Aleutian Islands. Puget Sound Navy Yard shipped M1 75mm pack howitzer No. 52 on M1 carriage No. 48, with all spare parts and accessories, to the Washington Navy Yard in early June 1941. The temptation to conclude that the large gunboats lost their 75mm landing howitzers at this time would be misplaced. The marines had been experiencing cracks in the recuperators and sleighs of their M1 75mm pack howitzers. The army had redesigned the carriage as the M1A1, and the shipment from Puget Sound likely points to the replacement of *Charleston*'s original landing howitzer with a better-designed and constructed piece.

Thereafter, the navy began the removal of other landing gun types from most of its warships, although there would remain several exceptions due to special circumstances. A complete survey of all battleships and heavy and light cruisers made in mid-August 1941 reflected the trend to discontinue the 37mm infantry guns and 81mm mortars as landing force weapons on these warships. When *Houston* had steamed west ten months before to relieve her sister *Augusta* as flagship of the Asiatic Fleet, she carried a substantial number of 37mm and 81mm rounds. On November

30, 1941, ten days before the Japanese bombed Cavite Navy Yard to destruction, *Houston* still had aboard a few hundred rounds for each landing piece, despite the August survey. Standard procedure called for the removal of ammunition when the guns firing such rounds were taken off—shipboard space was simply too valuable to spare. That the heavy cruiser still had her landing gun ammunition on board strongly suggests the continued presence of the guns as well. The prolongation of landing guns aboard *Houston* may have represented an exception permitted the Asiatic Fleet for its landing force equipment, as alluded to in an October 1941 letter from the chief of naval operations to the chief of the Bureau of Ordnance.

Other exceptions to the landing gun removals, as may be expected, lay in the navy's five oceangoing gunboats. *Sacramento*, *Asheville*, *Tulsa*, *Erie*, and *Charleston* each carried one 37mm gun and one 81mm mortar as landing pieces, with the last two warships named each still toting in addition one 75mm pack howitzer, probably the improved M1A1. The small Yangtze River gunboats, though frequently involved in putting bluejackets ashore along the river, normally carried nothing larger than machine guns as landing pieces. *Tutuila* was one of the two smallest such gunboats—a displacement of 395 long tons, a length just shy of 160 feet, and a draft of 5 feet 5 inches—and customarily operated in the upper reaches of the Yangtze. She remained in every sense on her lonesome, and for that reason she carried a pair of 81mm mortars in addition to her two 3-inch/23-caliber dual-purpose deck guns. War would catch *Tutuila* way upriver at Chungking, and eventually the American navy transferred her to the Chinese, likely with the two mortars still aboard.

The widespread removal of landing guns from the fleet did not end the presence of landing parties armed with lesser weapons—at least not yet. Many classes of ships still maintained a landing force aboard. They included aircraft carriers, the idea being that such parties on carriers would specialize in seizing air fields in order to prevent hostile forces from capturing them and using their facilities to attack the American task force. These landing forces were armed with the customary rifles, pistols, and light automatic weapons.

Within a day of each other in March 1941, Adm. Ernest King, commander-in-chief of the Atlantic Fleet, and Adm. Husband Kimmel, his counterpart commanding the Pacific Fleet, wrote to Adm. Harold Stark, chief of naval operations. Both men recommended that the landing forces be maintained as they were aboard battleships, cruisers, and destroyers, and agreed that the weight saved by the landing force equipment would be nugatory. They disagreed, however, on the need for a landing force in aircraft carriers, with only King desiring elimination in such ships. Kimmel's ocean was a great deal larger, with many more islands, and thus had potential for a lot more airstrips to endanger his task forces. In the following month, however, Adm. Royal Ingersoll, assistant chief of naval operations, directed the admirals commanding both the Pacific and Atlantic Fleets, with copies to each aircraft carrier in service, to eliminate the landing force from all carriers and to relinquish all of its equipment, except that for the marine detachment aboard. An amendment allowed the aircraft carriers to retain their steel helmets for crewmen whose battle stations lay topside, and their pistols to arm officers and aviators.

Destroyers for the moment kept their landing force, with the recommendation for every such warship in the Atlantic Fleet of thirty each .30-caliber rifles and .45-caliber pistols, three each .30-caliber automatic rifles and .30-caliber Lewis machine guns, and fifty allotments of infantry equipment. The recommended allowance for a destroyer tender was 140 rifles, fifty pistols, nine automatic rifles, three Lewis guns, and 175 sets of infantry equipment. In the autumn, the allowance for St. Louis-class light cruisers *St. Louis* and *Helena* and Brooklyn-class light cruisers was set at eighty rifles, eighteen pistols, fifteen automatic rifles, and six Browning machine guns. Battleships still had landing force gear aboard. In late November 1941, on the eve of war, the navy began the removal of the landing force equipment from its destroyers.

Sometime after their removal from warships, the Bureau of Ordnance had collected the Mark XI 3-inch landing guns into four major caches of such weapons. It did likewise, but to a far lesser extent, for the Mark VII 3-inch landing guns. In September 1941, the following facilities held significant numbers of these two landing gun marks, their limbers and accessories:

Washington Navy Yard: fourteen Mark XI landing guns
Bellevue Ordnance Annex (Washington, DC): two Mark XI landing guns; twenty-one Mark VII landing guns
Philadelphia Navy Yard: one Mark XI landing gun (removed from battleship *Texas*)
Mare Island Navy Yard: twelve Mark XI landing guns; one Mark VII landing gun
Puget Sound Navy Yard: ten Mark XI landing guns
Cavite Navy Yard: fourteen Mark XI landing guns; six Mark VII landing guns

A smattering of the former marine Mark VII landing guns could also be found here and there, but whereas nearly all of the fifty-five surviving Mark XI 3-inch landing guns were accounted for in the fall of 1941, just a few more than half of the Mark VII 3-inch landing guns were identified. Both marks of US naval guns would have a role, albeit a small one, in the war looming just over the horizon.

The Prewar Sale of Surplus Mark VII 3-inch Landing Guns

German forces overran the Netherlands in May 1940, surrogating the government of the Netherlands East Indies (NEI) to represent the largest Dutch population then remaining free. The occupation of the home country left the NEI as very tasty dangling fruit with enormous natural resources. The Japanese had long had their eyes on these mineral- and petroleum-rich islands, but now thanks to their German allies, such subjugation became that much easier. Acutely aware of that threat, the orphaned Dutch government set up the Netherlands Purchasing Commission (NPC) in New York City, whose goals included the procurement of ordnance. The President's Liaison Committee (PLC), loosely affiliated with the US government, acted to expedite the sales of both ordnance once standard but now obsolete, and modern armaments manufactured by private industry but not adopted by American forces. The artillery the Dutch commission found of particular interest included former navy deck guns that could be adapted as coast artillery, medium-caliber field artillery pieces, and small-caliber antitank guns.

A July 1940 memo between members of the PLC advised that fifteen 3-inch/23cal "old-type" field guns, with some shrapnel and common shell, but no high explosive rounds, might be made surplus. Their disposal, however, had to await the decision whether they remained necessary to the national defense. An internal note between PLC members Philip Young and Rear Adm. Ray Spear, USN, declared that the Dutch had since indicated that those fifteen 3-inch/23cal field guns—from their explicit, short barrel length almost certainly naval landing guns—were not suitable for their needs. Those guns were back on the table in June 1941, however, although the number now being considered had become one dozen. The US Navy had twenty-one of these former marine landing guns conveniently available at its Bellevue Annex of the Naval Gun Factory, and it is likely that the Dutch decided upon the reduced number of guns needed. The Dutch paid $85,000 to the US government in order to recondition the twelve landing guns. A separate contract covered the purchase of 5,000 3-inch rounds for $50,000.

With virtually no competitor for the task, the York Safe & Lock Company of York, Pennsylvania, became the prime contractor for reconditioning the landing guns. The twelve pieces were sent there by rail in late November 1941. It may not have been coincidental that the Martin Parry Company was also situated in York, because the two guns surviving today bear Martin Parry high-speed conversion rig that permitted towing by motor trucks. As devised by the company's chief design engineer, Adolph Buquor, the conversion kit consisted of two Budd steel disc wheels for pneumatic rubber tires attached to stub axles, which in turn were carried on cranked adapters that fitted over the ends of the original axletree. The high-speeded gun thus sat at the same height as the original one hauled by marines and permitted the use of the firing data in the existing range tables. The addition of radius rods and brake gear enabled towing speeds of 30 to 40 mph.

The tale becomes murky thereafter. By mid-January 1942, the Mare Island Ammunition Depot in California had assembled 5,004 3-inch/23cal rounds for the landing guns, of which the Dutch requested thirty to be sent at once for testing purposes. A few weeks later, apparently having heard nothing more, the navy queried the Netherlands Purchasing Commission whether it was able to receive the remaining 4,974 rounds. Postwar correspondence, however, shows that only those initial thirty rounds were ever delivered to the Dutch. In May 1942, two months after the surrender of the Netherlands East Indies, the chief of the Bureau of Ordnance informed the NPC that the York Safe & Lock Company had been able to refurbish the dozen landing guns for nearly $9,000 less than the $85,000 first estimated. That notification, coupled with the reality that today the few surviving Mark VII landing guns are to be found exclusively in Australia, bolsters the likelihood that the guns were shipped in advance of almost all of their ammunition.

The picture emerging is that the twelve Mark VII 3-inch landing guns departed the shores of the United States sometime after the Japanese attack on Pearl Harbor, during a period when their arrival in Java would still have been reasonably anticipated. Following a finite period of safe travel, Allied merchantmen crossing the Pacific by the relatively safe southern route soon found themselves either diverted to an Australian port or ordered back to the United States. Once in Australia, ordnance and other stranded military cargoes originally purchased by the Dutch were divided, for the most part equitably, among the armed forces of Australia, the United States, and the Netherlands (soon, however, the Dutch would transfer most of their equipment to Australian forces). Without their ammunition, these landing guns would have been of no use for beach defense, and as non-standard pieces, their employment in training would have been brief at best. It is likely they remained derelict for the war's duration, and of no military value.

Given that examples of the US Mark VII landing gun survive only Down Under, the irony is even greater that in this third decade of the twenty-first century, but few Australian artillery enthusiasts have an inkling of their wartime history.

The Prewar Transfer of Mark XI 3-inch Landing Guns

In late 1940, Maj. Gen. George Grunert, commanding officer of the Philippine Division, requested through channels that certain naval landing guns stored at Manila Bay, for which the navy did not anticipate future use, be transferred to the army. The 16th Naval District informed the Bureau of Ordnance that fourteen 3-inch/23cal landing guns remained stored at Cavite Navy Yard. The guns were practically unused and in good repair, but at that time had not been allocated a role in existing war plans. Also in the inventory were sixteen limbers, with an estimate of 8,000 assembled explosive shells and 15,000 unassembled shrapnel rounds. The commander-in-chief of the Asiatic Fleet (CinCAF) verbally approved the transfer. That inventory also comprised six Mark VII 3-inch landing guns, which at one time or another had seen service with the marines in Peking.

After the pot had simmered for five months, the bureau repeated its offer to transfer the fourteen Mark XI guns to the army, but now accompanied by 4,000 each shell and shrapnel rounds. The navy assured the army that no reimbursement would be necessary, other than the relatively inexpensive costs for packing and shipping. On that basis, Washington queried the commanding general of the Philippine Department if the army still wanted the naval landing guns, and Grunert answered in the affirmative. In early June 1941, the department stressed that these pieces would be a desirable asset in beach defense and identified two particular sectors—to receive eight and six guns respectively—where they would be of particular value in repelling hostile landings. In its turn, the War Department confirmed that need and the details of transfer previously proposed. At month's end, Brig. Gen. Eugene Reybold, acting chief of staff, underscored the immediate need for the landing guns, with Henry Stimson, secretary of war, following up at once with the approval of their swift transfer to the army.

Upon receipt, the Mark XI 3-inch landing guns, Mark VII limbers, more than 7,000 rounds of high explosive, and more than 4,900 rounds of shrapnel went to Fort Stotsenburg in Pampanga, Central Luzon, about 50 miles northwest of Manila. After the Philippine Department had had time to consider the matter, in early November

1941, it recommended sending all fourteen guns to the Visayan-Mindanao force. The intention was to assign eight pieces to the 61st Division and six to the 81st Division, but the need was stressed for panoramic sights in addition to the eight received with the guns. The Visayas and large Mindanao Island lay to the south-southeast of Manila Bay and represented a large insular area to defend, however many artillery pieces the department could afford to send there.

By late November, the department had reconsidered where to deploy the naval landing guns, now suggesting they go to Fort Mills, Corregidor Island. Such a move would have permitted the Model 1917 75mm field guns, also known as British 75s, then on the island to be freed up for use by the mobile field army on Luzon. It also represented a more sensible deployment of the naval landing guns, lacking the capability of high-speed towing as they did. In an assessment dated December 4, the department chief of staff and G-4 (logistics) agreed that the landing guns should not be sent to the Visayas, but rather the forty-year-old Vickers-Maxim 2.95-inch mountain guns would be assigned there.

There things stood when, four days later and 5,300 miles apart, the bombs began falling on Pearl Harbor, Oahu, and Clark Field and Fort Stotsenburg, Luzon.

War

Naval landing guns, whether in the hands of the US Navy or Army, played a very small role in the world war that began for Americans everywhere on December 7/8, 1941. Entirely lacking those small artillery pieces, the last naval landing forces of any substantial size participated in the amphibious landings on the coast of North Africa eleven months later. Sailors would thereafter fight and die on many a bloody beach, but not as members of a traditional landing party.

Four days after the onset of war in the Philippines, the commanding officer of the American field artillery, Brig. Gen. Edward King, now recommended moving the fourteen Mark XI 3-inch landing guns from Fort Stotsenburg to the Bataan peninsula for beach defense. At least one recent work (Gordon, 2011) makes the claim that the headquarters of the US Army Forces in the Far East (USAFFE) carried out the original plan to send these guns to the Visaya Islands. Information in *The Wainwright Papers*, however, effectively disputes that assertion. When the steamer SS *Corregidor* got underway for the Visayas late on December 16, she was badly overcrowded with well over 1,000 civilians fleeing to the islands to the south. Corresponding to the revised plan of December 4, she also had aboard some Vickers-Maxim 2.95-inch mountain guns intended for the military forces garrisoning those islands. Very early on the 17th, the vessel ran into the minefield near her eponym, the volcanic caldera of Corregidor, hit a mine, and sank quickly with a horrific loss of life. She also took the mountain guns down with her.

In accordance with the revised plan of December 12, as confirmed in more than one source, most of the Mark XI 3-inch landing guns went south to Bataan, and were allocated with other former naval guns to both I and II Philippine Corps for beach defense. There are accounts that some of these guns fought there, but they lack specificity as to type. In addition, conforming with the late November proposal, at least two made it to Corregidor. The defenders deployed one in the Middle Sector, between the tadpole head to the west and Malinta Hill to the east. It stood 200 yards north of the South Dock, likely in a beach defense role; Japanese shell fire destroyed it, date unknown. The other stood in the West Sector, 200 yards west of Geary Point, the conspicuous prominence on the south shore of the tadpole head. The artillerymen likely positioned it to defend the 60-inch searchlight on the point. Protected by its location, the landing gun survived the enemy's shells, only to be rolled over the cliff by its crew at around the time of the general surrender. It is within reason that additional Mark XI landing guns may have been sent to Corregidor and placed in reserve, and if so were likely destroyed before the surrender.

Apparently the six ex-marine Mark VII landing guns saw no action at all, assuming they survived the massive bombings of Cavite Navy Yard.

Although not a landing force, the substantial number of sailors left behind when the Asiatic Fleet moved south in the first days of the war became a naval battalion that saw action during the defense of Bataan. Cmdr. Francis Bridget commanded the unit. In late January 1942, the Japanese came down from Olongapo at the head of Subic Bay and landed two battalions of the 20th Infantry on the west coast of the peninsula, behind MacArthur's main line of defense. That shore, cut with numerous coves and inlets, and heavily forested, so confused the Japanese approach that its 2nd Battalion unintentionally split into two. One third of the battalion, about 300 men, landed on Longoskawayan Point, 3,000 yards west of the navy section base at Mariveles, and the remainder came ashore on Quinauan Point rather farther north. An observation post stood atop 617-foot Mount Pucot on Lapiay Point, just to the north of Longoskawayan Point, and the sailors and soldiers manning it spotted the Japanese landing on the morning of January 23. Given the perilous threat, the American and Filipino force had to react fast to avert disaster.

Any enemy landings in the vicinity of Mariveles became the responsibility of the navy to repulse, and Commander Bridget at once committed a force of 450 sailors and marines to counter it. He also requested the army for assistance. The bloody fight to win Longoskawayan Point will not be described in detail, because although the naval battalion took the offensive for this action, overall it remained part of the larger American force defending against an invading army. Such a distinction is an important one, because the action of a landing force is by its very nature offensive. Nonetheless, three relevant points are worthy of mention.

The heavy weapons used against the Japanese included two 81mm mortars and a machine gun platoon from the marines, with the army contributing a 2.95-inch mountain gun, a battery of 75mm field guns, and Battery Geary's 12-inch seacoast mortars firing from Corregidor. The huge 670-pound shells fired by the coast artillerymen, using point-detonating super-quick fuses, demoralized the enemy and did much to ensure the American victory there. In the end, skilled Philippine Scouts infantry finished off the Japanese landing force.

On a less fortunate note, it was 1914 all over again for the sailors going into battle. Despite the endless chat about providing more suitable uniforms for

Hauling Cole carts laden with the components of their machine guns, the marines on Luzon move up. The unit is specifically H Company, 2nd Battalion, 4th Marine Regiment (the famous China Marines), and in January 1942 it is heading for Longoskawayan Point on the west coast of Bataan in order to resist the Japanese landing there. (*USMC History Division*)

Although not a landing gun, the small Vickers-Maxim 75mm mountain gun was about the same size. The US Army acquired 120 of this British export piece as its 2.95-inch mountain gun, with many assigned to the Philippines and seeing action there in 1941–1942. One was sent to support the naval infantry fighting on Longoskawayan Point in early 1942. During the later campaign on Mindanao, another such obsolete gun knocked out a Japanese Type 95 Ha-Go light tank fording a stream. US Army photograph.

bluejackets fighting ashore, the navy had done little about those recommendations and endorsements. The service had accepted tacitly that the seamen's denim fatigue uniforms worn aboard ship constituted an acceptable substitute, being a deal less conspicuous than sailor whites. Nonetheless, for reasons not explained, the seamen at Longoskawayan Point fought in their white uniforms stained with coffee or other tropical dyes. A Japanese diary found after the battle reports strangely acting "suicide squads," wearing garish yellow uniforms and chattering loudly; in their eyes certainly intended to draw fire and thus give the Japanese positions away.

In spite of the disadvantages faced by the Americans, the attempts by the Japanese to defeat their enemy on Bataan by making repeated amphibious landings behind their line of defense, so successful against the British in Malaya, failed against the Americans. Instead, the Japanese lost about 1,800 men in those attempts with nothing to show for it.

Stateside, the dismantling of the fleet landing forces continued unabated with the coming of full-scale war. In late January, while the Americans of all three services battled to destroy the Japanese battalions that had come ashore in southern Bataan, the Naval Gun Factory notified the Bureau of Ordnance of eight Mark XI 3-inch landing guns that had fired a mere five rounds each. That uniformity suggests they had been in use at the naval academy, but were now present at the Washington Navy Yard. Although these pieces should have been in pristine condition, they exhibited pitting in the bore and/or chamber and the yard considered them unserviceable in that state. In a war in which the navy enjoyed almost unlimited funds, the $5,300 estimated to overhaul them as reserve artillery represented chicken feed. That the bureau followed through with its plan to refurbish them remains in doubt, given the competing need to produce current shipboard ordnance.

That belief is strengthened during the following month by the continued removal of all landing force equipment from the destroyers of the fleet, leaving a dozen rifles and pistols for the security of the ship. The directive underscored the widely understood doctrine that a landing force was intended to meet domestic emergencies in an otherwise peaceful period, with no intention to use such a force against strong enemy opposition in a time of war. Also, it reminded that landing forces must remain in an enhanced state of training, the possibility of which being prohibited by the state of war and the crews' obligations to their shipboard duties. The argument concluded that the space required for landing force equipment was not insubstantial and could be better used during wartime. The directive abided by the policy to leave aboard the steel helmets originally allocated to the landing force, now for the protection of the personnel exposed topside.

The next-to-last chapter in the tale of US Navy landing guns in World War II took place in March 1942, when Adm. Ernest King, commander-in-chief of the United States Fleet, ordered the removal of the 37mm infantry guns and 81mm mortars, along with their ammunition, from the battleships and the heavy and light cruisers that still carried them. There was some talk about developing more modern weapons for ships' landing forces, but all too soon, the need to prosecute the war consumed the navy's efforts and obviated other considerations.

The navy's need to deploy a trained landing force, however, had not quite come to an end. That finale would take place in November 1942 during the amphibious assaults on North Africa as part of Operation Torch, and would encompass four separate American naval landings. All four were quite extraordinary events, with one of them an utter disaster, and involve a multitude of armed forces: American sailors, marines, infantrymen, and rangers, and British bluejackets. At last for the American seamen, they wore uniforms appropriate for fighting ashore.

Operation Torch comprised a massive three-pronged undertaking, with landings intended to capture Casablanca, Morocco, on the Atlantic coast, and Oran and Algiers, Algeria, on the Mediterranean shore. Each landing was itself multipronged in order to capture an airfield or other satellite location, or to neutralize a potentially troublesome coast defense battery, in addition to the main landing.

The opposition to be encountered lay primarily in the French army and navy, with the army comprised as well of indigenous colonial units, such as the famous Chasseurs d'Afrique. That former horsed cavalry formation had been equipped with armored cars and tanks to fight a modern war. Both services were commanded by officers loyal, more or less, to the puppet government at Vichy, but whereas Allied intelligence considered many army units to be persuadable to the Allied cause, it remained a deal less sanguine about the navy. Proud, determined, courageous, and of excellent quality, both in its personnel and ships, la Marine Française was very nearly an entity unto itself, with its members' primary loyalty to that service. Moreover, it bore an abiding hatred toward its erstwhile ally, the Royal Navy, because the latter had bombarded French warships at Mers-el-Kébir, Algeria, in July 1940, to keep them out of German hands (despite the French solemn pledge to scuttle those ships to prevent such a seizure). In so doing, the British had killed more than 1,000 French sailors. Unlike in the United States and in the British Commonwealth, where army personnel manned the coast artillery, French harbor defense batteries had naval gun crews. Allied intelligence strongly held that whether aboard ship or ashore, the French navy would fight tenaciously. In that conviction, it was quite right.

Casablanca

Powerful surf presented one of the most troublesome obstacles encountered on the Atlantic coast of Africa, not surprisingly worsening as the winter months arrived. The Allied landings needed to move forward accordingly. On November 8, 1942, the Americans attacked at three places near Casablanca: Fédala and Mehdya to the northeast of the city and Safi to the southwest. In the early hours, the Clemson-class flushdeck destroyer-transport *Dallas* (six 3-inch guns), stripped down and with her mast shortened, took aboard an American army raider detachment of seventy-five men and the Free French river pilot René Malavergne. The raiders' mission was to take the airport at Port Lyautey, about 9 miles up the Sebou River from its mouth at Mehdya. *Dallas* attempted to force the defenses at the river mouth twice, just before and just after noon, but the Vichy French batteries rebuffed the ventures. Thereafter, light cruiser *Savannah* (five triple 6-inch turrets) shelled both the fixed harbor defenses and the mobile railway batteries.

During the night of November 9–10, a boat crew from the assault transport *George Clymer* made a second attempt to cut the cable bearing the net that extended across the river mouth. Despite being under constant small arms fire, including from machine guns, the sailors commanded by Lt. Mark Starkweather managed to sever the cable. Because the destroyer's bridge personnel encountered great difficulty in sighting the river jetties and negotiating the rough water at its mouth, *Dallas* did not enter the Sebou River until after 6 a.m. The experience and skill of pilot Malavergne proved immeasurable in getting the ship up the narrow and shallow channel. Making a mere 5 knots, *Dallas* continually scraped the bottom, but she did not ground. On the final leg to Port Lyautey, she had to thread the needle between two steamers scuttled as obstacles.

While putting the raiders ashore, *Dallas* came under fire from a battery of two 75mm guns and responded. The French battery certainly could shoot, putting more than one round close aboard the old destroyer. Fortunately, a Curtiss SOC-3 Seagull spotting plane off *Savannah* located the battery and dropped two impact-fused depth charges, which silenced the enemy guns. Although the raiders encountered automatic fire, they readily took the airport. Within a few hours, the first US combat aircraft flew in. For his cool precision on a dangerous mission, civilian volunteer René Malavergne received the Navy Cross.

Seen in this aerial photograph, the modified Clemson-class destroyer *Dallas* is anchored in the Sebou River, Morocco, with the essential and recently secured airport at Port Lyautey in the distance. She had wended her way up the river with some difficulty to put ashore seventy-five army raiders to capture the airport. In short order, American combat aircraft were making use of its runways. (*National Archives and Records Administration*)

To the southwest, at Safi, a similar adventure unfolded, this time involving two other flushdeck destroyer-transports, *Bernadou* and *Cole*, also rearmed with 3-inch main guns. This time, American bluejackets would join the armed force going ashore. The mast and part of the upper works of these Wickes-class destroyers had been cut down to reduce their silhouettes. *Bernadou* transported K Company and *Cole* carried HQ and L Companies, 47th Infantry; the former destroyer also had a naval landing party aboard. Both bodies of men had been specially trained as an assault force, but their respective missions differed. The seamen were to seize the dock facilities and the ships in the harbor, while the infantry occupied strategic points in a wider perimeter in order to protect the naval personnel in their task.

Bernadou departed from the transport area at a quarter to four in the morning of November 8 in order to find the harbor mouth. Shortly thereafter, *Cole* left to accompany the first wave of the main assault against Safi. Twenty-five minutes after departure, *Bernadou* approached the breakwater and there reciprocated a flashing light challenge, thus delaying fire from the harbor defense batteries. The respite did not last long, and soon enough a lively exchange of fire began. The accurate shooting of *Bernadou* and Gleaves-class destroyer *Mervine* (four 5-inch/38cal guns) silenced the hostile 75mm guns, but upon entering the inner harbor, the destroyer-transport came under fire from a quadruple 130mm gun coast defense battery. *Mervine* and then larger warships in the task force fired counterbattery, completely neutralizing the French battery.

Cole experienced difficulty in finding her way in, but received assistance from Ens. John Bell. That officer commanded the scout boat that had preceded the assault waves. By infrared lamp and radio, he directed both the landing craft and the warships accompanying them. An hour after leaving the transport area, the two destroyer-transports unloaded their landing forces on two separate Safi quays, thus far having suffered no casualties. Tanks landed from special lighters accompanied the assault infantry. Meanwhile, the naval landing party captured some tugs and secured the harbor cranes, all of which would prove useful to the Americans. *Bernadou* had grounded at the mole, though she lifted off with the tide. Later that morning, the straight-shooting *Cole* supported army troops in their capture of the main radio station in Safi.

Above and below: Wickes-class destroyers *Bernadou* and *Cole*, their masts removed and their upper works heavily cut down, modified, and strengthened, put ashore two rifle companies and the HQ company of the 47th Infantry Regiment, as well as a trained naval assault party, to capture and then protect the harbor at Safi, Morocco. Aboard *Bernadou*'s bridge, a signalman wigwags a message, while *Cole* steams close to light cruiser *Philadelphia*, the biplane wings of her Curtiss SOC Seagull floatplane partially seen. Both destroyers wear Measure 2, an early disruptive paint scheme. Photographs by US Navy and Eliot Elisofon, *Life*. (*National Archives and Records Administration*)

Oran

The Allied operations at Oran and the small harbor of Arzeu 25 miles to the east may be considered rather unusual because of the participation of American marines. To be sure, marine officers and enlisted men could be found in London, Rosneath, and Londonderry in the British Isles, in various headquarters and legations, and in detachments aboard capital ships—not to mention the exploits of the multilingual and multitalented Maj. Peter Ortiz, USMC, a member of the OSS, gaining fame postwar—but by interservice consensus, the marine amphibious landings so renowned today took place in the Pacific Theater.

Capt. Walter Ansel, USN, Lt. Col. Louis Plain, USMC, two other naval officers, and eleven enlisted marines constituted the headquarters and reserve element of an advance party. It worked in close conjunction with a British naval party and two companies of the 1st Ranger Battalion, all tasked with securing the port of Arzeu. At around 1 a.m., the rangers, in their first combat operation in strength, landed near Fort de la Pointe, at the landward end of the eastern breakwater, in order to neutralize the 75mm gun battery there. After a short battle, the rangers fired the prearranged Very flares signaling success, and Ansel ordered the landing craft assault (LCA) into the harbor to carry out their boarding mission on all ships there. With the dawn, the intensity of the fighting increased, mostly from the breakwater and the adjacent seaplane base. Just before 8 a.m., the rangers took the seaplane base and rounded up the French naval personnel and Senegalese soldiers, who had offered moderate to stiff resistance. The port town thereafter became a base assisting in the assault on the larger port of Oran to the west.

The great fortune thus far experienced by both the naval landing parties and their army counterparts conveyed in American naval vessels came to an end with the assault on Oran. For the members of the US Army, US Navy, US Marine Corps, and Royal Navy taken into the harbor in two former US Coast Guard cutters, the risky effort ended in bloody disaster. In this special action, Operation Reservist, the cutters HMS *Walney* (ex-USCGC *Sebago*) and HMS *Hartland* (ex-USCGC *Pontchartrain*) were to carry nearly 400 men of the 3rd Battalion, 6th Armored Infantry, twenty-six American bluejackets, six American marines, fifty-some British naval ratings, and twelve British demolition specialists into Oran harbor. Two motor launches would accompany the larger ships. There, it was anticipated that the assault infantry would take two coast defense batteries, while the naval contingent would capture and hold the wharves, as well as the ships tied up at the docks and anchored in the harbor. The sailors had been provided canoes to reach the more distant ships. Their essential job, as elsewhere, was to prevent any attempt to sabotage the wharves or scuttle the ships. The composite landing forces had been rigorously trained beforehand in the United Kingdom.

Capt. Frederick Peters, Royal Navy, the overall naval commander, had retired before the war, only to be recalled to service. To describe this officer as the perfect commander for a death or glory attack would be an understatement. *Walney*, the lead vessel, became his flagship. Lt. Col. George F. Marshall commanded the American infantrymen and Lt. Cmdr. George Dickey the American naval contingent. Unlike the operation at Safi, where *Bernadou* and *Cole* entered the harbor when the main assault began and thus enjoyed relative surprise, *Walney* and *Hartland* commenced their attempt to enter Oran harbor at 3 a.m. on November 8, after the balloon had gone up elsewhere and the element of surprise lost. The French sailors and soldiers defending Oran harbor were thus fully alerted and determined to retaliate for the earlier destruction suffered at Mers-el-Kébir. The cutters would now face an angry hornets' nest. Although some sources insist that the orders given to Peters required him to withdraw his ships if the French had been forewarned and their defensive fire too hot, the final decision was in fact left to his discretion.

A double boom protected the entrance to the east, between the land and the 3,000-yard-long northern breakwater. On the east, just inside the booms, a shorter jetty extended from the Môle Ravin Blanc, armed with a harbor defense battery. One of *Walney*'s tasks was to smash through the barrier, opening up the entrance for *Hartland* following. Not long after the cutters began their run, shrouded by smoke laid by one or both motor launches, the French detected them and turned on the harbor defense searchlights. Despite the smokescreen, the ships were bathed in bright light. *Walney* missed her mark, but Peters ordered her captain to try again. The cutter had to make a full circle to do so, this time smashing through the double booms, but now under heavy fire from the French batteries, both ashore and afloat.

Between Môle Millerand and Môle Giraud inside the harbor, also having batteries, *Walney* encountered the French sloop *La Surprise* steaming on the opposite course. Whether the cutter attempted to ram or avoid, the two ships scraped by, each firing at the other. *Surprise* gave more than she got, with *Walney* getting badly shot up. Below, with the casualties mounting, Marshall tried to rally his men. Some managed to get up on deck and fire back at their tormentors. At about this time, a shell hit the cutter's bridge, killing fifteen men, including her commanding officer, Lt. Cmdr. Peter Meyrick. Only Captain Peters survived the explosion, although partially blinded as a result. *Walney* thereafter became so much drifting wreckage, and soon capsized.

Hartland, dutifully following, also came under heavy fire, which partially blinded her commanding officer, Lt. Cmdr. Godfrey Billot. As had her sister, she missed the entrance on her first attempt, glancing off a jetty. *Hartland*'s second attempt succeeded, but nearly at once she passed by the French destroyer *Typhon*, tied up on the protected side of Môle Ravin Blanc. *Typhon* pounded the cutter at close range, inflicting serious structural damage and dreadful casualties, particularly among the gun crews and others exposed on deck. Now adrift and burning fiercely, *Hartland* eventually blew up and sank.

A relatively few men, armed and ready to fight, reached shore. Most survivors, sodden, dazed, and wounded to some degree, were simply glad to be back on dry land. Whatever the state of resolve of the Allied soldiers and sailors, the French soon rounded them up. Most of the prisoners were well treated, with the wounded seen to by French doctors and surgeons. When the Allied forces took possession of Oran, they released the survivors from their brief captivity. The butcher's bill had been terrible: Of the 393 officers and men of the American battalion, 189 were killed, including Lt. Col. George Marshall (awarded the Distinguished Service Cross posthumously), and 157

The Beginning of World War II and the End of US Navy Landing Guns   299

wounded. Of the twenty-six American sailors and six marines present, three seamen and two marines were killed, with a total of seven wounded. The casualties among the ships' companies and the British special landing parties aboard HMS *Hartland*, a total of about 257 officers and ratings, were 113 killed and 86 wounded. Adding the American and British casualties together, the force suffered 45 percent killed in action, died from their wounds, or drowned, and 37 percent wounded in action, thus 82 percent total casualties, indeed terrible by any measure.

Recognizing what the raid had been intended to achieve, the French added insult to injury by wrecking the harbor facilities and scuttling ships in a manner to block the harbor. With the element of surprise lost, Operation Reservist had been not only a disaster, but for nothing. In a final irony, the operation's naval commander, Capt. Frederick Peters, flew in an RAF aircraft to England, but it crashed in a fog, killing all aboard. He received the Victoria Cross posthumously.

The Allies undertook a similar effort to the east, Operation Terminal, with the objective of capturing the important Algiers harbor intact. Capt. Henry Fancourt, RN, commanded the force. The two British destroyers involved got roughly handled: HMS *Malcolm* suffered severe damage and had to withdraw, and HMS *Broke* eventually sank, after successfully delivering about half the men of the 3rd Battalion, US 135th Infantry, to a quay within the harbor, along with British naval personnel. The American troops managed to hold the harbor for several hours before surrendering—long enough to deprive the French of the opportunity to destroy its facilities. The defenders in turn soon capitulated to the larger Allied force entering the city. No US Navy landing contingent was involved in this operation, but given the results of the special operations at Oran and Algiers, the service needed no further reminder that a determined enemy armed with modern weapons had put an end to its landing parties of bygone years.

As the war wore on, American sailors participated in many additional missions ashore, on beaches or elsewhere, whether as boat handlers, scouts, raiders, or underwater demolition team members. Those operations often entailed great peril and required equally great physical courage and presence of mind, but they were no longer part of traditional landing parties. Similarly, the massive-scale marine amphibious assaults in the Pacific had evolved well past the corps' own landings earlier in the century. Such wartime undertakings thus lie beyond the scope of this work.

The US Coast Guard, separate but administered by the US Navy for the duration of World War II, played a large part in that conflict. Coast guardsmen crewed escort warships assigned to hunt U-boats and took the helms of assault landing craft to carry marines ashore onto disputed islands. At home, they patrolled beaches afoot and on horseback, and landed armed parties where the need arose. Here, eight two-legged and five four-legged members practice an armed landing on a South Carolina beach. The men carry both the M1903 and M1 .30-caliber rifles with the M1905 bayonet fixed. (*Office of US Coast Guard History*)

References Consulted

Ancheta, C. A., (ed.), *The Wainwright Papers*, Vol. I, pp. 21, 34–35, 83, "Artillery" (table and text), pp. 171–72; Vol. II, Exhibit G, "Beach Defense Artillery Tabulation," pp. 86–87 (Quezon City, Philippines: New Day Publishers, 1980).

Budanovic, N., "Operation Reservist: When the Allies Were Annihilated by Vichy French Soldiers," War History Online, September 22, 2016.

Dictionary of American Naval Fighting Ships, Naval History and Heritage Command (online). Consulted for each US naval warship and auxiliary mentioned in this chapter. Other sources were consulted for all British warships touched upon.

Gordon, J., *Fighting for MacArthur: The Navy and Marine Corps' Desperate Defense of the Philippines* (Annapolis: Naval Institute Press, 2011), pp. 73–75.

Howe, G. F., *Northwest Africa: Seizing the Initiative in the West*, Mediterranean Theater of Operations, U.S. Army in World War II, Center of Military History (Washington, DC: Govt. Print. Office, 1957), pp. 89–97, 102–07, 152, 158, 164–67, 170, 202–06, 241–44.

Mallonée, R. C., diarist, *The Naked Flagpole: Battle for Bataan* (San Rafael, CA: Presidio Press, 1980), pp. 100–01.

Miller, J. M., *From Shanghai to Corregidor: Marines in the Defense of the Philippines*, Marines in World War II Commemorative Series (Washington, DC: Marine Corps Historical Center, 1997), pp. 20–23.

Miskimon, C., "Daring Raid on Algiers Harbor," Warfare History Network, November 16, 2018.

Morton, L., *The Fall of the Philippines*, The War in the Pacific, United States Army in World War II, Center of Military History (Washington, DC: Govt. Print. Office, 1953), pp. 300–08.

US National Archives, Record Group (hereafter RG) 24, Records of the Bureau of Navigation, Entry 118, Ships' Log Books: deck log of *Houston* (CA 30), October 1940 and November 30, 1941, College Park, MD.

———, RG 74, Records of the Bureau of Ordnance; Washington, DC.

———, RG 80, General Records of the Navy Department, 1798–1947; College Park, MD.

———, RG 143, Records of the Bureau of Supplies and Accounts, College Park, MD.

———, RG 169, Records of the Foreign Economic Administration (including the President's Liaison Committee and the Office of Lend-Lease Administration); College Park, MD.

———, RG 225, Records of Joint Army and Navy Boards and Committees, College Park, MD.

———, RG 407, Records of the Office of the Adjutant General; College Park, MD.

———, RG 496, Records of General Headquarters, Entry 540, Records of Headquarters, US Army Forces in the Far East; College Park, MD.

US Navy, *The Landings in North Africa, November 1942,* Combat Narratives, Office of Naval Intelligence (Publications Branch, ONI: 1944), pp. 5–7, 19, 34–36, 41–43, 48–50, 65–69.

Whitman, J. W., *Bataan, Our Last Ditch: The Bataan Campaign 1942* (New York: Hippocrene Books, 1990), pp. 250–54, 257–65.

Retrospective

The landing party as carried out by the US Navy and other navies to further the national will by showing the flag, keeping the peace, safeguarding citizens abroad, and protecting national interests (usually commercial ones) stood as a timeworn practice. It was viewed as a right by the "civilized" nations plying the oceans of the earth to be conducted against the "uncivilized" or unstable ones generally restricted to the land, but rarely against each other. An exception can be found in the landings at Murmansk, Arkhangelsk, and Vladivostok by several nations allied against the Central Powers in World War I, when they became alarmed that the long-standing government of Russia would be overthrown by the Bolsheviks.

Three factors put an end to the practice, two of them tightly bound. The first lay in the increasing size and weight of the landing guns and other essential gear to be gotten ashore, a trend Dahlgren had warned against decades before. The second was the acquisition of modern arms by nations hitherto perceived as untamed and perilous. Armaments firms, pejoratively known as merchants of death, remained perfectly content to sell their wares to anyone waving about fistfuls of currency. The "wild and dark" continents became awash in mercenaries, soldiers of fortune, and military advisors—perhaps sullied and unwelcome in their homelands—who proved all too willing to instruct the "unruly philistines" in the fine arts of modern warfare. Bluejackets struggling to bring ashore landing guns, ammunition, and other cumbersome stuff now faced disciplined fire from small arms, machine guns, and perhaps light artillery, which inflicted heavier and increasingly unacceptable casualties.

The third factor can be ascribed to a farrago of inconvenience, expense, equivocal results, unwillingness of navies to take valuable time from shipboard duties to train seamen as naval infantry, and even morality. China in 1900, Veracruz in 1914, and the Russian ports in 1918–19 proved expensive, distasteful, and ambiguous. Two of them bordered on the disastrous. Large-scale landings by naval brigades vanished, while small-scale ones waited for World War II to see their end, at least in their long-traditional form.

As written in the preceding pages, each of the US Navy's breechloading landing guns had its limitations. The navy's trepidation and parsimony resulted in the 3-inch steel howitzer of the 1870s and 1880s not getting fixed metal cartridges, with the otherwise avoidable troubles in venting and priming. The adoption of the Fletcher Mark I field gun in the 1890s, when two far superior competing pieces were available, proved enormously expensive in money and time. Its poorly designed and weak recoil apparatus made the gun unreliable through its entire sorry service life. Its Mark I mod 1 successor represented a gratifying improvement, but the gun lacked an on-carriage recoil/counter-recoil system, and accordingly, it was obsolete upon arrival in the first decade of the twentieth century. The off-the-shelf Bethlehem Mark IV had the weakest breech mechanism of all landing guns and limited elevation as well, and remained an unpopular piece throughout its span of service.

The Ehrhardt Mark VII landing gun introduced at the end of the first decade, and given over to the marines, incorporated a well-designed sliding block breech mechanism, a good long-recoil system, and a panoramic sight, but again suffered from limited elevation. At a time when the marines fought in hilly and mountainous terrain, they needed an accompanying gun with a greater range and a substantially more curved trajectory. Also, the gun, field carriage, and limber were onerously heavy, needing animal drayage. Considering the time of development and the original input by the navy, the Bethlehem Mark XI should have been the best landing piece the navy ever had. Indeed, its excellent sliding block breech mechanism and split trail affording wide traverse and high elevation made it far better than its predecessors. It bore an Achilles heel, however, in that its recuperator leaked and required an accompanying compressed air bottle, hand in glove with empty motion in its elevating gear. Not to mention that in the steady progression in weight of the various components, the Mark XI landing gun and Mark VI field carriage were respectively the heaviest yet.

Insofar as seeing combat, the Dahlgren ML howitzers and the Mark I, Mark I mod I, Mark IV, and Mark VII 3-inch landing guns are confirmed as doing so. The 3-inch BL rifled howitzer and the Mark XI 3-inch landing gun certainly took part in either landings or campaigns, and likely were fired. Several of the numerous guns of lesser caliber, of either navy or army origin, also fought.

Ironically, time and technology aside, the navy never had a more satisfactory—or more frequently deployed—landing gun than the Dahlgren 12-pounder smoothbore, muzzleloading howitzer, in use before, during, and after the Civil War. Light, handy, simple, readily gotten ashore, popular, and accurate in the hands of experienced gunners, the US Navy's first purpose-built landing gun served as the ideal piece for a quarter-century.

APPENDIX
US Naval Landing Gun Survivors

The surviving examples of the several generations of American naval landing guns are numerous and widely distributed, not only within the United States but across the world. Not surprisingly, the US Navy and Marine Corps retain a goodly number of various types, both in their museum collections and on their active bases and establishments. The US Army Field Artillery Museum at Fort Sill, Oklahoma, has one example each of a naval landing gun and a naval mountain gun. Many such guns stand in city and town parks and on village greens; some remain in front of military fraternal posts. An unknown but seemingly large number are in the hands of private collectors, most of whom tend to be quite reticent about their holdings.

Dahlgren landing howitzers survive in great numbers, particularly on the east coast of the US (see page 31, lower, and page 34). Although the light, heavy, and rifled 12-pounders remain the most numerous among those enduring, there are examples of small 12-pounders, 24-pounders, and 20-pounder rifles (page 89, lower) to be found. The most careful listings of Dahlgren's "wicked devils" are found among the indices of Olmstead et al., *The Big Guns: Civil War Siege, Seacoast, and Naval Cannon* (Museum Restoration Service, 1997). The earnest gun searcher, however, should be aware that some of the pieces so identified have disappeared mysteriously or without explanation since that useful book was published a quarter-century ago.

Of the 1870s 3-inch breechloading rifles, most survivors are the smaller 350-pound type, the true landing howitzers, their barrels composed of either the earlier bronze (page 91) or the later steel (page 96, upper). A handful of 500-pound pieces provided largely to the Revenue Cutter Service also sailor on. Of importance to the ordnance student, the bigger brothers include those with trunnions cast with the gun, for example in the Washington Navy Yard, shown here, and at least one naval pair having the trunnions added with a shrunken-on band, displayed on the green in Chester, New Hampshire (page 96, lower).

The Fletcher Mark I 3-inch field gun has survived well, mostly equipped with the later Mark I mod 2 bronze or steel recoil cylinder having the characteristic expansion chamber atop its forward end (page 178). They represent popular pieces for military

A 500-pound 3-inch BL rifle mounted on a shipboard iron Marsilly carriage, Washington Navy Yard. (*Courtesy of Jim Bruns, National Museum of the US Navy*)

fraternal posts, and the Field Artillery Museum has one. A very few still display their original recoil sleeve lacking the expansion device. A battery of them were provided to the Michigan Naval Militia and one gun now stands in front of the Reese American Legion post, shown here. Another of the earlier variant is prominently displayed at the Deniz Müzesi, Istanbul's naval museum, likely from the Ottoman protected cruiser *Mecidiye* (or *Medjidiye*), completed in 1903 by William Cramp of Philadelphia. Not all such guns survive in good or complete condition.

Not faring as fortunately or numerously is the Naval Gun Factory Mark I mod 1 3-inch field gun, with only two found to date: one in front of the American Legion post in Montross, Virginia, shown here, and the other at Picatinny Arsenal, New Jersey, paired with a Mark I field gun.

The Bethlehem-designed Mark IV 3-inch landing gun has a few survivors, one once exhibited with the battleship USS *Texas* at San Jacinto, Texas, a pair in New Milford, Pennsylvania, displayed seasonally in its central park, shown here, and a pair in navy hands, currently in remote storage.

Unfortunately, it seems likely that none of the Ehrhardt-designed Mark VII 3-inch landing guns used by the Marine Corps remain within the United States. Only two are known for certain in their near entirety, both in Australia in private hands and both mounted on Martin Parry high-speeded field carriages. One is equipped with its original straight shield and the other with the modified hinged shield, the not-identical pair shown here. Efforts to interest the National Museum of the Marine Corps, Quantico, Virginia, in acquiring the latter piece have not been successful.

A handful of the Bethlehem-built Mark XI 3-inch landing guns survive in Virginia's Tidewater region. A trio are in navy hands: two of them in storage and one with its

Above left: A Mark I 3-inch field gun, American Legion Post 139, Reese, MI. (*Courtesy of Robert and Joyce Spiekerman*)

Above right: A Mark I mod 1 3-inch field gun, American Legion Post 252, Montross, VA. (*Nelson H. Lawry*)

Above left: A pair of Mark IV 3-inch landing guns, New Milford, PA. (*Nelson H. Lawry*)

Above right: The last known pair of Mark VII 3-inch landing guns, New South Wales, Australia. (*Courtesy of Damien Allan*)

Left: A Mark XI 3-inch landing gun and Mark VII limber, Norfolk Naval Shipyard, VA. (*Nelson H. Lawry*)

Mark VII limber displayed in Trophy Park, Norfolk Naval Shipyard, shown here. Another pair lacking wheels stand in front of the Tidewater American Legion post in Norfolk.

Only one Bethlehem-designed-and-built Mark XII naval mountain gun is known to survive; part of the large collection in the Field Artillery Museum, Fort Sill, and shown here.

Of the four British 12-pounder 8-hundredweight landing guns provided with the pair of Elswick-built protected cruisers in the late 1890s, three may be found today on the campus of Norwich University, Vermont. In addition, lots survive with naval reserve units just to the north in Canada, one of them assigned to HMCS *Unicorn*, Sasakatoon, Saskatchawan, shown here.

Regarding the smaller landing guns or those seeing brief or limited service, a number of Hotchkiss revolving cannon survive, but almost all appear on shipboard mounts. The navy has at least one 6-pounder in storage, and several collectors own long 1-pounders in handsome condition. Short 1-pounders survive in various places, and at least one pair is mounted on wheeled field carriages in the central square of Greenfield, Ohio, displayed here. Both navy and marine establishments possess examples of the Model 1916 37mm infantry gun mounted as a naval landing gun, with a pair shown in Crane, Indiana. The M1 75mm pack howitzer is a very plentiful survivor, mostly with steel wheels and pneumatic tires, as used by the army and marines during World War II and afterward. The piece mounted on either the M3 field carriage or the original M1 wooden-spoke-wheeled carriage issued as a prewar landing gun is uncommon, with one each of the latter at the US Army Field Artillery Museum, Fort Sill, and the National Museum of the Marine Corps, Quantico, whose piece is shown here.

Above left: The only known Mark XII 3-inch naval mountain gun, Fort Sill, OK. (*Courtesy of Gordon A. Blaker, U.S. Army Field Artillery Museum*)

Above right: An Elswick QF 12-pounder 8-hundredweight landing gun, HMCS *Unicorn* naval reserve establishment, Saskatoon, SK. (*Courtesy of Warren Noble, RCNR*)

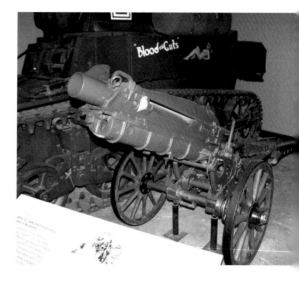

Above left: A pair of short Hotchkiss 1-pounder landing guns, Greenfield, OH. (*Nelson H. Lawry*)

Above right: An M1A1 75mm pack howitzer mounted on its M1 carriage, National Museum of the Marine Corps, Quantico, VA. (*Nelson H. Lawry*)

Below: A pair of Model 1916 37mm landing guns, naval reservation, Crane, IN. (*Courtesy of Jeffrey Nagan, US Naval Support Activity Crane*)

Index

ABCD ships, *also fig cap*
ABCD warships, *109*, 115, 147, *148*, 153, *154*, 157, 220–21, *see also* New Navy
Accles, James, 121; Gatling Gun ammunition drum invented by, 24, 121, 134
Aizpuru, Rafael, 105, 107–08
Allen, Seth, 82
Almy, John, 105
American military fraternal organisations and posts
  American Legion, Post 139, Reese, MI, 19, 305, *306*; Post 252, Montross, VA, 305, *306*; Tidewater Post 327, Norfolk, VA, 307
  Grand Army of the Republic (GAR), 132
  Veterans of Foreign Wars (VFW), Post 3438, West Carrollton, OH, *178*, 304–05 (by inference)
American steel industry, emergence of, 85–93, 108, 110, 136, 182–84
Ames, Adelbert, 66
Ames conundrum, 92
Anderson, Edwin, 233–34, *235*, 236; *see also* Medal of Honor
Ansel, Walter, 297
Anthracite coal, 90
Arias, Desiderio, 213–14, 216
Armstrong, James, 47
Armstrong, William, 142
Atwater, Charles, 156
Azueta, José, 225
Azueta, Manual, 225

Badger, Charles, 221, 232–33
Baker, Newton, 260
Bannerman, Francis, 193
Barclay, Charles, 156–57
Barnett, George, 198, 205
Barry, David, 258
Battles
  Assault Landing
    Algiers, 293, 299; Apia, 158, *159*; Bataan (Longoskawayan Point), 289–90, *291*, 292; Canton, 47, 269, *270*; Casablanca, 293 (Mehdya, 293–94; Port Lyautey, 293–94, 294; Safi, 295, *296*); Cavalla River, 42–43; Corregidor, 289–90; Faro de Las Cabezas de San Juan, 156–57; Fort Fisher, 46, 63, 64–66, 66; Fort Sumter, 63; Ganghwado/Shinmiyangyo, 74–75, 75–77, 78–79, 79–82, *81*; Guadalcanal, 280–81; Guánica, 155; Guantánamo Bay, 149, *150-51*, 151–53, *154*, 155; Murrells Inlet, 57; Oran, 293, 297–99 (Arzeu, 297); Roanoke Island, 53–54; *54*, 55, *55*, 66, *66*; Shanghai/Muddy Flat, 45; Strait of Shimonoseki, 70; Veracruz, 222–23, *224*, 225, 226–27, 227–28, 229–30, 231–34, *233*, *235*, 236–37, 237–38, 239
  Land
    Abu Klea, 118; Altamira, 216; Antietam/Sharpsburg, 55–56, *55*; Belleau Wood, 266; Coyotepe/La Barranca, 210–12; Cuzco Well, 152–53, *153–54*, 155; First Bull Run/First Manassas, 51–52; *52*, 56; Fredericksburg, *55*, 56; Guayacanas, 216, *216*; Harpers Ferry, 47; Honey Hill, 63, *64*; Hsiku Arsenal, 163–65, 168; Las Trencheras, 214, *215*; Llanos de Perez, 216; Monocacy, 59; Obozerakaya, 252; Peiho River, 162–165, *164*; Second Bull Run/Second Manassas, 55; Seletskoe, 251–52; Socorro (Phil.), 269; Suchan Valley, 256; Tagalii/Vailele, 158, *159*, 160; Tiagra, 252; Tientsin, 165–67 (East Arsenal, 165; West Arsenal, 165); Tulifinny Crossroads, 63, *64*; Wilson's Wharf, 37, 57–59
  Naval (including river actions)
    Albemarle Sound, *60*, 60–62; Bush River bridge, 59; Coronel, 164; Jutland, 165; Lake Onega, 254; Manila Bay, 147, 199, 249, *250*; Mers-el-Kébir, 293, 298; Mississippi River, 52–53; Rappahannock River (Port Conway, Port Royal), 56–57; Santiago de Cuba, 143, 149; Tai-O Bay, 46; Wilson's Wharf, *see* Land Battle
Bearss, Hiram, 216
Beato, Felice (Felix), 70, *81*
Belknap, George, 70
Bell, Henry, 70
Bell, John, 295
Bennett, Kenneth, 156
Berkeley, Randolph, 225, *226*
Bierer, Bion, 249, 254
Bigham, C. Clive, 163
Billot, Godfrey, 298
Bishop, C. S., 254
Blake, Homer, *75*, 76–77, 80
Bloch, Claude, 266
Boardman, William, 157
Boards
  Board of Organization, Tactics, and Drill (Greer Board), 1889, 197–98; Gun Foundry Board (Simpson Board), 1883, 113; other boards: selection of Remington BL carbine c1868, 72; selection of Fletcher Mark I field gun, 1895, 137–38; selection of Krag-Jørgensen rifle for all services, 1898, 134; examination of Mark II field carriage and Mark I limber, 1903, 183; selection of Mark XI landing gun, 1917, 247; consideration of equipment removal to permit mounting shipboard AA machine guns (*Texas* Board), 1936, 277
Boats, US Navy and US Coast Guard
  Armed motorsailers "Atlanta" and "Georgia", 254; canoes, 297; motor launches, 220, 225, 269; motor whaleboat, *300*; Picket Boat No. 1, 61–62, *62*; Picket Boat No. 2, 61; pulling boats (general), 24, 24, 46, *50*, 53, 70, 198, *226*, 282; pulling cutters, 23, *24*, 32, *44*, 57, *126*, 144,

149, 155; pulling launches, 24, 29, 32, *33*, 42–43, *44*, 51, 54, 57; pulling whaleboats, 149, 220; sailing launch, 144; steam cutters, 24, 73; 122, 144, *184*; steam launches, 24, *24*, 53, 73, 75–77, 108, 122, 149, 193, *226*, 228, *230;* use of pulling boats for lifesaving in rough water, 282

Boats, US Navy, national ensign flown on, *24*, *126*

Bormann, Charles; 39; fuse invented by, 39, *40*, 45, 59

Boxer, Edward, 99; fuse invented by, 99–100

Brady, Matthew, *44*, *50*

Bragg, Braxton, 63

Breese, James, 80

Breese, Kidder, 63

Bridges
Brooklyn, *173*; Bush River, *see* Naval Battle; Eads (Mississippi River), 90; Gunpowder River, 59; Iturbide (Carpentero Canal), 220; Long (Potomac River), 51; Rohrbach (Antietam Creek), 56; Warrenton Turnpike (Cub Run), 52

Bridget, Francis, 290

Broadwell, Lewis, 119; Gatling Gun ammunition drum invented by, 119–20

Brown, Charles, 82; *see also* Medal of Honor

Brown, Harvey, 51

Brown, John, 47

Brown, Robert, 80

Browning, John, 248; *see also* the automatic weapons bearing his name

Bruce, Lucien, 120–21; Gatling Gun ammunition feed invented by, *120*, 120–21, 134

Buchanan, Allen, 233

Buquor, Adolph, 287

Burnside, Ambrose, 51, 53, *54*, *55*, 55–56, 63

Butler, Smedley, 166, 210–11, 232, 236; *see also* Brevet Medal and Medal of Honor

Caceres, Ramon, 213

Campbell, Albert, 156

Campbell, Chandler, 214

Canada, William, 223, 231–32

Caperton, William, 213–14

Carranza, Venustiano, 219

Casey, Silas, 76–77, 80–81

Cash, John, 49

Cassel, Douglas, 76, 80–81

Castle, Guy, 227, 236

Catlin, Albertus, 233

Cervera y Topete, Pascual, 149

Chadwick, French, *195*

Chamorro, Emiliano, 210

Chao Yin-ho, 163–64

Christy, Harley, 241

Churches
Nuestra Señora de Guadalupe (Phil.), *see* Fort; Nuestra Señora de la Asunción, Veracruz, 231, 236

Churchill, William, 57

Cities and towns; *see also* Harbors and commercial ports

Cities
Fairfax, VA, *34*; Frederick, MD, *see* Land Battle of Monocacy; Garden City, KS, *89*; Greenfield, OH, 307, *308*; Harbin, Manchuria, 255; Prophetstown, IL, *91*; Saskatoon, SK, 307, *307*; York, PA, 287

Towns
Ashburnham, MA, *31*; Centreville, VA, 51; Chester, NH, *96*, 304; Coosawhatchie, SC, *64*; Crane, IN, 307, *308*; Dorado, PR, 143; Eddystone, PA, 241; Grahamville, SC, *64*; Matachin, Colombia/Panama, 107; Montross, VA, 305, *306*; New Milford, PA, 305, *306*; Nindiri, Nicaragua, 211; Reese, MI, 305, *306*; Sea Girt, NJ, *92*; Tejeria, México, 223, 228; West Carrollton, OH, *178*

Cleveland, Grover, 147

Cochrane, Henry, 151, 197

Coggins, Jack, 48–49

Cole, Edward, *248*, *266*; weapons and ammunition cart designed by, *248*, *266*, *267*, *275*, *278*, *280*, *291*

Cole, Eli, 205

Coleman, John, 82; *see also* Medal of Honor

Colt, Caldwell, 92–93

Colt, Samuel, 92

Communication and transportation companies

Communication
Central and South American Telegraph Co., 106–07

Transportation
Railroad: Charleston & Savannah RR, 63, *64*; Chinese Eastern Rwy, 255; Ferrocarril del Pacífico de Nicaragua, 210–11; Panama RR, 104–08; Philadelphia, Wilmington & Baltimore RR, 59; Trans-Baikal Rwy, 255; Trans-Siberian Rwy, 255–56

Steamship: Atlas Steamship Co., 106; Hamburg-America Line, 222; Mexican Navigation Co., 231; Pacific Mail Steamship Co., 105–08; Ward Line (New York and Cuba Mail Steamship Co.), 223, 225, 232

Cone, Hutchinson, 223

Conventions and treaties
Convention of Kanagawa, 1854, 43, *44*; Congress of St. Petersburg, 1868, 122; Tripartite Convention, 1899/1900, 160; Washington Convention, 1907, 210; Washington Naval Treaty, 1922, 268, 277

Cooke, James, *60*, 60–61

Cooper, George, *75*

Copp, Charles, 220

Courtney, Charles, *161*, 162

Cowles, John, 72

Cradock, Christopher, 164

Crane, Stephen, 152; *The Red Badge of Courage*, 152

Cresap, James, 225, 228

Crosley, Walter, 156

Curtiss SOC Seagull spotter floatplane:

aboard USS *Philadelphia*, *296*; off USS *Savannah*, 294

Cushing, William, 61–62, *62*, 65, 66

Dahlgren, John, 20, 27–28, *28*, 30, 32, *38*, 48, 51, 57, *58*, 63, 302; *see also* the guns designed by

Dale, George, 236

Daniels, Josephus, 223, 260

Davidson, William, 193

De Camp, John, 42–43

Denby, Edwin, 264

Deutsche Handels und Plantagen Gesellschaft (DHPG) plantation, Samoa, 158

Dewey, George, 249, *250*

Díaz, Adolfo, 210

Díaz, Porfirio, 219

Dickey, George, 297

Difficulty in or dissatisfaction with
Achieving breechloading and obturation, 85–86, *87*, 87–88, *89*, 94, *94*, 101, 104, 129, *189*, 190
Ammunition and fuses, 39, *40*, 45, 100, 152, 155, 166–68, 207, 209, 211–12
Bethlehem Steel Co. in meeting its contractual deadlines, 190, 260–62, *263*, 264, *265*
Field uniforms of; marines, *150*, 153, 153–54, 155, 167, *200*, *206*; sailors, 71, 229–30, *233*, 234, 237, 282, 290, 292–93
Fore River Ship & Engine Co. in completing its contract for steel forgings, 182–84; Fore River Shipbuilding Co., 203
Mark I (Fletcher) field gun, 137, *139*, 141, *141*, 168, 171–80, *178*
Mark IV (Bethlehem) landing gun, *189*, *191*, 189–92, 244, 246, 273, 303
Mark XI (Bethlehem/US Navy) gear system and lost motion, 262, 266, 268, 303; hydropneumatic recuperator, 261–62, 269, 303
Model 1895, Colt-Browning machine gun, 134, *135*, 152, 158, 163–64, 168
Model 1895 Winchester-Lee infantry rifle, *133*, 133–34, 155, 162–63, 167
Poor workmanship by gun makers: American & British Manufacturing Co., 203; Bethlehem Steel Co., 260–62
Producing steel of adequate quality and strength, 85–86, 88, 90–91, *91*, 92, 136, 182–84, 203
Recoil and recuperation, *189*, 189–90, 192, 261–62, 269
Recording & Computing Machines Co. in failing to produce panoramic sights, 261–62, *263*
Steady weight increase in successive landing guns, 15, 28, 30, 88, 93, 98, 103, 123, 185, *186*, 187, 189–90, 207, 212, 239, *245*–46, 246, 262, 280, 302–03; the necessity for animal haulage, 123, 187, 207, 212, 246, *246*, 262, 303
Venting, axial versus radial, 88, 95, 96, 97, 99, 103

Diplomatic representation abroad
    International: consulates, 103, 223,
        255; legations, 145, 160, *161*, 162,
        163, 165, 167
    United States: consulates, 104–05,
        147, 157, *157, 159*, 195, 223, 225,
        228, 231–32, 255–56, 269, *270*;
        legations, 72, 147, 157, *157*, *161*,
        201, 207, 210, 273, 297
Dorward, Arthur, 166
Dougherty, James, 82; *see also* Medal
    of Honor
Drake, Franklin, 80
Dresel, Herman, 156
Drew, Edward, 72, 76
Driggs, William, 138; *see also* the guns
    and gun company associated with
Dunlap, Robert, 166, 205
Durham, Charles, 211
Dutton, Robert, 157

Earle, Ralph, 220, 247, 260–61
Early, Jubal, 59
Edged and pointed weapons
    Arrows, 72; bayonets, 24, 55, 72,
        133, *133*, *148*, *154*, *157*, *159*, 166,
        167, *200*, 214, 229, 252, 282, *300*;
        cutlasses, 63, 64–65, 75, *124*; spears,
        72, 82, 162; swords, 58, 72, *131*,
        162 (Mameluke, *150*)
Edsall, Norman, 160
Elliott, George, 152–53, 199
Elliott, Gilbert, 60
Ellis, Augustus V. H., 51
Ellis, Earl (Pete), 205
Ellyson, Theodore G., 234, 236
Eo Jae-sun, 82
Eo Jae-yeon, 81–82
Eshleman, Benjamin, 56
Estrada, José, 210
Evans, Robley, 197

Fagan, Louis, 57
Fall, Albert, 264
Fancourt, Henry, 299
Farragut, David, *65*, 109
Fellowes, William, 46
Fitzgerald, John, 152; *see also* Medal
    of Honor
Fletcher, Frank F., 137–38, 222–23, 225,
    227, 231–33, 236–37
Fletcher, Frank J., 232
Floyd, Henry, 249, 251
Flusser, Charles, 60
Folger, William, 86, 121–22, 132
Foote, Andrew, 47, 53
Foreign nations
    China, 45, *46*, 47, 70, 72, 104, 122,
        145, 157, *157*, 185, 193, 199,
        268–69, *270*, 274, 278, *291*, 302;
        see also Boxer Rebellion under Wars
    Dominican Republic, 213–14, *214*–16,
        216
    France, 69, 70, 72, 86, 106, 113, 122,
        129, 219, 241–42, *248*, 249, 252;
        see also Assault Landing Battles at
        Algiers, Casablanca, and Oran, and
        Operation Torch
    Germany, 95, 113, 122, 129, 158,
        160, 163, 166, 195, 199, 202, 205,
        241–42, 244, 249, 286, 293

Great Britain (England), 45–46, 70,
    132, 147, 158, 160, 163, 165,
    195, 199, 223, 225, 249, 282,
    292–93; ordnance matters, 27,
    35, 57, 95, 99–100, 113, 118, 129,
    134-35, 142–44, *143*–44, 162–63,
    167, 248, 289, *291*, 307; *see
    also* cooperation with US Navy in
    various Assault Landing, Land, and
    Naval Battles during the nineteenth
    and twentieth centuries
Japan, 70; Commodore Perry's visits,
    43, *44*, 45, *46*, 69; cooperation with
    other powers to suppress Boxer
    Rebellion, 162–63, 165, 167; war
    scares with United States in 1907
    and 1913, 187, 199, 205; in conflict
    with US during Russian intervention
    and World War II, 167, 255, 258,
    280, 285–87, 290, *291*, 292
Korea, intervention into, 69–73,
    74–75, 75–77, 78–79, 79–81,
    81, 82–83, Pyonginyangyo, 69,
    Shinmiyangyo, 69
Mexico, 27, 271; political turmoil,
    219–20; incident at Tampico,
    220–22, *221*; occupation of
    Veracruz, 127, *206*, 216, 219,
    222–23, *224*, 225, *226*–27, 227–28,
    229–30, 231–34, *233*, *235*, 236–37,
    237–38, 239, 242, 244, *246*, 302
Nicaragua, 110, 194, 209–12, 219,
    268–69; *see also* Land Battles of Las
    Trencheras and Guayacanas
Philippines, 163, *186*, 187, 192–94,
    194, 199, 205, 207, 237, 256, 269,
    289, *291*; *see also* Islands of Luzon
    and Mindanao, and Assault Landing
    Battles of Bataan and Corregidor
Russia, 86, 113, 122, 129, *267*, 302;
    cooperation with other powers
    to suppress the Boxer Rebellion,
    163–64, *164*, 165, 167; Allied
    interventions in North Russia and
    Siberia during and after World War
    I, *221*, 249–51, *250*–51, 252, *253*,
    254–56, *257*, 258
Soviet Union, 258
Spain, 129, 143, 192–93, *194*, 223,
    225, *226*, 269; *Virginius* crisis, 87;
    see also Spanish-American War
Other Nations
    Algeria, 293; *see also* Assault
        Landing Battles; Argentina,
        122; Australia, 287–88, 305
        (New South Wales, 306); Brazil,
        122, 143; Canada, 307, *307*;
        Colombia, 104–05, 107; *see also*
        Panama; Cuba, 143, *144*, 179,
        see also Assault Landing Battle
        at Guantánamo Bay; East Africa
        (regional), *144*; Finland, 249;
        Formosa, 70–71; Greece, 242;
        Haiti, 236, *245*, 269; Hawaiian
        Kingdom, 147, *148*; Honduras,
        269; Ivory Coast, 42; Liberia, 42;
        Loo-Choo, 45 (Great Loo-Choo,
        Okinawa, 45); Morocco,, 195,
        293; *see also* Assault Landing
        Battles at Casablanca; North
        Africa (regional), 289; Panama,

104–09, *106*, 149, *151*, 161, 211;
    Portugal 108; Sudan, 118; Turkey/
    Ottoman Empire, 242, 305; West
    Africa (regional), 42, 160
Forney, James, 59
Forts and fortified places
    Chinese
        Barrier (Pearl River), 47; Taku (Hai
            River), 162
    Confederate
        Caswell, 66; Defiance/Three-gun
            Battery, 53–55; Fisher, *see* Assault
            Landing Battle; Sumter, *see*
            Assault Landing Battle [*see also*
            United States Fort]
    French
        de la Pointe, 297
    Korean
        Chojidondae/Marine Redoubt,
            74, 77, 78; Chojijin, *74*, 77;
            Deokjindondae/Fort Monocacy,
            74, *78*, 79; Deokpojin/Fort Palos,
            74, 80, 82; Sondolmokdondae/
            High Citadel/Fort McKee, *74*, 78,
            80, *81*, 81–82; Yongdudondae/
            Elbow Fort, 73, *74*, 80
    Mexican
        Castle San Juan de Ulua, 223, *224*
    Philippine
        Nuestra Señora de Guadalupe
            (church of), 193
    United States
        Adams, 49; Camp McCalla, *151*,
            151–53, *153*, *154*; Douglas, 278;
            Lafayette, 244; Mifflin, 191; Mills,
            289–90 (Battery Geary, 290);
            Pickens, 49, 51; Pocahontas, 58;
            Powhatan, 57–58; Sill, 304, 307,
            307; Stotsenburg, 288–89, air
            attack on,, 289; Sumter, firing on
            and surrender of, 28, 49, *49*; *see
            also* Assault Landing Battle
Fortson, Eugene, 213–14
Foster, John, 54–55
Foster, Paul, 228
Fox, Gustavus, 48
Freeman, Angel H., 158, 160
Freeman, Charles, 281–83
Fullam, William, 197, 199
Fuller, Ben, 166–67, 205
Funston, Frederick, 237
Furlong, William, 281

Gardner, Alexander, 56
Gatling, Richard, 118; *see also* his
    machine gun
Geographic features
    Land features
        Beach
            Orchard, Bronx, NY, *132*
        Capes
            Fear, *see* Cape Fear River;
                Hatteras, 53; San Juan, 155–57,
                see also Assault Landing Battle
        Islands
            Aleutians, 284; Assateague, 109,
                see also Lighthouse; Blanquilla
                reef, 271; Camiguin, 193; Corfu,
                108, *109*; Culebra, 172, 205,
                206; Fishers, *126*; Florida Keys,
                87 (Key West, 87, 143, 149,

150); Ganghwado, *see* Assault Landing Battle; Guam, *186*, 187, 205, 207; Hawaiian, 147 (Ford, 147, Oahu, 201, 277, 282, 289); League, *131*, 176, 205, *see also* Philadelphia Navy Yard; Long, 241; Nantucket, 199; Netherlands East Indies, 286–87 (Java, 287); Outer Banks, 53, *54*; Philippines, 163, *186*, 187, 199, 205, 207, 238, 288–89, *see also* Philippine-American War and World War II (Corregidor, 289–290, *see also* Assault Landing Battle; Luzon, 193, 256, 288–89, 291; Mindanao, 289; for both islands, *see also* World War II); Puerto Rico, 143, 155–57, 172; Roanoke, *see* Assault Landing Battle; Roncador Cay, 110; Russky, 256, 258; Ryukyu, 45 (Okinawa, 45); Samoa, 101, 102, 109, 145, 158 (Upolu, 158), *see also* Assault Landing and Landing Battle; Santa Rosa, *see* Fort Pickens; Shameen 269; Vieques, 205; Visayas, 289

Isthmus
    Panama, 104–08, 149, 161; *see also* Panama Canal, Panama Railroad, and Panama expedition

Peninsulas
    Bataan, 289–90, *see also* Assault Landing Battle; Federal Point, 63; Gimpo, 72; Kowloon, *see* Seaport

Points
    Federal, *see* Peninsula; Fishermans, 151, 153; Geary, 290; Lapiay (Mount Pucot), Longoskawayan, Quinauan, for all three *see* Assault Landing Battle of Bataan

Watercourses and water bodies
    Bays
        Buzzards, 132; Chesapeake, 59, 271; Edo/Tokyo/Yedo, 43; Florida, 87; Guantánamo, 149, 151, 153, 161, *see also* Assault Landing Battle; Manila, 163, 199, 201, 288–89, *see also* Naval Battle; Panama, 105, 108; Pensacola, 49, 51; Scapa Flow, 249; Subic, *186*, 187, 290; Tai-O, *see* Naval Battle; Tyutuke, 256; Zaliv Nakhodka/Amerika Bay; Zolotoy Rog/Golden Horn Bay, 255–56

    Canals
        Carpentero, 220; Panama, 106, 158

    Creeks
        Antietam, 56; *see also* Land Battle; Bull Run, 55 *see also* Land Battles; Cub Run, 52, *see also* Land Battle of First Bull Run; Soochow, 45

    Gulfs
        Guinea, 42; Mexico, 220, 234

    Harbors, *see* Harbors and commercial ports
    Inlets
        Kola, 249; Murrells, *see* Land Battle
    Lakes
        Baikal, *255*; Michigan, *126*; Onega, *see* Naval Battle
    Rivers
        Broad, *64*; Bush, *see* Naval Battle; Cape Fear, 61, 63, 66; Cavalla, 42; Cumberland, 53; Dvina, 251–52, 254; East, *173*; Hai, 162; Han, 72; Gunpowder, *see* Bridge; Hudson, 244; James, *50*, *57*, *59*; Kusnetchikha, 254; Makyoung, 269; Mississippi, 52–53, *65*, 90; Pánuco, 220; Pasig, 193; Pearl, 47; Peiho, 157, 161–62, *164*, *see also* Land Battle; Potomac, 51, *59*; Rappahannock, 56; *see also* Naval Battle; Roanoke, 60–61, *62*; Salée (misconceived; *see* Yeomha Strait); Sebou, 293–94, *294*; Severn, 26; Susquehanna, *59*; Taedong, 69; Tennessee, 53; Yangtze, 268–69, 285
    Roads
        Hampton, 149, 266, 269, 271; Roze, 72
    Sounds
        Albemarle, 53, *54*, 60, *60*; Croatan, 53, *54*; Pamlico, 53, *54*; Roanoke, 53, *54*
    Seas
        Aegean, 242; Barents, 251; Caribbean, 110, 143, 281; East China, 70; Ionian, 108, *109*, 242; Mediterranean, 108, 242, 293, *see also* Assault Landing Battles of Algiers and Oran; White, 251; Yellow, 70, 162
    Straits
        Shimonoseki, *see* Assault Landing Battle; Sondolmok, 73, *74*, 80, 82; Yeomha, 72–73, *74*, 76–77, 80

Gibson, James, 56
Gibson, John, 157
Gillespie, Archibald, 197
Gilmor, Harry, 59
Gisburne, Edward, 227
Goldsborough, Louis, 53
Gomer, Louis-Gabriel de; 30; the gun chamber of his design, 30, *31*
González, Pablo, 220
Gordon, John, 289
Grace, Patrick, 83; *see also* Medal of Honor
Grady, John, 233, 236–37
Graeme, Joseph, 140, 181
Graham, William, 28
Grant, Frank, 193
Graves, William, 256
Green, Joseph, 57
Greene, Israel, 47
Greer, James, 197–98
Grimes, Cary, 56
Grove, Conrad, 255–56
Grunert, George, 288

Guangxu (Emperor), 157
Guest, John, 45
Gulick, Charles, 188
Gun component or carriage component makers
    Archibald Wheel, 261; Ajax Manganese, 177; Cleveland Welding & Mfg., 261; I. G. Johnson, 175–76; Martin Parry, 287, 305; Parsons Manganese, 177; Paul Reeves & Son Manganese, 177; Railway Steel Spring, 176–77; Recording & Computing Machines, 261–62; Republic Iron & Steel, 261; York Safe & Lock, 287
Gun makers
    Cannon and gun makers, *see also* individual guns
        Aboukhoff (Obukhov), 129; Cyrus Alger, 48; American & British Mfg., 188, 201–04, 246–47; American Ordnance, 138, 140, 142, 188; Ames Mfg., 48, 92; Armstrong (William), 129, *see also* Elswick Ordnance; Armstrong-Whitworth, 143; Atlantic Steel Works, 88, 90; Baldwin Locomotive Works, 241; Bethlehem Iron (later Steel), 140, 142, 185, 189, 247, 260–62, 264, 269; Black Diamond (Parks Bros.), 88, 90; Chrome Steel, 86; Crucible Steel, 202–03; Driggs Ordnance, 129, 137–38; Elswick Ordnance (EOC), 95, 142–43, *144*; Fore River Ship & Engine, 182–84; Fore River Shipbuilding, 184, 203; Gruson EGMF, 129; Hontoria, 129; Hotchkiss et Cie, 85, 87–88, 110, 129, 131, 135, 137–38, 248, 307; Krupp AG, 85, 87–88, 95, 113, 129, 165; Krupp Grusonwerk, 202; Meudon Ordnance Works, 87; Midvale Steel, 88, 90–91, 136, 184, 247, 260; Nashua Iron & Steel, 88, 90; Naval Gun Factory/Washington Navy Yard, 27–28, 47–48, 51, *58*, 86–88, 90, 92–93, 97, 113, 127, 132, 135, 137, 142, 168, 172, 175, 177–79, *182*, 182–83, 185–87, 203, 241, 261, 266, 268, 272; Norway Steel & Iron (Naylor & Co.), 91–92; Otis Iron & Steel, 90; Pratt & Whitney, 122, 138; Rheinische Metallwaaren und Maschinenfabrik (Ehrhardt), 201–03; Schneider-Creusot, 129, 247; Skoda Works, 129; St. Chamond, 129; Tredegar Iron Works, 48, 52; Watervliet Arsenal, 113; Whitworth (Joseph), 129; Wiard Iron Works, 48
    Machine gun and small arms makers, *see also* individual weapons
        Colt's Patent Fire Arms, 92–93, 117–18, 122; Hotchkiss et Cie, 114–15, 138, *216*, 248; Maxim, 122, 134, 168, 248; New England Westinghouse, 250; Remington Arms, 71–72, 113, 115, 117,

Index 313

250; Sharps Rifle, 115; Smith & Wesson, 117; Springfield Armory, 114, 129; Vickers, Son & Maxim, 129, 248; Winchester Repeating Arms, 114, 133

Guns
  Austrian export
    Mannlicher 8mm infantry rifle, 163
  British Army
    Enfield .58-cal infantry rifle, 129
    Maxim machine gun (also widely exported), 134, 163, 168, 248
    Vickers Mark I machine gun, see US Model 1915
  British export
    18-pounder field gun, 251
    Maxim 1-pounder pompom, 103
    Maxim-Nordenfelt 7mm volley gun, 143–44
    Vavasseur 12-pounder BL rifle (French service), 86
    Vickers-Maxim 75mm mountain gun, see US Army, 2.95-inch mountain gun
    Whitworth 12-pounder BL rifle (Confederate service), 57
  Confederate manufacture or service
    Brooke 6.4-inch (100-pounder) RML, 60, 60–62, *62*
    Dahlgren landing howitzer, copies, 48; captured, 47, 52, 55, 56
  French army
    Mle 27 Brandt 81mm mortar, 275, 277
    Mle 1873 Reffye 75mm field gun and breech mechanism, 86–88, *87*, *89*
    Mle 1897 75mm field gun, 142
    Mle 1914 Hotchkiss heavy machine gun, 248, see also Mexican and US services
    Mle 1915 Chauchat machine rifle, see US Navy and US Marine Corps
    Mle 1916 37mm infantry gun, see US Army, US Navy, and US Mariine Corps
  French navy
    75mm BL, 294–95, *297*
    130mm BL, 295
    194mm BL, 249
  German export
    Krupp artillrery, 129, 165–66
    Mauser infantry rifle, 144, *152*, 204, see also Mexican service
  Japanese army and navy
    Murata infantry rifle, 167
    Type 95 Ha-Go light tank, *291*
  Korean army
    32-pounder, 77, 79
    gingal, 72
    matchlock, 72
  Mexican army and navy
    1-pounder, 228, *230*, 234
    Hotchkiss machine gun, 225
    Mauser infantry rifle, 227
  Royal Navy
    Landing guns
      7-pounder RML, 142, *143*, 158
      9-pounder, 6cwt RML, 142, 162, 166

Elswick/Armstrong-Whitworth QF 12-pounder, 8cwt, 142, *144*, 166, see also US Navy

Shipboard guns
  6-inch/100-pounder BL, 158
  6-inch BL, 225
  9.2-inch, 166
  12-inch BL, 249

Russian army
  Model 1891 Mosin-Nagant 7.62mm infantry rifle, *164*, 250–51, see also US Navy

Spanish navy
  280mm BL, 225

US Army, Union Army, and National Guard artillery
  Coast artillery
    12-inch seacoast mortar, 290
  Field guns
    2.38-inch BL, experimental, 136
    3-inch, 350-pound BL rifle, landing pattern, 132, *132*
    3-inch, 500-pound BL rifle, landing pattern, 92
    Dahlgren 12-pounder landing howitzer, 47, 51–52, *52*, 56
    Model 1841 12-pounder mountain howitzer, *37*
    Model 1885 3.2-inch field gun, *132*
    Model 1902 3-inch field gun, 202; Model 1902 field caisson, 212
    Model 1911 3-inch mountain howitzer, 213
    Model 1916 37mm infantry gun, *278*, see also US Navy and US Marine Corps
    Model 1923-E2, M1, M1A1 75mm pack howitzer, 269, 277, 279, 284, 307, see also US Navy and US Marine Corps
    Vickers-Maxim 2.95-inch mountain gun, 213, 256, 289–90, *291*

Landing gun
  Wiard puddled wrought-iron 3.4-inch RML, 48, 53

US Navy and US Marine Corps
  Boat, field, and deck carriages
    Dahlgren, boat, 30, 32, *33*, 34, 43, 51, 61, 62; field, 28, 30, *31*, 32, *34*, 34–35, 38, 43, *44*, 49–50, 51, 52, 55, 57, 71, 77, 104, 121; pivot, 34, *65*
    Hotchkiss, boat, 125, 127, 131, *230*; field, *123–24*, 123–25, *126*, 127, 131, 147, *308*; shipboard, 124, 125
    Howitzer, 3-inch, boat, 24, 97; field, 23, 25, *91*, 97–98, *102*, 132, *132*; Marsilly, 88, *305*
    Mark I, 133, *139*, *141*, 140–42, *154*, *169*, 171, *173*, 174, 176, 178–80, 182, 275, 303, *306*
    Mark I mod 2, actually recoil mechanism, 174, *178*, 178–80, 201
    Mark II, 142, 180–85, *181–82*, *184*, 233, 243, 267, 303, *306*
    Mark III, to mount shipboard 3-inch/50cal gun, 185–87, *186*, 205, 273, 280

Mark IV (Bethlehem), *189*, *191*, 189–92, *226*, 230, *238*, 244, *306*
Mark V (Ehrhardt), 202–04, *206*, 207, *208–09*, 211–12, *214–15*, *238*, 239, *245–46*, 271–72, *274*, 305, *306*; shield conversion, 244, *245*, *306*
Mark VI (Bethlehem), 247, 260–62, *263*, 266, 268, 271, 276, 277, 279, *280*, 303, *306*
Mark VII mountain (Bethlehem), 247, 264, *265*, *307*
M1 pack and landing howitzer, 279, 284, *308*

Field limbers
  General, 103, 168, 246
  Hotchkiss, 123–24
  Mark I, 180, 182–83, 185
  Mark II, 185, 187, 205, 273, 280
  Mark III, 189, *226*, *238*
  Mark IV, 202, 207, *208–09*, 212, 215, 239, 244, *245–46*, 271–72, 286, 303
  Marks V and VI/US Army Model 1902, 212, 239, 244, *245*, 286
  Mark VII, 247, 261, 266, 268, 277, *280*, 286, 288, *306*, 307

Landing guns
  Dahlgren muzzleloading boat and landing howitzers
    3.4-inch (12-pounder) rifle, 29–30, 36–37, 48, *50*, 56, 61, 63, 71, 76, 81, 83, 85, 303, 304
    4-inch (20-pounder) rifle, see Shipboard guns
    12-pounder SB, general,, 27–30, *31*, 32, *33*, 34–37, *37–38*, 38–39, 43, *44*, 45–49, *49–50*, *52*, 51–63, 62, 66, 66. 71, 73, 75–77, 79–83, 85–86, 88, 100–01, 103–05, 121, 131, 303; light, 29, 30, 62, 63, 71, 76, 83, 131–32, 304; medium/heavy, 28, 29, 30, *31*, 33–34, 42, 63, 71, 76, 83, 131–32, 304; small, 29, 30, 32, 71, 304
    24-pounder, see Shipboard guns
  Light breechloading landing guns
    1-pounder: American Ordnance, 127; Driggs Ordnance, 127; Hotchkiss, 144; long 125, *126*, 127, 131, 307; short 125, *126*, 127, 307, *308*; Naval Gun Factory, 127; Pratt & Whitney, 125, 127; see also shipboard guns
    6-pounder, *119*, 135, *136*, 307; Driggs-Schroeder, 135, 168; Hotchkiss, 135, 168, see also shipboard guns
    9-pounder: American & British Mfr. (tested), 188
    Hotchkiss revolving cannon, 37mm, 103, 107, *123*, 123–24, 147, see also shipboard service
    Model 1916 37mm infantry and landing gun, 275, 277, *278*, 281, 284–85, 292, 307, *308*

Standard breechloading landing guns
 3-inch, 350-pound rifled howitzer, bronze, 86–88, 91, *91*, *93*, *99*, *101*, *102*, *109*, *132*, 136, 304; steel, 23–26, *24*, 86–88, 90–94, *96*, *99*, *101*, *102*, 103, 107–110, 122, *132*, 168, 303, 304
 3-inch, 500-pound rifled howitzer, bronze, 86, 88, 93, 99, 101, 109; steel, 86, 88, 90–93, 96, 97, 99, 304, *305*
 Driggs-Schroeder 3-inch landing gun (tested), 137–38, 168
 Elswick/Armstrong-Whitworth 12-pounder, 8cwt, 142–44, *144*, 242, 307, *307*
 Hotchkiss 10-pounder landing gun (tested), 137–38, 168
 M1, M1A1 75mm pack and landing howitzer, 269, 277, *279*, 284–85, 307, *308*, *see also* US Army
 Mark I (Fletcher) 3-inch field gun, *133*, 135, 137–39, *139*, *141*, 140–42, 148–49, 152–53, *154*, 155, 161–68, *169*, 171–72, *173*, 174–78, *178*, 179–80, 192, 199, 201, 207, 273, *275*, 303–05, *306*, *see also* Difficulty in or dissatisfaction with
 Mark I mod 1 (US Navy) 3-inch field gun, 180–85, *181–82*, *184*, 192, 211, 222, *233*, *243*, 266, *267*, 269, 270, 271–72, *173*, 273, *275*, 303, 305, *306*
 Mark IV (Bethlehem) 3-inch landing gun, *189*, *191*, 189–92, 222, 225, *226*, 227–28, *230*, 232–33, 236–37, *237–38*, 244, 246, 272–73, 303, 305, *306*, *see also* Difficulty in or dissatisfaction with
 Mark VII (Ehrhardt) 3-inch landing gun, 192, 201–05, *206*, 207, *208–09*, 209–14, *214–15*, 222–23, *238*, 239, *245–46*, 246–47, 256, 266, 271–73, *274*, 286–88, 290, 303, 305, *306*
 Mark XI (Bethlehem/US Navy) 3-inch landing gun, 244, 246–48, 260–62, *263*, 264, 266, 268–69, 271, 275, 276, 277, 279–80, 280, 286, 288–90, 292, 303, 305, *306*; *also see* Difficulty in or dissatisfaction with
 Mark XII (Bethlehem) 3-inch mountain gun, 247–48, 260–61, 264, *265*, 269, 307, *307*
Mortars
 Brandt 81mm, infantry, 277
 M1 81mm, infantry, 266, 275, 277, 278, 281, 284–85, 290, 292
 Stokes 3-inch, trench, 266, 275, 277, 278

Machine guns
 Gardner (tested), 118
 Gatling, various models and calibers, *24*, 103, 107, 110, 118–22, *119–20*, *123*, 131, 134, 147, 157, *157*, 204; Accles feed drum, *24*, 121, 134; Broadwell feed drum, 119–20; Bruce feed device, *120*, 120–21, 134; Bulldog gun, 118–19, *120*
 Lowell battery (tested), 118
 Mark VI Lewis gun .30 cal, *248*, 248–50, 254, 266, 282, 286
 Model 1895 Colt-Browning ("potato digger") 6mm, 134, *135*, 144–45, 149, 153, *153*, 155–56, 158, *159*, 161, 162–63, 165, 168, 193; .30 cal, 134, 205, 216, 222, *229*, 236, *238*, 250
 Model 1909 Benét-Marcié .30 cal, 216, *216*, 222, 236
 Model 1914 Hotchkiss heavy 8mm, 248
 Model 1915 Chauchat light 8mm, 249
 Model 1915 Colt-Vickers .30 cal, 248
 Model 1917 Browning water-cooled .30 cal, 266, 286
 Model 1918 Browning automatic rifle (BAR) .30 cal, 266, 282, 286
Personal arms
 Shoulder weapons
  M1 Garand .30 cal, 280–81, *300*
  Models 1861 and 1863 Springfield .58 cal muskets, 72, 76, 116, 129
  Model 1867 Remington cal .50-45 rolling block carbine, 71–72, 76, *78*, 82–83, 113, 116
  Model 1870 Remington cal .50-70 rolling block rifle, 71–72, 113, *126*
  Model 1891 Mosin-Nagant 7.62mm, US manufacture, 250–51, *253*
  Model 1895 Winchester-Lee 6mm, *133*, 133–34, 144, 149, *153–54*, 155–56, *157*, *159*, 162–63. 167, 193, *200*
  Models 1896 and 1898 Krag-Jørgensen cal .30-40, 134, 167, 192–93, *200*, 204, *206*
  Model 1903 Springfield .30 cal, 193, 204, *206*, 215, *229–30*, *238*, 250, 280, 282, *300*
  Models 1921 and 1928 Thompson SMG .45 cal, 268, *274*, 275
  Remington-Keene cal .45-70, 115, *116*
  Remington-Lee cal .45-70, 115–16, *116*, 131, 133, *148*
  Springfield rifle cal .45-70 trapdoor, various models, 103, *106*, 116, *117*, *148*
  Whitney (Plymouth) .69 cal musket, 72, 116

  Winchester-Hotchkiss cal .45-70, 104, *106*, *114*, 114–115, *116*, *119*, *126*
 Pistols
  Colt
   Models 1851 and 1861 Navy .36 cal revolvers, 71, 117
   Model 1860 Army .44 cal revolver, *28*, 117
   Model 1871 .38 cal conversion revolver, 117
   Model 1872 .44 cal conversion revolver, 117
   Model 1889 New Navy .38 cal DA revolver, 117
   Model 1911 .45 cal semiautomatic pistol, 204, *209*, *230*, *237*, 250, 282
  Remington
   Cap and ball Navy .36 cal revolver, various, 71
   Model 1867 cal .50-25 rolling block pistol, 71, 76
   New Model Navy .38 cal conversion, 117
  Smith and Wesson revolvers, 117
Shipboard guns
 1-pounder: 24, 125, 156, 193, *228*, *230*, 254; Driggs-Schroeder, long, *131*
 3-pounder, *24*, 125, 155, 193, 254
 6-pounder, 125, 155–56, 185, 193–94, 220
 3-inch/23cal, 244; 3-inch/50cal, *186*, 187, 205, 221, 222, 228, 234, 236, 273, 280, 293, *295*, *296*
 4-inch BL, 115, *154*, 156, 163, 185, 210
 4.7-inch QF, 143–44
 5-inch/38cal, *295*; 5-inch/40cal, 156–57, 176, 222; 5-inch/50cal, 194, 210, *221*, 222, 254, 256, 271; 5-inch/51cal,, 249, *250*
 6-inch BL 115, 156–58, *159*, 161, 185, 294
 6-inch QF, 143–44
 8-inch BL 115, 147, 157, 165, *184*, 185, *191*, 193, 195, 210, 241, 255, 258, 268
 8-inch ML, 75, 115
 10-inch BL, 115, 148, 156
 12-inch BL, 115, 152, 156, *173*, 185, *191*, 222, 232–34, *235*, 242
 13-inch BL, 155, 161, 172, 178
 14-inch BL, as railway guns, 241
 15-inch pneumatic, 115
 32-pounder ML, 59, 132
 Dahlgren 4-inch (20-pounder) RML, 29–30, *34*, 37, 48, *65*, 75, 88, *305*
 Dahlgren 4-inch (20-pounder) rifled BL *65*, 88, *89*, 304
 Dahlgren 9-inch smoothbore ML *50*, 75
 Dahlgren 24-pounder smoothbore ML, 29, 34–36, 43, 48, 53, 55, 61, 71, 75, 131, 304
 Dahlgren 50-pounder RML, *58*

Hotchkiss revolving cannon, 37mm, 122, *124*, 124–25, 307, *see also* field service; 47mm and 53mm, 115, 122–23, 125, 307
Parrott RML: 20-pounder, 59; 30-pounder, 70; 60-pounder, 75; 100-pounder, *50*, 59
Y-gun for depth charges, 244

Haggerty, Daniel, 227
Hall, Newt, 161, *161*, 197
Halsey, William, 194
Hammes, R. B., 236
Hammond, J. B., 54–55
Hand grenades, 61, 266, 282
Hanrahan, Dennis, 81
Harbors and commercial ports; *see also* Cities and towns
  Lake ports
    Cleveland, 90, 261; Granada, 210–11; Managua, 210–11
  River ports
    Albany/Watervliet, 113; Canton, 47, 269, 270, *see also* Assault Landing Battle; Chicopee Falls, 48, 92; Düsseldorf, 201; Hartford, 122; Irkutsk, 255; Khartoum, *see* Sudan, Foreign Nations; New Orleans, 106; Philadelphia (Nicetown), 88, 90, 176–77, 191, 207, 210, 305; Plymouth, NC, 60; Port Conway and Port Royal, *see* Naval Battle on the Rappahannock River; Port Lyautey, *see* Assault Landing Battle of Casablanca; Pratt's Landing, 57; Pyongyang, 69; Tampico/Madero), 220–23, *221*, 225, 232, incident at, 220–21, 239; Tientsin, 157, *157*, 161–65, *161*, *164*, *see also* Land Battle; Trenton, 48; Washington, 51, 59, 107, 172, 266, 271, 286, 288, *see also* Navy Yard; Wilson's Wharf, *see* Land Battle; Yeung Kong, 269
  Seaports
    Algiers, *see* Assault Landing Battle; Amerikanka, 256; Annapolis, 86. 99, 179, *see also* Naval Academy; Apia, 101, 109, 158, *159*, *see also* Land Battle of Tagalii; Arkhangelsk, *221*, 249, *251*, 251–52, 254, 302 (Bakharitza, 251, *253*; Solombola, 254); Ashby Harbor, 53, *54*, *see also* Assault Landing Battle of Roanoke Island; Baltimore, 59; Bluefields, 110, 210; Boston, 48, 91, 130, 132 (Dorchester, 130); Bridgeport, 138, 188, 201–02; Casablanca, *see* Assault Landing Battle; Charleston, 63; Chefoo, 104; Cherbourg, 110; Colón/Aspinwall, 104–09, *106*; Corinto, 194, 210–11; Fajardo, 156; Guánica, 155–56; Havana, 138, 148; Havre de Grace, 59; Istanbul, 305; La Playa, PR, 155; Honolulu, 147, *148*, *see also* US Naval Station; Kowloon, 46, *46*; Monte Cristi, 214, *215*; Murmansk, 249, 251, 254, 302; Nagasaki, 70; New London, 205; Newport, *49*, 179, 205; New York, *50*, 88, 104, 106, 108, 143, 176, 193, 241, 244, 286 (Bronx, *132*, 175; Brooklyn, 86, 148, *150*), *see also* New York Navy Yard; Norfolk, 143; Oran, *see* Assault Landing Battle; Panama City, 104–05, 107–08; Pearl Harbor, *see* US Navy Stations and Yards; Pensacola, 46, 49, 51; Petrograd/St. Petersburg, 249; Ponce (La Playa), 155; Puerto México/Coatzoacalcos, 239; Puerto Plata, 214; San Juan, 155; Santo Domingo, 213, *214*; Santiago de Cuba, 143, 149; Savannah, 63, *64*; Seattle, 284; Shanghai, 45, *270*, *see also* Assault Landing Battle; Tangier, 195; Tongku, 161, 165; Tsingtao, *270*; Venice, *250*; Veracruz, *206*, 219–20, 222–23, *224*, *see also* Assault Landing Battle; Vladivostok, 249, 252, 254–56, 257, 258, *258*, 302; Wilmington, NC, 63, 66, *see also* Assault Landing Battle of Fort Fisher; Yokohama, 43, *44*, 165, *see also* Convention of Kanagawa

Hardaway, Corbin, 252; *see also* Navy Cross
Hardaway, Robert, 57
Harding, Warren, 264
Harrington, Francis, 148, *154*
Harris, Thomas, 59
Harris, Townsend, 45
Harrison, William, 232
Haselden, [Lieutenant Colonel], 251–52
Hatch, John, 63
Hay, John, 195
Hébert, Louis, 61
Henderson, Alexander, 76
Henderson, Archibald, 198
Herbert, Hilary, 138
Heyerman, Oscar, 80
Heywood, Charles, 107, 198
Hicks, Donald, 251, *251*, 252, *253*, 254; *see also* Navy Cross
Higginson, Francis, 155
Hinojosa, Rámon, 220
Hoke, Robert, 63
Hontoria, José González; guns designed by, 129
Hoover, Herbert, 166
Horne, Frederick, 194
Hotchkiss, Benjamin, 122, 138; *see also* guns built by his company
Houston, J. B., 106–07
Houston, Nelson, 76, 81
Howard, Archie, 255–56
Huerta, Victoriano, 219–22; Hueristas, 220, 239
Hulbert, Henry, 160; *see also* Medal of Honor and Navy Cross
Huntington, Robert, 148–49, *150*, 151–53, 155, 197–98
Hurricane, *see* Typhoon
Huse, Harry (Henry), 155, 223, 231
Hutchins, Charles, 103, 116, 134,

Indirect laying, 201, 239, 244, *245*, 246, *263*, *265*; *see also* Panoramic sight
Influenza 1918, 252
Ingersoll, Royal, 285

Jackaway, Joseph, 59
Jeffers, William, 85–88, 91, *91*, 92, 97, 130
Jellicoe, John, 165
Jenkins, Friend, 137–38
Jewell, Theodore, 195
Jewett, Henry, *161*, 162
Jimenez, Juan, 213
Johnson, Bradley, 59
Johnston, Rufus, 236
Johnstone, James, 163
Jordan, Leland, 225
Jouett, James, 105–08

Kalmikoff, Ivan, 255
Kane, Theodore, 105
Kautz, Albert, 158
Keene, John, 115; *see also* the rifle invented by
Kelly, John, 45
Kelly, Patrick, 61
Kemp, Thomas, 249
Keranen, Emil, 252; *see also* Navy Cross
Kerensky, Alexander, 255
Kerrick, Charles, 194
Kimball, William, 107
Kimberly, Lewis, *75*, 76–77, 80
Kimmel, Husband, 285
King, Edward, 289
King, Ernest, 285, 292
Kipling, Rudyard, 107
Knight, Austin, 255–56

Lamb, William, 63, 66
Lansdale, Philip, 158, 160
Larimer, Edgar, 273, 275
Leahy, William, 194
Lee, Fitzhugh, 59
Lee, Robert E., 47, 55
Lee, Samuel, 57, 63
Lejeune, John, 157, 199, 205, 213, 222, 234, 237
Lesseps, Ferdinand de, 106
Leutze, Eugene, 187
Lighthouses
  Assateague, 91; *see also* Island; Faro Benito Juárez, 228; Faro de Las Cabezas de San Juan, 156, *see also* Assault Landing Battle; Faro Venustiano Carranza, 234
Lili'uokalani (Queen), 147, *148*
Lincoln, Abraham, 51
Lippett, Thomas, *161*
Liscum, Emerson, 165, 167
Lodge, Henry Cabot, 221
Lôme, Enrique Dupuy de, 148
Long, Charles, 166
Long, John D., 149
Longnecker, Edwin, 107
Low, Frederick, 70, 72–73, 83
Lowry, George, 225, 228, 231
Lung, George, 160
Lyon, Henry, 103

Maass, Gustavo, 223, 231
MacKenzie, Alexander S., 71

Madero, Francisco, 219
Maffitt, John, 61
Magill, Louis, *123*, 153; *see also* Brevet Medal
Mahan, Alfred T., 101
Malavergne, René, 293–94; *see also* Navy Cross
Malietoa Laupepe (King), 158
Mallory, Stephen, 48
Manney, Henry, 174
Marshall, George F., 297–98; *see also* Distinguished Service Cross
Marvin, Joseph, 86
Mason, Newton, 137
Mason, Theodorus, 101
Mata'afa Iosefo (King), 158
Matthews, Edmund, 63
Maxim, Hiram, 122, 134, 248, *see also* Machine gun and other guns
Mayo, Henry, 220–23, 225
Mayo, Isaac, 42
McCain, John Sr., 194, *194*
McCain, John III, 194
McCalla, Bowman, 105–08, 149, *151*, 151–52, *161,* 161–62, 165
McCawley, Charles, 198
McCloy, John, 228, *230*; *see also* Medal of Honor
McCluney, William, 46
McCrea, Edward, 75, *75*
McCully, Newton, 254
McDonnell, Edward, 225, 227
McIlvaine, Bloomfield, 80
McKee, Hugh, *74*, 78, 80, 82
McKelvy, William, 210
McKenna, Richard, 269
McKenzie, Alexander, 82; *see also* Medal of Honor
McKinley, William, 148
McLean, Thomas, 80, 82
Mead, William, 76–77, 80–81, 83
Meade, Robert, 165–68
Medals awarded
  Brevet Medal, 153, 166; Distinguished Service Cross, 298; Medal of Honor, 82, 83, 152, 160, 166, 228, *235*, 236; Navy Cross, 160, 252, 254, 294; Victoria Cross, 70, 299
Meigs, Montgomery, 51
Mena, Luis, 210–11
Merchant steamers and sailing vessels
  Acapulco, 106–08; *Antonio López*, 143; *Baltimore*, 144; *City of New York*, 53; *City of Para*, 106–07; *Colón*, 105; *Corregidor*, 289; *Corsair II*, 155, *see also* as renamed, USS *Gloucester*; *Dauntless*, 92; *Esperanza*, 223, 232; *General Sherman*, 69–70; *J. S. Lowell*, 109; *Mexico*, 223; *Pocahontas*, 53; *Rose Standish*, 45; *Rover*, 70; *Sonora*, 231, 236; *Surprise*, 69; *Venezuela*, 149, *see also* as renamed, USS *Panther*; *Ypiranga*, 222–23, 227, 231, 239
Metals
  For cartridge cases
    brass, 190; coiled brass, 124–25; copper, 83, 87; drawn brass, 118, 125, 129; steel, 179; tin, 35
  For gun components
    bronze, 28–29, *33*, 48, 63, 79, 87, *89*, 93, 97–98, 103, 118, 120, 131–32, 136, 180, 183, 304; hardened bronze, 85–86, 88, *91*, *91*, 93; manganese bronze, 172, *173*, 174–75, 177, 179, 183
    copper, 28, 30
    iron, 27–28; cast iron, 85; puddled wrought iron, 48, 53; wrought iron, 85
    steel, 85–88, 90, *91*, 92–93, 103, 110; alloy steel, 129, 136 (chrome steel, 86; nickel steel, 129, 136, 182–84, 201, 203, 247); carbon steel, 85, 129, 175, 183; cast steel, 85; crucible steel, 85; gun steel, 183, 203; semi-steel, *see* puddled wrought iron; soft steel, 93
    tin, 28
    zinc, 28
  For primers, vents, and fuses
    brass, 39, 100, 125; bronze, 100; copper, 30, 93, 95, 96, 99, 100; lead, 39, 100, 125; tin, 39
  For rotating bands
    copper, 88, 99; soft brass, 124; soft lead, 37, 38, 88
Metcalf, Victor, 199
Metford, William, his rifling pattern, 134
Meyrick, Peter, 298
Miles, Nelson, 155–56
Military formations and units
  British: Royal Marines, 45–46, 158, 160; Royal Welsh Fusiliers, 165
  Confederate Army
    Formations: Army of Northern Virginia, 55, 63; Citadel Corps of Cadets, 63; Hoke's Division, 63; Fitzhugh Lee's cavalry division, 59; Hampton Legion, 52
    Units: Portsmouth Battery/Grimes' Battery, 56; Rowan Artillery/Reilly's Battery, 52; Washington Artillery/Eschleman's Battery, 56
  French Army: Chasseurs d'Afrique, 293; Senegalese troops, 297
  Imperial Japanese Army: 2nd Battalion, 20th Infantry, 290, 292; other, 167, 255, 280
  International, mixed, or irregular formations and units:
    Czechoslovak Legion, 255, 256; Escuadra de Guantánamo, 152; Honolulu Rifles, 147
  Mexican Army: Veracruz Military District, 223
  Philippines Military
    Philippine Army: 61st and 81st Divisions, 289; Philippine Constabulary, 269
  Russian Military
    Bolsheviks, 251–52, *251*, *253*, 254, 302
    Imperial Russian Army, 163–65, *164*, 167, 251 (Cossack cavalry, 163, *164*)
    Red Army, 254–55, 258
    White Russian Army, 254, 258
  Union Army
    Formations
      Army of the James, 58; Army of the Potomac, *55*
      IX Corps (Burnside's Corps), 56; XXIV Corps, 66; 2nd Division, 51; Coast Division, 53–55
      Regiments: 1st New York Marine Artillery, 53; 1st Regiment US Colored Troops, 58–59; 8th Connecticut Volunteer Infantry, 56; 9th New Jersey Volunteer Infantry, 54–55; 9th New York Volunteer Infantry (Hawkins' Zouaves), 55; 10th Regiment US Colored Troops, 58–59; 99th New York Volunteer Infantry (Union Coast Guard), 54–55
    Units
      I Company, 71st New York State Volunteer Infantry, 51–52, *52*; K Company (Whiting's Battery), 9th New York Volunteer Infantry, 56
  United States Army
    Formations
      I and II Philippine Corps, 289–90; American Expeditionary Force Siberia, 256; American North Russia Expeditionary Force, 252, *253*; US Army Forces in the Far East (USAFFE), 289–90 (Philippine Department, 288–89; Philippine Division, 288)
      5th Brigade, 237
      Philippine Scouts, 290
      Regiments: 9th Infantry, 165–67; 27th Infantry, 256; 31st Infantry, 256, *258*; 339th Infantry, 252, *253*
    Units
      1st Ranger Battalion, 297; 3rd Battalion, 6th Armored Infantry, 297–99; 3rd Battalion, 135th Infantry, 299; HQ, K, and L Companies, 47th Infantry, 295, *296*
  United States Marine Corps
    Advance base force, 187, 198–99, 205, *206*, 207, 273, *see also* Advance Base Brigade and Advance Base School
    Formations
      Advance Base Brigade, 205, 222–23, 225, *226*, 233–34, 237; 1st Marine Brigade, 201; 4th Marine Brigade, 248, 249
      1st Marine Regiment, 163–67; 1st Provisional Marine Regiment, 210; 4th Marine Regiment, 214, 215, 216, 291; 10th Marine Regiment, 277
    Units
      First Marine Battalion (Huntington's Battalion), 148–49, *150*–51, 151–55, *153*–54; 6th Machine Gun Battalion, 248, 249; B Company/39th Company (Peking Legation Guard), 273, *274*; D Company, 1st Marine Regiment, 166; E Company, 1st Provisional Marine Regiment,

210–11; F Company, First
  Marine Battalion, 148, 151–53,
  154; H Company, 4th Marine
  Regiment, *291*
Military and Naval Museums
  Turkey
    Deniz Müzesi, Istanbul, 305
  United States
    National Museum of the Marine
      Corps, Quantico, VA, 305, 307,
      308
    US Army Field Artillery Museum,
      Fort Sill, OK, 307, *307*
Minié, Claude-Étienne, his musket ball,
  72, 129
Moffett, William, 232
Monaghan, John, 160
Monffort, Aurelio, *227*
Morgan, J. Pierpont, 155
Morris, Richard, 56
Motion pictures depicting US naval
  landing parties
  The Sand Pebbles (20th Century
    Fox, 1966), based on the Richard
    McKenna novel, 269
  The Wind and the Lion
    (Metro-Goldwyn-Mayer, 1975), 195
Murders: of American survivors from
  the bark *Rover*, 70; of the crew of the
  merchant steamer *General Sherman*,
  69; of US Navy bluejackets from the
  brig USS *Perry*, 57
Murphy, Paul, *123*
Myers, John (Jack), *123*, *161*, 161–62,
  195, 197

Nalty, Bernard, 47
Naval academies, other schools, and
  training establishments
  Massachusetts: Nautical Training
    School (Massachusetts Maritime
    Academy), 132
  Mexico: Escuela Naval Militar,
    Veracruz, 225, 228, *230*, 231, 234,
    236
  United States Navy and Marine Corps
    Advance Base School, 205; Naval
    Academy, 23, 25–26, 49, 86–87,
    91, *91*, 101, *102*, 109, 121,
    132, 136, *136*, 169, 179, 188,
    197, 199, 207, 228, 234, 269,
    276, 280, *280*, 284, 292; Naval
    War College, 205; School of
    Application, 168
  Marine Corps Recruit Depot,
    64; Naval Training Ship
    Monongahela, *114*, 179; Naval
    Training Station, Great Lakes,
    126; Naval Training Station,
    Hampton Roads, 268–69; Naval
    Training Station, Newport 179,
    268–69
Naval brigade/fleet brigade/landing
  brigade, 62–63, *64*–65, 66, 70, 83,
  101, 103, 105, 130, 164–67, 223, 225,
  228, 233, 254, 302
Naval bureaus, echelons, fleets,
  squadrons, flotillas, and special units
  French Navy: Far Eastern Squadron,
    69; main battle fleet, Mers-el-Kébir,
    293, 298; coast artillery gunners

defending Casablanca, Oran, and
  Algiers harbors, see Assault Landing
  Battles
United States Navy
  Bureaus: Construction and Repair,
    277; Navigation, 107; Ordnance,
    28, *57*, *58*, *65*, 85–86, *87*, 88,
    89, *91*, 92, 95, 99, 121, 130–32,
    137, 140, 144, 168, 172, 174–80,
    182–93, 199, 201–04, 207, 209,
    213, 220, 239, 244, 246–47,
    256, 260–62, 266, 268, 272–73,
    275, 277, 280–81, 284–88, 292;
    Ordnance and Hydrography, 27;
    Supplies and Accounts, 177
  District: 16th Naval, 288
  Fleets: Asiatic, *235*, 255, 267,
    268–69, *270*, 274, 275, 284–85,
    288, 290; Atlantic, 220, 221, *221*,
    222-23, 231–32, 242, 285–86;
    Pacific, *279*, 282, 285; United
    States, 292
  Flotillas: Potomac, 56; Albemarle
    Sound, 60–61
  Squadrons: African, 42; Asiatic, 70,
    72, 79, 201; East India, 43, 45;
    European, 195; Flying, 143, 149;
    Mississippi River, 52–53, *65*; North
    Atlantic, 105–06, 149; North
    Atlantic Blockading, 57, 63, *65*;
    Pacific, 158; South Atlantic, 195;
    South Atlantic Blockading, 57, *58*,
    62, *64*; South Pacific, 104; Special
    Service, 266, 271, 281–83; White/
    Squadron of Evolution, 108, *109*
Neville, Wendell, 223, 227, 233, 236
Newbolt, Henry, 118
New Navy, 86, *109*, 113, 115, *154*, 197,
  see also ABCD ships
Newspapers and periodicals
  Chicago Tribune, 152; *McClure's
    Magazine*, 152; *New York Times*,
    138; *New York World*, 152
Niblack, Albert, 197
Nichols, Edward, *75*
Nichols, Samuel, 197
Nieh Shih-ch'eng, 163
Nimitz, Chester, 194, *194*
Norton, Charles, 107

O'Bannon, Presley, 197
Obturating, firing, and detonating
  devices
  Breech mechanisms
    Bethlehem Mark IV, 190, 244, 246,
      273, 303; Bethlehem Mark XI,
      247, *263*, 303; Driggs-Schroeder,
      135; Ehrhardt, 201, *209*, *215*;
      Elswick QF, 142; Fletcher, 137–39,
      139, *178*, 180, *306*; Hotchkiss, 87,
      125, 135, 137; Krupp, 87; Reffye,
      86–88, *87*, *89*; 3-inch howitzer,
      91, *91*, 93–94, *94*
  Firing locks
    Badger, 140, 180–81; Dahlgren, 30,
      31, *34*, 38–39; Gatling, 118–19,
      120; Graeme, 140, 181; Hotchkiss
      RC, 122–23; Tasker/United States
      Ordnance Co., 180–81, 188
  Firing devices
    3-inch howitzer, 95, 96, 97

Fuses
  Bormann (time), 39, *40*, 45, 59, 83;
    Boxer (time), 99–100; Hotchkiss
    (percussion), 124–25; navy
    percussion, 39, 59, 83; navy time,
    31, 39; other or in general, 26,
    35–36, 99, 131, 142, 155, 209,
    212, 290, 294
Gas checks
  Broadwell ring, 93, 95; cup gas
    check, 95, 97, 103. fixed metal
    round, 88, 95, 97
Primers and primer vents
  Primer: friction, 39, *40*, 95, 99;
    percussion, 83, 115, 118, 131,
    140, 181; quill percussion, 30,
    38–39, *40*
  Primer vent: axial, 88, 95, *96*;
    radial, 30, *31*, *96*, 97, 99, 103
O'Callaghan, George, 45
O'Kane, James, 63
Olmstead, Edwin et alia, 304
O'Neil, Charles, 188
Operation Torch, 293–99 (Operation
  Reservist, 297–99; Operation
  Terminal, 299)
Ordnance Department, US Army, 129,
  186, 204, 213
Ortiz, Peter, 297
Osborn, Luther Wood, *159*
Otis, Elwell, 193

Panoramic sights, 186, 190, 201, 204,
  207, *245*, 246–47, 261–62, *263*, 264,
  265, 289, 303; see also Indirect laying
Parker, Foxhall A. , 87
Pattison, Robert, 132
Pearson, Frederick, 70
Pearson, R. H., 45
Pegram, Robert, 46
Pelham, John, 57
Pendleton, Edwin, 176, 186, 190
Pendleton, Joseph, *123*, 205, 210–11,
  214, 216
Perdicaris, Ion, 195
Perkins, Constantine, 160
Perry, Matthew C., 43, *44*, 45, *46*, see
  also Convention of Kanagawa
Perschke, George, 252
Peters, Frederick, 297–99; see also
  Victoria Cross
Picking, Henry, 76–77
Pillsbury, John, *75*, 80, 199
Pino, Pedro del, 156
Plain, Louis, 297
Plunkett, Charles, 241–42
Poe, Edgar Allen Jr., 258
Poole, Frederick, 249
Porter, Benjamin, 54–55, *64*–66, 66
Porter, David, *65*
Porter, David D., *46*, 51–53, 63, *65*–66
Potts, Templin, 137
Powell, William, 164
Preble, George, 63
Prestan, Pedro, 105, 107
Preston, Samuel, *64*–65
Projectiles
  Canister, 29, 35, 37–38, 45, 55, 57,
    59, 62, *62*, 76, 82–83, 101, 103,
    124–25, 127, 155; grape, 29, 37, 55,
    57, 59; common shell, 21, 23, 26,

29, 35–37, *37, 38*, 39, 45, 55, 57, 59, 60, 76, 81, 83, 88, 99, 124–25, 127, 209, 287–88; high explosive shell, 21, 124, 143, 149, 156, 164, 182, 228, 236, 242, 254, 290, 294, 298 (lyddite/fused picric acid, 166–67); solid shot, 29, 37; shrapnel/spherical case, 21, 23, 26, 29, *31*, 35–36, *37*, 38–39, 42, 45, 47, 55, 76, 82–83, 99, 103, 135–36, 142, 149, 152, 155, 162, 166–68, 172, 182, 204, 207, 209, 211–12, 287–88
Propellants
  Gunpowder/black powder, 29–30, *31*, 35–37, 39, *40*, 71–72, 85, 88, *94*, 95, 97, 99, 114, 124, 129, 134; nitro-based powder/smokeless, 129, 134, 140, 190–91
Purchase of surplus Mark VII landing guns
  Netherlands Purchasing Commission, 286–87; President's Liaison Committee, 286–87
Purvis, Hugh, 82; *see also* Medal of Honor
Pyrotechnics, 42, 57

Quick, John, 152–53; *see also* Medal of Honor
Quinn, [Mate], 76, 79

Ragsdale, James, *157*
Raisuli, Mulai Ahmed er, 195
Ramsay, Francis, 86
Reed, Allen, 193–94
Reffye, Jean-Baptiste Verchère de, 86, *87*; *see also* the gun and breech mechanism of his design
Reiter, George, 149, 151
Reybold, Eugene, 288
Rockwell, Charles, *75*, 75
Rodgers, Frederick, 156–57
Rodgers, John, 70–73, *75*, 76,
Rodgers, William, 103
Roosevelt, Franklin D., 247, 264
Roosevelt, Theodore, 195, 199, 210
Roosevelt, Theodore Jr., 262, 264
Roxborough, George, 90
Roze, Pierre-Gustave, 69; *see also* Roze Roads
Rush, William, 223, 225, 227–28, 234
Russell, John H., *123*, 205

Sampson, Harold, *169*
Sampson, William, 137, 149, *169*
Schley, Winfield Scott, 76, 82, 143, 149
Schoonmaker, Cornelius, 110
Schroeder, Seaton, 73, 81–82, 138, 197; *see also* the guns bearing his name
Scott, Winfield, 222
Scribner, Edward, 183–84
Scrugham, James, 281
Seeley, William H. H., 70; *see also* Victoria Cross
Selfridge, James, 156,
Semyonov, Grigori, 255
Seymour, Edward, 162; relief expedition commanded by, *151*, 162–65, 168
Shankland, William, 56–57
Sherman, William T., 63, *64*
Sherinsky, [Colonel], 164

Shoup, David, *274*
Shrapnel, Henry, 35; *see* Projectiles for the round he devised
Sicard, Montgomery, 86–88, 92, *94*, 131
Simmons, John, 59
Simpson, Edward, 113 , *see also* Board, Gun Foundry
Sims, William, 199
Snow, Albert, 76, 80–81, 83
Society of Righteous Harmonious Fists, 160, *also see* Boxer Rebellion under Wars
Soley, John, 101, 103, 130–31
Southerland, William, 211
Sparrow, Herbert, 271
Spear, Ray, 287
Spotts, James, 57
Standley, William, 281
Stark, Harold, 285
Starkweather, Mark, 294
State militia and national guard (army)
  New Jersey, 92, 132, *132*; Pennsylvania, 132; Utah (Volunteer Light Infantry), 193
State national guard armories housing naval militia and naval reserve units
  Randolph Street Armory, Chicago (6th Battalion, US Naval Reserve), *275*; Seattle Armory (1st Artillery Section, 1st Division, Washington Naval Militia), *243*; South Armory, Boston (Dorchester Yacht Club, later Massachusetts Yacht Club, Massachusetts Naval Militia), 130
State naval militia, 127, 130–32, 184, 192, 241–42; by state:
  California, 116, 130; Connecticut, 130, 132; Massachusetts, 116, 130; Michigan, 130, 305, *306*; New York, *126*, 130, *131*; Rhode Island, 130; Washington, *243*
Steedman, Charles, 104–05
Steele, George, 211
Stephenson, Lowry, *274*
Stevens, John, 147
Stickney, Herman, 223
Stimson, Henry, 288
Stirling, Yates Jr. , 194
Stockholm, Andrew, 61
Stoddard, George, 63
Stuart, James E. B. , 47
Suárez, José Pino, 219
Swain, Julius, 59
Swinburne, William, 147
Swindler, Leland, 256

Taft, William H. , 210, 219
Tanumafili (Prince), 158
Tauscher, Hans, 201–02
Taussig, Joseph, 163, 165
Taylor, Acting Ens., 57
Teapost Dome and Elk Hills scandals, 264
Terry, Alfred, 63
Tilton, McLane, 72, 76–77, 79–80, 82–83, 197
Tomas, Enrique, 152
Totten, George, 80
Tracy, Benjamin, 197–98
Tribes, clans, and military-political factions

Barbo, 42; Berber, 195; Bolshevik/Red Russian, 250–52, *251*, *253*, 254, 256, 258; Boxer, *see* Boxer Rebellion; Colorum, 269; Cossack 163, *164*, 255; Formosan, 70–71; Gribo, 42; Mataafan, 158, 160; Moro, 192; Red Guard (Canton, 1927), 269; Tiger Hunter, 81, *81*; White Finn, 249; White Russian, 254, 258
Twining, Nathan, 203
Typhoon (hurricane), gales, and destructive surf
  Apia, 101, *102*, 109–10, 158; Assateague Island, 109; Atlantic coast of Africa, 42–43, 293; Cape Hatteras/Outer Banks, 53; Chesapeake Bay, 271; South Carolina coast, 57
Tzu Hsi (Dowager Empress), 157, 160

Underwood, Robert, 207, 209–12
United States federal executive departments
  Interior, 264; Navy, 108, 122, 202, 262, 281; State, 256; Treasury, 88, 92; War, 193, 260, 280, 288
United States foreign policies
  Dollar diplomacy, 210; Monroe Doctrine, 210
US Military and Naval Installations (*see also* Forts, United States)
  US Army Armories and Arsenals
    Benicia Arsenal, 193; Frankford Arsenal, 90, 212; Harpers Ferry Armory and Arsenal, 47; Manila Ordnance Depot, 193; Picatinny Arsenal, 305; Springfield Armory and Arsenal, 114, *see also* the shoulder weapons named for this armory; Watervliet Arsenal, 113
  US Marine Corps Barracks
    Boston, 106, 207; Cavite, 205, 207 273; Charleston, 207; Guam, 207; Guantánamo Bay, 207; Honolulu, 201; Mare Island, 207; New York, 106, 148, 150, 207, *208*, Norfolk, 207; Olongapo, 205, 207, 256; Pearl Harbor, 273; Peking, 207, 273; Philadelphia, 106, 207, *209*, 210; Port Royal/Parris Island, 207; Portsmouth, 207; Puget Sound, 207; Quantico, *245*, *248*; San Diego, 205; Sitka, 207; US Naval Academy, 188, 207, 212; Washington, *123*, 207
  US Navy Land Establishments (other than Naval Academy, schools, and training stations, *see*)
  Hospitals
    Mare Island, 167; Russky Island, 256; Yokohama, 165
  Ordnance establishments and storage facilities
    Bellevue Ordnance Annex, 286–87; Cavite, *see* Stations and Yards; Fort Lafayette Naval Ammunition Depot, 244; Fort Mifflin Naval Magazine/Ammunition Depot, 191; Iona Island Naval Ammunition Depot,

244; Mare Island, 287, *see also* Yards; Naval Gun Factory, 93, 113, 127, 132, 135, 137, 140, 142, 168, 171–72, 175–79, 182–83, *182*, 185–87, 190, 192, 203, 241, 261, 266, 268, 272–73, 287; Navy Operating Base Hampton Roads, 266; Olongapo, *see* Stations; Pearl Harbor, *see* Yards; Philadelphia, *see* Yards; Puget Sound, *see* Yards; Washington Navy Yard (pre-Naval Gun Factory), 27–28, 47–48, 51, *58*, 86–88, 90, 92–93, 97, 131–32, *see also* Yards

Stations
 Cavite, 163, 172, 179, 194, 201, 205, 207; seized from Spain, 199; *see also* Yards; Ford Island, Pearl Harbor, 147, *see also* Yards; Guam, *186*, 187 205,, 207; Guantánamo Bay, 179, 205, 207; Honolulu, 179; Mariveles, 290; Olongapo/Subic Bay, 187, 290; Port Royal, SC, 207

Testing facilities
 Experimental Battery, Annapolis, 86, 90, 99; Naval Proving Ground, Dahlgren, VA, 277; Naval Proving Ground, Indian Head, MD, 137, 140–41, 171–72, 174–77, 179, 183, 185, *186*, 186–88, 190–91, 202–04, 261–62, *also* naval use of the US Army's Sandy Hook Proving Ground, NJ, 204

Yards
 Boston, 131, 179, 207; Cavite, 268,, 273, 286, 288; air attack on, 285, 290; *see also* Stations; Charleston, 207; Mare Island, 179, 205, 207, 286; New York (Brooklyn), 61, 109, 144, *150*, *173*, 179, 207, 266; Norfolk/Gosport, 48, 179, 187, 205, 207, 306, 307; Pearl Harbor, 277; air attack on, *279*, 287, 289; Philadelphia/League Island, 59, *131*, 176, 187, 205, 207, 209, 266, 272–73, 280, 286; Portsmouth (Kittery, ME), 109, 179, 207; Puget Sound, 179, 207, 284, 286; Washington 51, *58*, 131, 168, 204, 207, 244, 280, 284, 286, 292, 304, *305*, *see also* Ordnance Establishments

US National Archives, 168
US Naval commissioned officer rank structure:
 Before 1860, 28, *44*; from 1862, 14, 15, 52–53, *58*, 65, *66*, *75*; from 1882, 16, 26, *49*, *161*, 162–63; from 1883, 15; from 1902, 15, 16, 26, 193–94; from 1912, 15, 26
US Naval Institute: lectures/papers, 87, 101, 103, 130–31, 197, 199; periodical, 87, 131, 199; prize essay, 116
Uribe, Virgilio, 228

Valle, Pedro del, 281–83
Vandegrift, Alexander, *274*
Varley, Cromwell, 195

Wainwright, Jonathan, by inference from *The Wainwright Papers*, 289
Wainwright, Richard, Jr., 155, 197–98
Wainwright, Richard III, 225
Walker, John, 107–08
Wallace, Lew, 59
Waller, Littleton, 163–65, 168, *238*, 239
Warley, Alexander, 62
Warrington, Lewis, 27
Wars
 American Civil War/War of Rebellion, 27, *28*, *33*, 36, 42, 46, 46, 47–49, *49–50*, 51–63, *52*, *54–55*, *58*, 60, 62, 64–66, 66, 70–71, 76. 85, 88, 90, 104, 110, 113, 129, *131*, 152, 165, 179, 198, 304; Banana Wars, 194, 205, 209–214, *214–16*, 216, 219, 236, *245*, 268–69; Boer War, 103, 142; Boxer Rebellion, 133, 145, 157, *157*, 160–68, *161*, *164*, 193; Crimean War, 129; First Barbary War, *150*; Franco-Prussian War, 127, 222; Mexican-American War, 27, 222; Philippine-American War/Philippine Insurrection, 192–94, *194*, 237; Second Barbary War, *44*; Spanish-American War, 76, 110, *131–32*, 142–43, *144*, 145, 148–49, *150–51*, 151–53, *153–54*, 155–57, 163, 168, 192, 198–99, 225; Taiping Civil War, 45; War of 1812, *44*; World War I, 125, 127, 133, 142, *144*, 160, *206*, 241–42, *243*, 244, *245*, *248*, 248–49, *250*, 252, 255, *263*, 266, 267, 269, 282, 302 (North Sea Barrage, 241; Western Front, 241–42, 249, *278*); World War II, 160, 165, 167, 194, 205, *206*, 222, 232, 242, 258, 266, 276, 283, 289–90, *291*, 292–95, *294*, 296, 297–99, *300*, 302, 307
Warships and naval auxiliary vessels
Brazilian navy
 Amazonas, 143, *see also* USS *New Orleans*, New Navy; *Almirante Abreu*, 143, *see also* USS *Albany*, New Navy
Confederate States Navy
 Alabama, 76, 110; *Albemarle*, 60–62, *60*, *62*; *Atlanta*, *58*; *Young America*, 58
French Navy (la Marine française)
 Amiral Aube, 249; *La Surprise*, 298; *Typhon*, 298
Greek navy
 Kilkis, 242; *Lemnos*, 242
Imperial German Navy (Kaiserliche Marine)
 U-156, 241
Royal Canadian Navy
 Unicorn, Saskatoon, SK (land establishment named by tradition as a ship), 307, *307*
Royal Navy
 Broke, 299; *Camperdown*, *143*; *Eaglet*, 46; *Encounter*, 45; *Essex*, 225; *Glory*, 249; *Grecian*, 45; *Hartland* (ex-USCGC *Pontchartrain*), 297–99; landing craft assault (LCA), 297; *Malcolm*, 299; motor launches, 297–98; *Rattler*, 46; *Royalist*, 158, 160; *Suffolk*, 257; *Terrible*, 166; *Victoria*, *143*; *Walney* (ex-USCGC *Sebago*), 297–98
Russian Navy
 Amerika, 256; destroyer at Murmansk, 254
Spanish navy (Armada Española)
 Arayat, *see* US Navy; *Callao*, *see* US Navy; *Emperador Carlos V*, 225, *226*; *Pampanga*, *see* US Navy; *Paragua*, *see* US Navy; *Samar*, *see* US Navy
Turkish (Ottoman) Navy
 Mecidiye/Medjidiye, 305
United States Army
 Bombshell, 61; *Grapeshot*, 53; *Laguna de Bay*, 193; *Thomas*, 256; *Zouave*, 53
United States Coast Guard/Revenue Cutter Service
 McCulloch, 193; *Pontchartrain*, *see* HMS *Hartland*; *Sebago*, *see* HMS *Walney*
United States Navy
 Warship classes
  Arayat (gunboats), 193; *Brooklyn* (light cruisers), 275, 286; *Clemson* (destroyers), 293, *294*; *Connecticut* (pre-dreadnought battleships), *184*, *235*, 242; 272; *Denver* (protected cruisers), 266, 271; *Gleaves* (destroyers), 295; *Kearsarge* (pre-dreadnought battleships), 21; *Michigan* or *South Carolina* (dreadnought battleships), 201; *Mississippi* (pre-dreadnought battleships), 242; *New Orleans* (protected cruisers), 143–44, *144*, 242, 256; *S-boats*, *270*; *St. Louis* (semi-armored cruisers), *185*, 193; *St. Louis* (light cruisers), 286; *Tennessee* (super-dreadnought battleships), 279; *Virginia* (pre-dreadnought battleships), 21, *191*; *Wickes* (destroyers), 295, *296*
 Warships, sail
  Constitution, 42; *Jamestown*, 70; *John Adams*, 27–28; *Levant*, 47; *Monongahela* (converted), 114, *114*, 179; *Perry*, 57; *Plymouth*, 43, 45; *Portsmouth*, 47; *Saratoga*, 43, 45
 Warships, sidewheel
  Cœur de Lion, 56–57; *Daffodil*, 57; *Hatteras*, 76; *Hunchback*, *50*; *Jacob Bell*, 56–57; *Malvern*, 65–66; *Miami*, 60; *Michigan*, 115; *Mississippi*, 43, 45; *Monocacy* (oceangoing gunboat), 70, 73, *74–75*, 75, 77, 79, 80–82, 114, 162–63; *Otsego*, 61; *Philadelphia*, 51; *Powhatan*, 45–46, *46*, 51, 115; *Sassacus*, 60–61; *Southfield*, 60;

Susquehanna, 43; *Tallapoosa,* 109; *Wyalusing,* 61
Warships, steam screw (Old Navy) *Alaska,* 70, 75, 76, 114; *Alliance,* 105, 107; *Anacostia,* 56; *Benicia,* 70, 73, 75, 76–77, 81, 105; *Canandaigua,* 57; *Colorado,* 70, 72, 75, 75–76; *Currituck,* 56; *Dawn,* 37, 57, 59; *Despatch,* 91, 109; *Enterprise,* 132; *Fuchsia,* 59; *Galena,* 105, *106,* 114; *Hartford,* 70, 122; *Kearsarge,* 110; *Lancaster,* 114; *Mary Sanford,* 57; *Monitor,* 85; *Nahant, 131; Nipsic* 57, 110; *Palos* (armed tug), 70, 73, 75, 75–77, 114; *Penobscot,* 104; *Pensacola* (screw steamer), 104–05; *Pequot,* 57; *Picket Boat No. 1* and *Picket Boat No. 2* (torpedo boats, see Boats); *San Jacinto,* 47; *Shenandoah,* 107–08, 114; *Swatara,* 104, 106, 114, 121, 130; *Ta-Kiang,* 70; *Tennessee,* 106–07, 115; *Thetis,* 115; *Trenton,* 91, 101, *102,* 109–10, 115; *Tuscarora,* 104–05; *Vandalia,* 108–10; *Wachusett,* 114; *Wyoming,* 70; *Yantic,* 114

Warships. steam screw (New Navy) *Albany,* 143–44, 256, 258, *258; Amphitrite,* 156–57; *Annapolis,* 210; *Arayat* (ex-Spanish), 193; *Atlanta,* 109, 115; *Baltimore,* 157; *Boston,* 109, 115, 147, *148,* 157; *Brooklyn,* 165, 195, 255–56, *257,* 266, 267; *Callao* (ex-Spanish), 193; *Castine,* 115; *Charleston* (protected cruiser), 193; *Chicago,* 109; *Cincinnati,* 156–57; *Concord,* 115; *Detroit,* 115; *Dolphin,* 153, *154,* 220–22, 281; *Gloucester,* 155, 225; *Hancock,* 222, 234, 237; *Hannibal,* 156; *Iowa, 173; Illinois, 172; Leyden,* 156–57; *Maine, 117,* 138, 148; *Marblehead, 149, 151,* 152; *Massachusetts,* 155, 172, 174, 179; *Monterey,* 115; *Nashville,* 163, 168; *Newark,* 161, 164, 168; *New Orleans,* 143–44, *144,* 256, *258; New York, 133, 139,* 169, *200,* 266 (later *Saratoga,* 266; later *Rochester,* 266); *Olympia,* 249, *250,* 252, 254; *Pampanga* (ex-Spanish), 193, 269; *Panay* (ex-Spanish), *194, 194; Panther, 149, 150,* 151, 153, 176–77; *Paragua* (ex-Spanish), 193–94;
Philadelphia (protected cruiser), 115, 122, 158, *159; Prairie,* 210, 213, *214,* 222–23, *225, 226, 228, 231, 236; Puritan,* 156; *Raleigh,* 157; *Samar* (ex-Spanish), 193; *San Francisco,* 222–23, 232–34, 236; *Solace,* 163, 225, 234; *Texas* (pre-dreadnought battleship), 152–53; *Vesuvius,* 115; *Yankton,* 254; *Yorktown,* 109

Warships, twentieth century *Alabama,* 178–79; *Arkansas,* 232–33, 236; *Asheville,* 185, 266, 269, 275, 281, 285; "Atlanta" (armed motorsailer, see Boats); *Augusta,* 268, 275, 284; *Bernadou,* 295, 296, 298; *Birmingham,* 266, 281; *Black Hawk,* 268; *California* (armored cruiser), 210–11 (*see also* as renamed *San Diego*); *California* (battleship), *279; Charleston* (semi-armored cruiser), 185; *Charleston* (oceangoing gunboat), 275, 279, 281, 284–85; *Chester* (scout cruiser), 222–23 232–33, 236; *Chester* (heavy cruiser), 268; *Chicago,* 268; *Cleveland,* 266; *Cole,* 295, 296, 298; *Connecticut,* 222; *Dallas,* 293–94, *294; Decatur, 194; Denver, 194,* 210, 266; *Des Moines, 221,* 222, 254, 266, 281; *Edsall* (successively two ships, destroyer and destroyer escort), 160; *Erie,* 275, 279, 281, 285; *Florida,* 222–23, 225, 228, 231, 233, 236; *Galveston,* 254, 266, 281; *George Clymer,* 294; *Georgia* (battleship), *191,* 200, *243;* "Georgia" (armed motorsailer, see Boats); *Helena,* 286; *Houston,* 268, 275, 284–85; *Hulbert,* 160; *Huron* (armored cruiser, ex-*South Dakota*), 258, 266, 268; *Idaho,* 242; *J. Fred Talbott,* 281; *Jacob Jones,* 281; *Kansas, 184,* 272; *Kewaydin,* 271; *Lansdale* (successively destroyers in two classes), 160; *Louisiana,* 234, 237; *Louisville,* 268, 277; *Luzon,* 268; *Mervine,* 295; *Michigan,* 234, 237, *238; Milwaukee,* 185; *Mindanao,* 268; *Minnesota,* 222, 234, 237, 272; *Mississippi,* 242; *Monaghan* (successively destroyers in two classes), 160; *Monocacy* (river gunboat), 185, 269; *New Hampshire,* 233–34, *235,* 236–37; *New*
Jersey, 185, 233; *Northampton,* 268; *Oahu,* 268; *Omaha,* 281; *Orion,* 234; *Palos* (river gunboat), 185, 269; *Panay* (river gunboat), 268; *Penguin,* 269; *Pensacola* (heavy cruiser), 268; *Perry,* 281; *Philadelphia* (light cruiser), 296; *Pigeon,* 269; *Pittsburgh* (armored cruiser, ex-*Pennsylvania*), 266, 270; *Quincy,* 275; *Rhode Island,* 185; *Richmond,* 281; *Sacramento,* 185, 258, 269, 270, 275, 281, 285; *Salt Lake City,* 268; *San Diego* (armored cruiser, ex-*California*), 241; *Savannah,* 294; *South Carolina,* 232–34, 236; *St. Louis* (semi-armored cruiser), 185; *St. Louis* (light cruiser), 286; *Tacoma,* 210, 271, 281; *Tennessee, 279; Texas* (super-dreadnought battleship), 277, 286, 305; *Trenton,* 281; *Tulsa,* 266, 268, 281, 285; *Utah,* 222–23, 225, 227–28, 232–34, 236; *Vermont,* 233, 237; *Vincennes,* 275; *Virginia,* 185, *243; Wichita,* 275; *Zane,* 281

Washington, George, *132*
Water bucket, 171, *184*
Welles, Gideon, 48
Wells, Charles, 76, 82
Wertman, [Gunners Mate], 228
Wheeler, William, *75,* 76, 80, 82
Whigham, Henry J., 152–53
Whiting, James, 56; Whiting's Battery, 56.
Whiting, William, 63, 66
Whitney, William, 105–08
Wilkinson, Theodore, 225
Williams, Dion, 199, 205, 207
Williamson, James, 252
Wilson, Henry Lane, 219
Wilson, Woodrow, 219, 221–22, 256
Wiltse, Gilbert, 147
Wise, Edward, 58
Wise, Frederic, 213
Wood, Leonard, 194
Wood, [Lieutenant], 155
Wood, Thomas, *123*
Woodman, John, 62
Woodward, Douglas, 254; *see also* Navy Cross
Wright, Robert, 105–06
Wurtsbaugh, Daniel, 162
Wynne, Robert, 165

Young, Lucien, *148*
Young, Philip, 287

Zaragoza, Ignacio Morelos, 220
Zelaya, José Santos, 210
Zeledón, Benjamin, 211